THE CERTIFIED
QUALITY PROCESS
ANALYST HANDBOOK

ASQ Quality Press
Milwaukee, Wisconsin

NEW AGE

Table of Contents

List of Figures and Tables

Introduction

This handbook is designed as a reference for the Certified Quality Process Analyst Body of Knowledge (BoK), providing the basic information needed to prepare for the CQPA examination.

There are four main sections in the CQPA Body of Knowledge, further subdivided into related subsections. These sections are:

- Quality Basics

- Problem Solving and Improvement

- Data Analysis

- Customer–Supplier Relations

The first part of the handbook, Quality Basics, begins with the quality principles embodied by the ASQ Code of Ethics. The fundamental elements of a quality system are described in this section.

The second part focuses on problem solving and improvement, including such tools as Pareto charts, scatter diagrams, the plan–do–check–act (PDCA) cycle, quality management, project management, the Taguchi loss function, and Taguchi's signal-to-noise ratios, concluding with lean tools.

To support any effective problem solving and improvement endeavors Part III, Data Analysis, provides the analytical methods to interpret and compare data sets and model processes. Basic statistics, probability, sampling methods, statistical process control (SPC), basic statistical decision tools, regression and correlation, and design of experiments (DOE) are explained in this section.

Finally, any successful enterprise must understand its Customer–Supplier Relations, the last section of the BoK. Effective customer–supplier relations are key to high customer satisfaction, the ultimate measure of a company's worth.

The relevant portion of the CQPA Body of Knowledge is excerpted at the beginning of each handbook chapter as a guide for the reader.

Part I
Quality Basics

Part I

1

Chapter 1

A. ASQ Code of Ethics

> Identify appropriate behaviors for situations requiring ethical decisions. (Apply)
>
> **Body of Knowledge I.A**

CODE OF ETHICS

Fundamental Principles

ASQ requires its members and certification holders to conduct themselves ethically by:

I. Being honest and impartial in serving the public, their employers, customers, and clients.

II. Striving to increase the competence and prestige of the quality profession, and

III. Using their knowledge and skill for the enhancement of human welfare.

Members and certification holders are required to observe the tenets set forth below:

Relations with the Public

Article 1—Hold paramount the safety, health, and welfare of the public in the performance of their professional duties.

Relations with Employers and Clients

Article 2—Perform services only in their areas of competence.

Article 3—Continue their professional development throughout their careers and provide opportunities for the professional and ethical development of others.

Article 4—Act in a professional manner in dealings with ASQ staff and each employer, customer, or client.

Article 5—Act as faithful agents or trustees and avoid conflict of interest and the appearance of conflicts of interest.

Relations with Peers

Article 6—Build their professional reputation on the merit of their services and not compete unfairly with others.

Article 7—Assure that credit for the work of others is given to those to whom it is due.

As a Certified Quality Process Analyst, you will be expected to perform your process analysis duties as a professional, which includes acting in an ethical manner. The ASQ Code of Ethics has been carefully drafted to serve as a guide for ethical behavior of its members and those individuals who hold certifications in its various disciplines.

Specifically, the Code of Ethics requires that CQPAs are honest and impartial and serve the interests of their employers, clients, and the public with dedication. CQPAs should join with other quality process analysts to increase the competence and prestige of the profession. CQPAs should use their knowledge and skill for the advancement of human welfare and to promote the safety and reliability of products for public use. They should earnestly strive to aid the work of the American Society for Quality in its efforts to advance the interests of the quality process analyst profession.

Performing your duties in an ethical manner is not always easy. Ethical choices are often not clear-cut and the right course of action is not always obvious. The ASQ Code of Ethics is not meant to be the one final answer to all ethical issues you may encounter. In fact, the Code has been revised and will continue to be revised from time to time in order to keep it current as new perspectives and challenging dilemmas arise.

To improve relations with the public, CQPAs should do whatever they can to promote the reliability and safety of all the products that fall within their jurisdiction. They should endeavor to extend public knowledge of the work of ASQ and its members as it relates to the public welfare. They should be dignified and modest in explaining quality process analysis work and its merit. CQPAs should preface any public statements they make by clearly indicating on whose behalf they are made.

An example of the CQPA's obligation to serve the public at large arose when a process analyst found that her manufacturing facility was disposing of hazardous waste materials in an illegal fashion. Naturally, she wrote up the violation and sent her report to the director of manufacturing. Her manager told her a few days later to "just forget about" the report she had written. This CQPA considered the incident a significant violation of her personal ethics and she told her boss that she felt the larger public interest outweighed the company's practice in this case.

Her arguments were persuasive and eventually the company reversed its former practice of disposing of the hazardous materials in an illegal manner.

To advance relations with employers and clients, CQPAs should act in professional matters as faithful agents or trustees of each employer or client. They should inform each client or employer of any business connections, interests, and affiliations that might influence the CQPA's judgment or impair the equitable character of his or her services. CQPAs should indicate to his or her employer or client the adverse consequences to be expected if his or her judgment is overruled. CQPAs should not disclose information concerning the business affairs or technical processes of any present or former employer or client without formal consent. CQPAs should not accept compensation from more than one party for the same service without the consent of all parties. For example, if the CQPA is employed, he or she should not engage in supplementary employment in consulting practice without first gaining the consent of the employer.

Another example of an ethical issue concerning loyalty to employers and clients occurred when a quality consultant was asked to give another client in a competing firm a short description of the first client's product. The information that the second company was requesting was readily available on the Internet, but this CQPA felt that he should not be the source of information about a competitor's project since that information was obtained while engaged by the competitor.

To improve relations with peers, CQPAs should give credit for the work of others to those to whom it is due. CQPAs should endeavor to aid the professional development and advancement of those in his or her employ or under his or her supervision. Finally, CQPAs must not compete unfairly with others. They should extend friendship and confidence to all associates and to those with whom he or she has business relations.

An example of a potential violation of the ASQ Code of Ethics as it relates to dealing with peers occurred when a consulting quality process analyst was asked to do a job that required more work than she could do on her own. She properly enlisted the support of a peer, a highly qualified and very senior Certified Quality Engineer. The CQE so impressed the client that they asked him to take the lead on the engagement. The CQE, however, chose not to violate the Code of Ethics. Instead, he discussed the situation openly with the CQPA and the client and it was agreed to keep the leadership for this engagement in the CQPA's domain.

Chapter 2

B. Quality Planning

> Define a quality plan, understand its purpose for the organization as a whole and who in the organization contributes to its development. (Understand)
>
> **Body of Knowledge I.B**

To get things done, an organization needs a plan. Plans may be very simple with merely a goal to be reached and a list of the steps necessary to achieve the goal. But effective organizational plans have all of the following:

1. Responsibilities of all the people associated with the project

2. Clear goals or objectives

3. Logical steps to reach the goals or objectives

4. Identification of necessary resources

5. An implementation approach

    a. Deployment of the goals or objectives

    b. Convey priorities and rationale to stakeholders and performers

    c. Measures of progress

    d. Milestones (checkpoints in the schedule to check progress)

6. Measures of success

RESPONSIBILITIES

Top-level management in an organization is responsible for establishing the firm's commitment to quality. It is this commitment from top management that constitutes total quality management. Feigenbaum[1] defined total quality control as follows:

Total quality control's organizationwide impact involves the managerial and technical implementation of customer-oriented quality activities as a prime responsibility of general management and of the mainline operations of marketing, engineering, production, industrial relations, finance, and service as well as of the quality control function itself.

Kaoru Ishikawa defined companywide quality control (CWQC) thusly:[2]

To practice quality control is to develop, design, produce, and service a quality product which is most economical, most useful, and always satisfactory to the customer.

In addition to top management endorsing its commitment to quality, most organizations appoint a group to develop guidelines, measure progress, and assist in implementing the quality objectives. This group may be formally called the *quality council* although many organizations call it by different names or merely include the quality council function in the roles of the executive committee. The council, whatever it is called, acts as a steering committee for the corporate quality initiative. In addition to ensuring that the quality initiative is implemented, the quality council must incorporate total quality strategies into the strategic business plan. Some of the specific responsibilities of the quality council may include:[3]

- *Developing an educational module: The modules may be different in content and duration for different organizational levels.*

- *Defining quality objectives for each division of the organization: This encompasses what improvement methods to employ, who is responsible, and what measures will be used.*

- *Developing and helping implement the company improvement strategy: Roadblocks and system problems must be resolved and the quality council is in a unique position to influence these decisions.*

- *Determining and reporting quality cost: There is no better way to encourage quality improvement than to show that poor quality costs a great deal. Evaluating the dollar value associated with poor quality in terms of prevention, appraisal, and failure costs helps management to measure quality using an easily understood denominator.*

- *Developing and maintaining an awareness program: The awareness program should be built slowly. It shouldn't be a surge of slogans, posters, and banners that will subsequently fade away. Continuous emphasis on excellence, sharing of success stories, recognition, and setting standards are all part of this process.*

Whenever possible, the planning process should include the people who will ultimately implement the plan. There are two reasons for this:

1. The people who will ultimately implement the plan probably have more realistic expectations of what is feasible.

2. People who feel that they participated in planning the work to be done are often more committed to performing it later.

SET GOALS/OBJECTIVES

Quality planning always begins with establishing what needs to be accomplished or what products and services need to be provided. This is often not an easy task since the various stakeholders in any process often have significantly different perspectives of what needs to be accomplished. Therefore, it is almost always necessary to negotiate the goals that will be the objectives of the quality project or effort.

Quality goals should be set for the organization and for each quality project to be performed. The quality council may establish these for the organization, but corporate quality goals are most often derived from the corporation's strategic mission, which is influenced by customers and top management. The quality goals must be consistent with the organization's mission and should be included in the corporate strategic plan. Quality goals, like all strategic goals, should be SMART:

S = Specific

M = Measurable

A = Achievable

R = Realistic

T = Traceable

To be SMART, the goal has to identify who is responsible, what must be accomplished, where the goal is to be achieved, and when the goal is to be accomplished.

To be *measurable,* the goal must be quantified. Goals that are not quantified are not measurable and are not SMART.

To be *attainable,* the goal must be achievable. It must be possible to reach this goal, at least in theory. If the goal is not feasible, it is not reasonable to expect people to try to achieve it. The goal should stretch the organization slightly so that everyone reaches for higher levels of performance but it must not be beyond the organization's ability to reach. Breaking difficult long-term goals into simpler, attainable steps so the managers and workforce believe that what they are doing is possible is a good management practice.

To be *realistic,* a goal must represent an objective that is both important and *feasible,* given the capabilities of the organization. In establishing goals, it is necessary to consider the abilities of the individuals who will achieve them and to consider the capability of the organization's processes. If the organization does not have the skills, materials, tools, training, processes, and knowledge to achieve the goal, it is unrealistic to expect that the organization can achieve it.

To be *traceable,* the goal must clearly be derived from the corporate mission. If the goal is not clearly in line with the overall strategic corporate vision it will be impossible to keep the organization focused on achieving it.

An example of a customer satisfaction goal that is specific, measurable, attainable, realistic, and traceable is:

To ensure that we are satisfying our customers, 90 percent of our customers will give us 5's or 6's on our customer satisfaction survey by January, 2008.

According to Juran,[4] total quality control should be a structured process and should include the following:

- A quality council
- Quality policies
- Strategic quality goals
- Deployment of quality goals
- Resources for control
- Measurement of performance
- Quality audits

Quality goals must be linked to product or service satisfaction, customer satisfaction, or cost of quality.

LOGICAL STEPS TO REACH THE GOALS

Stated goals alone are not sufficient for a plan. The plan must provide a road map for reaching the goals. The job must be broken into specific tasks that, when performed in sequence, will result in the goals being met. The project management tools and practices described in Chapter 12 of this handbook will aid in breaking the quality initiative into specific, interrelated tasks.

RESOURCES

All the personnel, equipment, materials, facilities, and processes necessary to perform the tasks must be identified in the plan. The company's total quality processes must include the means for identifying and procuring the resources necessary.

Any people who are involved with the quality programs should be identified in the corporate quality plan. Individuals must be assigned specific tasks to plan, implement, measure, and improve the quality program in the organization. It is imperative that those assigned specific responsibilities be held accountable for performing the assigned tasks. Holding individuals responsible for their actions is the role of the individual's supervisors or manager although this process may be monitored by the quality council and senior management.

AN IMPLEMENTATION APPROACH

For a plan to be of any real value to an organization, it is essential that the plan be implemented. "Deployment of the plan" means informing the individuals who will execute the plan of their responsibilities and holding them accountable for implementing the plan. Plan deployment is necessary or the plan is merely a document that has no impact on the organization's activities.

MEASURES OF SUCCESS

Finally, for a plan to be effective, it must include measures of success. Some measures of the success of corporate quality planning include:

- Are products delivered on time?
- Are costs of developing/delivering the products less than the price charged?
- Are the products produced all sold (overproduction)?
- Are the customer's requirements and expectations met?
- Are the customers satisfied with the service provided?
- Was documentation provided (if expected)

Chapter 3

C. Cost of Quality (COQ)

> Describe and distinguish the classic COQ
> categories (prevention, appraisal, internal
> failure, external failure) and apply COQ
> concepts. (Apply)
>
> **Body of Knowledge I.C**

THE COST OF QUALITY

It's a term that's widely used—and widely misunderstood.

The "cost of quality"[1] isn't the price of creating a quality product or service. It's the cost of *not* creating a quality product or service.

Every time work is redone, the cost of quality increases. Obvious examples include:

- The reworking of a manufactured item

- The retesting of an assembly

- The rebuilding of a tool

- The correction of a bank statement

- The reworking of a service, such as the reprocessing of a loan
  operation or the replacement of a food order in a restaurant

In short, any cost that would not have been expended if quality were perfect contributes to the cost of quality.

Total Quality Costs

As Figure 3.1 shows, quality costs are the total of the costs incurred by:

- Investing in the prevention of nonconformance to requirements

- Appraising a product or service for conformance to requirements

- Failing to meet requirements

Prevention Costs

The costs of all activities specifically designed to prevent poor quality in products or services.
Examples are the costs of:

- New product review
- Quality planning
- Supplier capability surveys
- Process capability evaluations
- Quality improvement team meetings
- Quality improvement projects
- Quality education and training

Appraisal Costs

The costs associated with measuring, evaluating, or auditing products or services to assure conformance to quality standards and performance requirements.
These include the costs of:

- Incoming and source inspection/test of purchased material
- In-process and final inspection/test
- Product, process, or service audits
- Calibration of measuring and test equipment
- Associated supplies and materials

Failure Costs

The costs resulting from products or services not conforming to requirements or customer/user needs. Failure costs are divided into internal and external failure categories.

Internal Failure Costs

Failure costs occurring prior to delivery or shipment of the product, or the furnishing of a service, to the customer.
Examples are the costs of:

- Scrap
- Rework
- Reinspection
- Retesting
- Material review
- Downgrading

External Failure Costs

Failure costs occurring after delivery or shipment of the product—and during or after furnishing of a service—to the customer.

Figure 3.1   Quality costs—general description.

Examples are the costs of:

- Processing customer complaints
- Customer returns
- Warranty claims
- Product recalls

Total Quality Costs

The sum of the above costs. This represents the difference between the actual cost of a product or service and what the reduced cost would be if there were no possibility of substandard service, failure of products, or defects in their manufacture.

Figure 3.1   *Continued.*

Chapter 4

D. Quality Standards, Requirements, and Specifications

> Define and distinguish between quality
> standards, requirements, and specifications.
> (Understand)
>
> **Body of Knowledge I.D**

The quality assurance (QA) function may include design, development, production, installation, servicing, and documentation. The phrases "fit for use" and "do it right the first time" arose from the quality movement in Asia and the United States. The quality goal is to assure quality at every step of a process from raw materials, assemblies, products and components, services related to production, and management, production, and inspection processes, through delivery to the customer, including a feedback loop at each step for continuous improvement.

One of the most widely used paradigms for QA management is the plan–do–check–act (PDCA) approach, also known as the Shewhart cycle. The main goal of QA is to ensure that the product fulfills or exceeds customer expectations, also known as customer *requirements*.

A companywide quality approach places an emphasis on four aspects:

- Infrastructure (as it enhances or limits functionality)

- Elements such as controls, job management, adequate processes, performance and integrity criteria, and identification of records

- Competence such as knowledge, skills, experience, and qualifications

- Soft elements, such as personnel integrity, confidence, organizational culture, motivation, team spirit, and quality relationships.

The quality of the outputs is at risk if any of these four aspects are deficient in any way.

In manufacturing and construction activities, these business practices can be equated to the models for quality assurance defined by the international standards contained in the ISO 9000 series for quality systems. From the onset of the Industrial Revolution through World War II, companies' quality systems relied primarily on shop floor inspection or final lot inspection. This quality philosophy lacked a proactive approach and held the potential that if a product made it to final lot inspection and defects were found, the entire lot was rejected, sometimes at considerable cost and waste of personnel and material resources. The quality problem was not addressed and corrected during manufacture. With the pressures of raw material costs and limited resources, most businesses adopted the more recent quality philosophy to limit adding value to a product or activity as soon as poor quality is observed, and to implement check steps along the process to observe and correct as early as possible. This led to the *quality assurance* or *total quality control* philosophy.

The key to any process, product, or service is to understand and provide what the customer wants. The major process, product, or service characteristics are:

- Reliability

- Maintainability

- Safety

Various industries have lost market share due to a failure to adequately understand *customer requirements* and failure to translate those requirements into *specifications* that a company would need in order to be able to satisfy the customer requirements.

Specifications are typically an engineering function and are critical in the product design, planning, and realization phases.

In addition to physical, measurable specifications, there may be other customer expectations, such as appearance. This may be included as a quality requirement that the quality function would establish, document, and implement.

For instance, parameters for a pressure vessel must include not only the material and dimensions but operating, environmental, safety, reliability, and maintainability requirements. Moreover, external standards outside of the company and the customer may dictate further requirements that the product must meet. They may be published federal environmental standards or statistical sampling standards, such as ASQ Z1.4-2003, or performance and testing method standards published by a professional association such as the American Society for Testing and Materials (ASTM).[1]

Chapter 5

E. Documentation Systems

> Identify and describe common elements and
> different types of documentation systems
> such as configuration management, quality
> manual, document control, etc. (Understand)
>
> **Body of Knowledge I.E**

To the quality professional—and other business support personnel—documentation systems are essential although often underappreciated tools. There are three purposes for a documentation system in an organization. These are to:

1. Guide individuals in the performance of their duties

2. Standardize the work processes throughout the organization

3. Provide a source of evidence regarding practices

New employees, people recently reassigned to fulfill new responsibilities, and candidates for appointment to new jobs need a reference for the policies, materials, tools, resources, and practices necessary to do their new job. Documentation serves as the basis for training new employees in both formal training settings and on-the-job training. Documentation also provides more seasoned individuals a reference for the work they are doing.

For an organization to function effectively, the practices of all of the individuals must be standardized so that others will know what to expect from each person. Failure to standardize is one of the most prevalent reasons for confusion and friction in processes involving multiple departments, functions, or teams. As will be stressed below in the discussion of configuration management, the documentation system must permit changes so that improvements can be made and to serve the unique needs of specific customers, but the documentation system must standardize the common elements of the work being done by everyone.

Finally, the documentation system serves as a source of evidence of practices in which the organization is engaged. Documents must be available to support appropriate responses under the following conditions:

1. When legal or regulatory challenges are made

2. When marketing and sales requires proof of the organization's capabilities

3. When senior managers and project managers need to plan and estimate future performance

4. When quality personnel need to spot opportunities for improvement, sources of waste, and problems

Nearly all quality certification programs require the organization being certified to demonstrate that it employs a documentation system to control its documents. For example, ISO 9001:2000[1] criteria include a requirement that registered organizations possess an extensive documentation system. ISO 9001:2000 requires that the following documents and records be maintained:

4.2.3 Control of Documents

The QMS [Quality Management System] documents, including quality records, must be controlled. A documented procedure must be established to approve, review, update, and identify the current revision level of QMS documents. The documented procedure must ensure that relevant versions are available at points of use. Documents must be legible, readily identifiable, and retrievable. Documents of external origin must be identified and controlled; and prevent unintended use of obsolete documents.

4.2.4 Control of Quality Records

The organization must control and maintain QMS records to provide evidence of conformance to requirements and of effective QMS operation. A documented procedure must be established for the identification, storage, retrieval, protection, retention time, and disposition.

NEED FOR A FLEXIBLE AND CURRENT DOCUMENTATION SYSTEM

No documentation system is perfect. While every effort must be made to ensure that the documents in the system are complete, the manner in which jobs are performed changes over time and the documents describing the work are seldom 100 percent up to date. Accordingly, the documentation system must be flexible.

When individuals retrieve a document and discover that it is out of date, they will be reluctant to retrieve documents from the documentation system in the future. It is essential that the documents in the system are kept current even though it is often an expensive and difficult activity to keep the documents up to date.

LIBRARY

Every organization needs to have a library of documents necessary for the organization to perform its functions. ISO 9001:2000 has several documentation requirements including a quality manual. Items typically contained in the quality manual or in the organization's library may include:

- Policies and procedures
- Work instructions
- Contracts
- Design inputs
- Process details
- Engineering changes
- Final inspection data
- Warranty changes
- Legal agreements
- History of defects
- Disposition of customer complaints
- Warranties
- Document descriptions and blueprints
- Histories of supplier performance

There is an expense associated with archiving and maintaining the library. The cost for building and maintaining the library must be borne by the organization. Companies that have tried to eliminate the costs associated with the documentation system have often discovered that this was a serious mistake when they were later unable tc respond to an environmental agency challenge or were sued and unable to produce evidence of their practices in court.

CONFIGURATION MANAGEMENT

Organizations that produce products (hardware or software) control their documents with a configuration management (CM) program. Examples of organizations that produce products include:

- A manufacturing firm that produces electronic components
- A software firm that makes and sells computer games
- A consulting firm that produces training materials for courses to prepare candidates for ASQ certifications

The organization's documentation system must be flexible to permit changes to be made to documents when changes are made to products or their design. To allow documents to be modified in a systematic and intelligent fashion, most product-producing organizations adopt configuration management. The CM system

Table 5.1   Example document traceability matrix.

CPI Program 7267.39 Requirements						
Control number	Requirement	ORD	SS	CPCI specification	VCRM	Test document
CN 301	Sense change signal	3.1.7	2.1.2	3.4.1.2	4.1.2	4.1.2
CN 302	Assemble control data	3.1.8	2.1.5	3.1.3	4.1.2.1	4.1.2.1
CN 303	Transfer data to external module	3.1.9	2.7.6	3.4.1.3	4.1.3	4.1.3
CN 304	Calculate new delay time	3.1.10	2.1.4	3.7.9	4.1.4	4.1.4
CN 305	Inform control module	3.1.11	2.1.7	3.4.1.4	4.1.5	4.1.5
CN 306	Return to search mode	3.1.12	2.1.8	3.4.1.5	4.1.5.1	4.1.5.1

ensures that all the documentation describing the product reflects all of the changes that have been made to it during design, development, or installation.

For example, if an electronic component manufacturing firm designs a capacitor to satisfy a customer requirement for handling 12.5 amps but later is informed that this requirement must be increased to 13.0 amps, the customer and the company must be certain that this change in the requirement has been reflected everywhere that it appears. Of course the requirement will be changed in the operational requirement document (ORD) and the system specification (SS). It must also be changed in the component specification, in the installation instructions, and in any training documents that have been generated. The purpose of a configuration management system is to make sure the requirements are changed in all the documents, not just some of them. Failure to make the changes universally will result in confusion and perhaps serious failure.

Configuration management is an established discipline. Most professional organizations have adopted configuration management to control their policies, standards, practices, and manuals. For example, ASQ practices CM to control its handbooks, certification requirements, bodies of knowledge, and other documents. The IEEE also has a standard, 729 1983, for software CM.[2]

CM is often administered by a *configuration control board* in the organization. This is a team of personnel responsible for ensuring that document revisions are universally incorporated throughout all the organization's documents. At a minimum there must be a *document traceability matrix,* a spreadsheet that traces each requirement. An example of a simplified requirements document traceability matrix is shown in Table 5.1.

DOCUMENTATION CONTROL

All organizations should employ some version of a documentation control (DC) system. The DC system is very much like the CM system that manufacturers and product providers require. But even if the organization does not provide customers with products, it is highly desirable that the organization's documents all be organized and kept current and available. Revising the documents should follow a simple procedure that ensures that all of the agencies and individuals who are affected by any changes proposed have an opportunity to review the revisions before they are made, and that the changes, once they are made, are reflected universally in all relevant documents. ISO 9001:2000[3] requires that revisions to quality documents be identified along with the revision level. Change markings can occur throughout the document. For example, additions may be shown underlined and deleted words can be shown as lined through—features all word processors provide.

A word of caution: Individuals should be discouraged from making hard copies of documents to keep handy since these documents will not reflect the latest changes that have been made. Once a documentation control system is in place, it should be adhered to by everyone in the organization, and if it has been automated, the continued use of hard copies should be discouraged.

Chapter 6

F. Audits

1. AUDIT TYPES

> Define and describe various audit types:
> internal, external, system, product, and
> process. (Understand)
>
> **Body of Knowledge I.F.1**

An audit is an evaluation of an organization, its processes, or its products. Audits are performed by an independent and objective person or team who are called auditors. The purpose of an audit is to provide decision makers such as senior managers with unbiased information on which decisions can be made. Audits can be concerned with a single aspect of the business, such as financial status, or they can assess performance in many or all of the functions engaged in a business process, such as in an audit of a supplier's operation to determine if a contract award should be made. Audits collect data from the auditors' observations and data that is maintained by the audited organization, and this data is compared to standards, regulations, or practices that have been established for the purpose of assessing performance.

There are two major classifications of audits:

- *Internal audits* are entirely under the purview of the company being audited. The auditors may be specially trained members of the audited firm's staff or they may be external consultants or contracted personnel who are capable of collecting and analyzing the necessary audit data. The results of an internal audit are delivered to the audited company's management so that these managers can make decisions or direct changes based on the audit results. Nearly always if a firm is anticipating an external audit it will arrange for an internal audit to prepare for the external one. Internal auditors should be independent of the function that is being audited to ensure that the audit is unbiased.

- *External audits* are performed by and are entirely under the purview of an agency outside the audited company, such as a government agency or an independent auditing firm. There are regional, national, and international accounting firms and auditing corporations who are capable of performing extensive and thorough audits of finance, quality, safety, and compliance with regulations such as environmental and hazardous waste disposal restrictions. The board of directors of a company may initiate an external audit or an external agency may require an external audit to grant a certification or for some other reason. If the external agency requiring an audit contracts with an outside source—neither the audited firm nor the requesting agency—the audit is referred to as a *third-party audit*. (The *first party* in this case is the auditee; the *second party* is the audit authority, usually the entity requiring the audit be performed.) Third-party audits are performed by auditors who are independent of both the auditee and the requesting agency. An ISO 9001:2000 audit is a third-party audit performed by a registrar.

Auditing is an integral part of any organization's quality control program. ISO 9001:2000 requires external audits to register a company as compliant with its standards, and a fundamental requirement of ISO 9001:2000 is that the candidate firm demonstrate that it conducts routine internal audits.

Ordinarily, the auditee is informed that the audit will occur and when. This is necessary since the auditors will need to interview workers, staff, and managers in the audited organization and will need to look at data that may not be immediately available. In order to minimize the time spent auditing (which impacts routine performance of the audited organization), it is nearly always necessary to notify the auditee in advance. However, sometimes an unannounced audit is appropriate. A third-party or external audit team can arrive and state that they are present to conduct an unannounced audit. Firms that are subject to such unannounced audits must be prepared at all times for such an interruption of their normal duties. They should perform frequent, routine internal audits and have sufficient margin in their schedules and budgets to accommodate these otherwise disruptive audits.

There are three types of audits:

1. Product audit

2. Process audit

3. System audit

The *product audit* assesses the final product or service to determine its quality before it is used by the customer. Product audits may be performed by the customer. Like other audits, product audits may be performed as internal or external audits. Product audits usually take less time than a process or system audit since they assess a single product at a time.

The process audit is an assessment of a single process. It is, accordingly, more complex than a product audit but ordinarily less difficult and time-consuming than a system audit. The process audit may be performed as an internal or external audit. The focus of a process audit is a single function such as all of the steps in manufacturing a product or all of the steps in servicing a type of customer.

The system audit is an assessment of the entire system or all of the functions of an organization. These audits typically cannot be conducted rapidly. It is not uncommon for a full system audit to take from two to five days and include several auditors who engage dozens of individuals from the audited organization's staff in interviews. Two examples of system audits are (1) a formal ISO 9001:2000 registration audit and (2) a pre-award audit for a potential new supplier. System audits are often performed by third parties and are usually ultimately external audits. However, as has already been suggested, organizations often perform internal audits even of the entire system in order to better prepare for an external full-system audit.

For more information about quality audits, consult *The ASQ Auditing Handbook.*[1]

2. AUDIT PROCESS

> Describe various elements, including audit preparation, performance, record keeping, and closure. (Understand)
>
> **Body of Knowledge I.F.2**

The audit process consists of three phases:

1. Preparation
2. Execution
3. Closure

Preparation

When preparing for any quality audit, the following actions should be taken:

Formal Authorization. The first step in the audit process is obtaining formal authorization from someone in a leadership position, usually someone in upper management, for the audit. This authorization provides the audit team leader the authority to execute the audit, the audited organization the permission to cooperate with the auditors, and the organization the resources necessary to plan and execute the audit.

Establish the Purpose of the Audit. Audits are conducted to improve processes, implement a continuous improvement program, take corrective action, and for many other reasons. Ultimately, the audit purpose must be made clear to the auditors and auditees alike.

Determine the Audit Type. There are three audit types: product audits, process audits, and system audits. In addition, audits may be customized to specific auditing needs.

Establish the Scope of the Audit. To avoid the audit's scope growing over time ("scope creep"), the scope should be clearly defined during the preparation phase. Once the audit is under way, the scope may change, but a clear understanding of the scope that was agreed to at the beginning of the project will provide a basis for orderly and systematic change rather than chaos.

This task of establishing the scope of the audit is not always easy since most people are not comfortable estimating or predicting how long it will take to perform a job in the future. The scope of the new audit should be based on past experiences—consult the records of past audits to determine how long it took to perform them and then adjust the estimates derived from the actual time and effort required to perform them in the past based on what is known about the new audit. For example, if a similar audit was performed in the past by three people in three days, but the three people felt that was not sufficient time, the new audit may be scheduled to take three people four days.

Identify the Necessary Resources. The personnel, equipment, materials, documentation, training, tools, and procedures that are required to perform the audit must be stated.

Form a Team. The auditors should be identified and then provided an opportunity to develop into a team before the audit commences. It may be useful to employ a facilitator to expedite the team-building process.

Assign Team Roles and Responsibilities. Once the auditors have been identified, it is necessary to assign them to specific roles and to identify each of their responsibilities. The team leader is identified and given the responsibility for leading the team, facilitating the audit process, and constituting the remainder of the audit team. Other audit team members are normally assigned responsibilities within their areas of expertise in auditing methods and practices or in specific technical areas. The audit team leader should attempt to assign individuals to duties they are capable of performing and arrange for instruction or training in areas that the team members will need during the audit.

Identify the Audit Requirements. Specific audit activities and deliverables, including audit forms to be completed, working papers, detailed questions to be asked during interviews, debriefing activities, and reports, constitute the audit requirements.

Schedule the Audit Activities. A final component of the audit plan is a detailed audit time schedule. This schedule must include not only dates and times for the delivery of required documents and reports but must also identify milestones. Milestones are checkpoints in the schedule when the status of the audit activities can be assessed to determine progress.

Execution

The performance of the audit constitutes the execution phase. The following activities are required during this phase:

- Managing and administering the audit process
  - Managing scope creep (the growth of the scope of the audit from what was originally agreed to) with continuous communication to all involved parties
  - Conducting all necessary meetings and individual interviews
- Creating working papers
- Conducting a kick-off meeting
- Collecting data
  - Interviews
  - Reviewing documents maintained by the audited organization
- Analyzing data to find patterns and trends and to create findings
- Conducting an exit meeting

The documentation created during the execution phase includes:

- Working papers
- Status reports
- Preliminary findings
- Results
- Debriefings
- Closure reports
- Corrective action requests
- Follow-up report forms
- Forms for corrective actions

Closure

An audit is considered closed when both parties involved in the audit are satisfied that all corrective actions have been successfully completed. The following reports are generated during the closure phase:

- Final report
- Final report with exceptions
- Terms and conditions for closure (agreements between the agent authorizing the audit and the auditee regarding what constitutes closure)

3. ROLES AND RESPONSIBILITIES

> Identify and define roles and responsibilities of audit participants (lead auditor, audit team member, client, and auditee). (Understand)
>
> **Body of Knowledge I.F.3**

The roles and responsibilities for each participant in an audit are listed in this section.

Agent Authorizing the Audit

This may be an external agency, registrar, certifying body, board of directors, customer purchasing department, process manager, or product manager.

- Authorize the audit
- State the audit purpose
- Determine the audit type
- Establish the scope of the audit
- Approve the schedule
- Approve the closure terms and conditions

Senior Management

- Authorize resources
- Direct the organization to cooperate with and assist the auditors

Audit Team Leader

- Create the audit plan for approval by senior management and the agent authorizing the audit
- Manage and administer the audit process
- Form the team
- Assign the team members' roles and responsibilities
- State the audit requirements
- Schedule the audit activities, deliverables, and milestones
- Administer documents during the execution phase
- Call and lead the kickoff, progress, and exit meetings

Auditor

- Participate in planning the audit
- Participate in team-building activities
- Collect data
- Analyze data
- Prepare working papers
- Attend meetings
- Give reports
- Document findings

Auditee

- Be available when the audit is scheduled
- Provide all evidence requested by the auditor such as documents, work in progress, and computer data normally employed in the performance of the auditee's duties
- Answer all questions posed by the auditor (normally, auditees should not volunteer information not requested by the auditor)

Chapter 7

G. Teams

1. TYPES OF TEAMS

> Distinguish between various types of teams
> such as process improvement, work group,
> self-managed, temporary/ad hoc, cellular, etc.
> (Analyze)
>
> **Body of Knowledge I.G.1**

A team is a group of people who perform interdependent tasks to work toward a common mission.

Some teams have a limited life, for example, a design team developing a new product or a process improvement team organized to solve a particular problem. Others are ongoing, such as a department team that meets regularly to review goals, activities, and performance.

Understanding the many interrelationships that exist between organizational units and processes, and the impact of these relationships on quality, productivity, and cost, makes the value of teams apparent.

Many of today's team concepts originated in the United States during the 1970s through the use of quality circles or employee involvement initiatives. But the initiatives were often seen as separate from normal work activities, not as being integrated with normal work activites.

Team design has since evolved into a broader concept that includes many types of teams formed for different purposes.[1]

Although they may take different names in different industries or organizations, this text presents seven types of teams:

- Process improvement teams
- Self-managed teams
- Temporary/ad hoc teams
- Work groups

- Cellular teams

- Special project teams

- Virtual teams

Process Improvement Teams. These are project teams that focus on improving or developing specific business processes and are more likely to be trying to accomplish breakthrough-level improvement (as explained in Chapter 10). Process improvement teams:

- Are typically cross-functional—different functions/skills related to the process

- Have a management sponsor who:

  - Charters the team

  - Ensures that the team has appropriate resources

  - Ensures that the team has organizational support

- Achieve a specific goal

- Are guided by a well-defined project plan

- Have a negotiated beginning and end

The leader of a process improvement team is usually selected by the project's sponsor, and the team meets on a regular basis to:

- Plan activities that will occur outside the meeting

- Review information gathered since the previous meeting

- Make decisions to implement process changes

In cases where a team member does not have the expertise, an independent facilitator (not associated with the specific process) may be asked to work with the team.

There are many organizations who have implemented the *kaizen* approach for productivity improvement with its main goal of eliminating waste (kaizen is discussed further in Chapter 10). An organization taking the kaizen approach may also use *kaizen events* or *kaizen blitzes* to focus on process improvement through an accelerated team process that focuses on a narrow project implementing the recommended changes within a three- to five-day period. In these cases, the team facilitator typically has more authority than with most teams.[2]

A quality process analyst (QPA) will often be a member of a process improvement team. Because it's challenging enough to support the development and deployment of organizational change, it's always good to first ensure that the mission has been accepted and approved by upper management and that it's purpose is well defined and measurable. Some of the challenges a QPA may experience include dedicating enough time to work on a specific project and employees being reluctant to provide data or information. Once a solution is provided, employees

• Have a well-thought-out vision of how the team will fit into the entire organization	• Set performance expectations for the team
• Ensure that the organizational culture can and will support the team	• Develop a method for providing feedback to the team
• Ensure that the organization has the resources to commit to this type of change	• Set boundaries in which the team will be allowed to operate
• Train team members in skills that will allow them to function together.	• Recognize that the self-managed team is not leaderless and that they may need management intervention

Figure 7.1   Guidelines for successful self-managed teams.

Source: ASQ's Foundations in Quality Self-Directed Learning Series, Certified Quality Manager, Module 1 (Milwaukee: ASQ Quality Press, 2001): 1-58.

may be unwilling to embrace the change that needs to happen in order for the solution to be effective for quality improvement. To promote the change process, the QPA can communicate the positive benefits of the change to their peers. By including and encouraging fellow employees to participate when possible, they will be more inclined to support the mission.

Self-Managed Teams. Also called self-directed teams or high-performance teams because of their broader scope of responsibilities. They are groups of employees involved in directly managing the day-to-day operation of their particular process or department. Some of their responsibilities are those that are traditionally held for managers. Self-managed teams:

- Are authorized to make decisions on a wide range of issues (for example, safety, quality, maintenance, scheduling, and personnel)
- Set goals
- Allocate assignments
- Resolve conflict[2]

Because these employees have broader responsibilities, they assume ownership of a work process. The leader of a self-managed team is usually selected by the team members and in many cases the role is rotated among the members over time. Other positions entailing scheduling or safety, for example, are also often rotated among the team members. Self-managed teams require up-front planning, support structures, and nurturing. Figure 7.1 highlights a number of guidelines that can increase the success of self-managed teams.

With the various roles that self-managed teams take on, training is necessary to move from a traditional work environment to a self-managed work environment. Often even the managers need training to transition from their current role to the new support role. For self-managed teams to be successful, they will need training. For example, one Fortune 500 company provides training in the following subjects:

- How to maintain focus on the customer

- How to develop a vision and mission that are integrated with the larger organizational mission

- Understanding roles and operating guidelines

- Skills required for working together to make decisions, plan work, resolve differences, and conduct meetings

- The concepts and strategies for empowerment

- Setting goals (objectives) and solving problems for continuous improvement[3]

Attempts to transition from a traditional work environment to a self-managed work environment require culture change. This culture change can be very dramatic and is best approached by first implementing cross-functional process improvement teams and/or work teams as learning mechanisms. Self-managed teams are more likely to be successful as part of a start-up of a new facility.

Temporary/Ad Hoc Teams. These types of teams are often initiated to address a specific problem or situation. The team would be formed based on the work to be performed. They may not use the usual formal structure (for example, agenda, regular meeting frequency), but the same general principles and processes still apply. Organizations open to using ad hoc teams must be flexible.

An empowered organization will use temporary teams when deemed useful to carry out a short-term mission, for example conducting internal audits. Some companies, for example engineering design companies, may form ad hoc teams with other engineering, design, and/or manufacturing experts to meet their customers' needs. Various situations can arise that require immediate and dedicated attention, in which case ad hoc teams would have to be formed, such as:

- Preparing for a major customer's audit team coming to do an assessment on the adequacy of your quality management system

- Disaster relief—fire occurring and the process needs to relocate

- Recovery from your organization's information management computer system being compromised by an outside virus

Usually, management will designate a person to:

- Form a small, cross-functional team

- Address the situation to the team

- Call upon other technical expertise as needed

Work Groups. Also called *natural teams*, work groups bring together employees in a participative environment. Work groups have an ongoing charter to continually improve the work processes over which they have direct responsibility and ownership. The team leader is usually the individual responsible for the work area, such as a department supervisor.

Work groups function similarly to quality circles, in which department personnel meet weekly or monthly to review performance of their process. They may:

- Monitor feedback from customers

- Track internal performance measures of processes

- Track internal performance measures of suppliers

Work groups are similar to process improvement teams, with the key differences being that they are neither cross-functional nor temporary. Should they need additional support a facilitator is usually available, and outside resources may be brought in on a temporary basis when necessary.

Since work groups are an ongoing organizational structure, it is critical that the organizational systems and values support the effort. Certain basic elements should be considered when an organization is attempting to initiate work groups:

- Top management support

- Clear communication

- Improvement objectives and expectations

- Team training

- Appropriate competencies

- Supportive compensation and performance appraisal systems

Other issues to consider include:

- Team's scope of responsibility and authority

- Degree of autonomy

- Information needed by team and where obtained

- Decision-making process

- Performance measures and success factors

- Recognition and rewards for performance

- Competencies that must be developed

- Selection process for leaders and facilitators

- Risk management process

An implementation plan including the necessary support systems should be defined before initiating such groups. If work groups are being implemented in an existing organization where a participative management style has not been used before, a pilot program in a department where they are likely to succeed is highly recommended.[4]

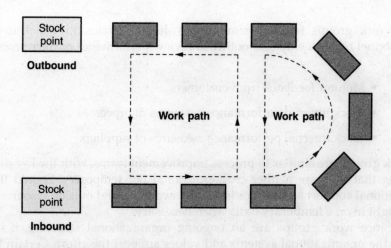

Figure 7.2   Typical U–shaped cell layout.

Source: R. T. Westcott, *The Certified Manager of Quality/Organizational Excellence Handbook,* 3rd ed. (Milwaukee: ASQ Quality Press, 2006): 403.

Cellular Teams. Processes can be organized into *cellular teams,* where the layout of workstations and/or machines used in a designated part of a process is typically arranged in a U-shaped configuration (see Figure 7.2). This allows operators to proceed with a part from machine to machine to perform a series of sequential steps to produce a whole product or a major subassembly.

The team that operates the cell is usually totally cross-trained in every step in the series. Team effectiveness depends on coordination, timing, and cooperation. Team members' competencies must be as closely matched as possible to maintain a reasonable and consistent work pace.

Cell team members usually assume ownership and responsibility for establishing and improving the work processes. Leadership of the cellular team may be by a person designated as lead operator or a similar title. In some cases, the role of lead operator may be rotated among the members. The lead operator is usually responsible for training new team members, reporting team output, and balancing the flow of work through the process. A cellular team is a specialized form of a self-directed work group.[5]

Special Project Teams. This type of team is used when a need develops to form a long-duration, totally dedicated project team to implement a major new product, system, or process.

Some examples are:

- Reengineering an entire process

- Relocating a major segment of an operation

- Mergers, acquisitions, or divestitures

- Replacing a major information system

- Entering a new market with a new product
- Preparing to apply for the Baldrige Award

Such special project teams may operate as a parallel organization during their existence. They may even be located away from the main organization or exempt from some of the constraints of the main organization. The core team members are usually drafted for the duration of the project. Persons with additional expertise may be called into the team on a temporary, as-needed basis. Usually the project is headed by an experienced project manager. Depending on the nature of the team's objectives, external specialists or consultants may be retained to augment the core competencies of the team members.[6]

Virtual Teams. In today's global and electronic business environment, it may be expedient to have team members who do not work in the same geographic location. These virtual teams also require many of the same roles and processes, but the substitution of electronic communications (for example, e-mail, videoconferencing) for face-to-face meetings provides an additional challenge to team leadership. A key competency of team members is the ability and motivation to self-manage their time and priorities.

Virtual teams have special needs, some of which are:

- Telephones, pagers, local area networks, satellites, videoconferencing
- Computers, high-speed and wireless connections, Internet technology, e-mail, Web-based chat rooms/forums
- Software for communications, meeting facilitation, time management, project management, groupware, access to online databases

The benefits of virtual teams are:

- Team members can work from anywhere, at any time of day
- Team members are selected for their competence, not for their geographic location
- Expenses may be reduced or eliminated, such as travel, lodging, parking, and renting/leasing or owning a building[7]

2. TEAM-BUILDING TECHNIQUES

> Define basic steps in team-building such as introductory meeting for team members to share information about themselves, the use of ice-breaker activities to enhance team membership, the need for developing a common vision and agreement on team objectives, etc. (Apply)
>
> **Body of Knowledge I.G.2**

In *The Certified Manager of Quality/Organizational Excellence Handbook,* Russ Westcott explains that team processes can be considered in two major groups of components: task-type and maintenance-type. Team-building techniques enhance both groups. *Task-type* processes keep teams focused and on the path toward meeting their goals. *Maintenance-type* processes help maintain or protect the well-being of the relationships among team members.

Key task-type components include:

- Documenting and reviewing the team's objective

- Having and staying on an agenda for every team meeting

- Defining or selecting and following a technical process that fits the particular project mission

- Using decision-making techniques appropriate to the situation (explained further in this section)

- Defining action items, responsibilities, and timing, and holding team members accountable[8]

Maintenance-type components also have a dramatic impact on team performance. The dynamics of the interactions between team members are just as critical as the tasks that keep the team on the path toward meeting its objectives. Team maintenance helps alleviate the problems created by the fact that each team member has his or her own perspectives and priorities. These are shaped by individual personalities, current issues in each member's personal life, and the attitudes of both the formal and informal groups within the organization to which the team member may belong. Although the team may have specific objectives and agendas, each individual may interpret them differently.

The first team meeting can set the tone for the entire team effort. With good planning, the first meeting will address:

- Team member introductions where members share information about themselves

- Team members getting acquainted

- Having the sponsor attend the meeting and emphasize the importance of the meeting

- The team's mission and scope

- Defining the structure for future meetings

- Defining or selecting the technical process improvement methodology/model to be used

- Drafting the team's objectives

- Defining/reviewing the project plan and schedule

- Setting the meeting ground rules (norms)

- Working out decision-making issues

- Clarifying team members' roles[9]

Handling Team Introductions

There are various methods of helping team members approach their initial meeting to ease the anxiety they might be having about working with one another. The manager can conduct team introductions by asking team members to share basic information with the group, such as name and job, thing they like best about their job, or an interest outside of work.

Conducting special exercises, or ice-breakers, during the first meeting, makes it easier for the team members to feel more comfortable in their new environment. They promote spontaneous interaction and conversation. There are plenty of training materials to help in the area of ice-breakers.

Another idea is to initiate building camaraderie among the team members by coming up with a team 'name.' Allowing the team members to share their individual good or bad experiences of working on other teams is a great way to expose their personalities and values. This kind of sharing can bring the team members closer together.

Setting Team Goals and Objectives

"Team goals must flow from organizational goals and departmental goals to ensure that team members and the entire organization are moving in the same direction.

Objectives are defined as quantitative statements of future expectations that include a deadline for completion. Team objectives should be approached the S.M.A.R.T. W.A.Y.,"[10] as described in Figure 7.3.

Preventing Problems

Team members need basic guidelines to add predictability to their work environment along with creating a sense of safety around team interactions. Having ground rules or norms, working out decision-making issues, and having assigned roles are common ways of preventing team dynamic problems.

Specific:	Focus on *specific* needs and opportunities
Measurable:	Make each objective *measurable*
Achievable:	Ensure that objectives are challenging yet *achievable*
Realistic:	Stretching is fine, but the objective should be *realistic*
Time frame:	Every objective should indicate a *time frame* for achievement
Worthwhile:	Every objective should be *worth doing*
Assigned:	*Assign* responsibility for each objective
Yield results:	Ensure that all objectives *yield* desired results

Figure 7.3 Objectives made the S.M.A.R.T. W.A.Y.

Source: ASQ's Foundations in Quality Self-Directed Learning Series, Certified Quality Manager, Module 1 (Milwaukee: ASQ Quality Press, 2001): 1-66. Reproduced with permission from R. T. Westcott & Associates.

Table 7.1 Meeting ground rules.

Area	Topics to consider
Logistics	• Regular meeting time and place • Procedure for notifying members of meetings • Responsibilities for taking notes, setting up the room, and so on
Attendance	• Legitimate reasons for missing a meeting • Procedure for informing the team leader when a meeting will be missed • Procedure for bringing absent members up to speed
Promptness	• Acceptable definition of "on time" • Value of promptness • Ways to encourage promptness
Participation	• Basic conversational courtesies (for example, listening attentively, holding one conversation at a time) • Tolerable interruptions (for example, phone calls, operational emergencies) • Confidentiality guidelines • Value of timely task completion

Source: ASQ's Foundations in Quality Self-Directed Learning Series, Certified Quality Manager, Module 1 (Milwaukee: ASQ Quality Press, 2001): 1-65.

Ground Rules

These are group norms regarding how meetings will be run, how team members will interact, and what kind of behavior is acceptable. Each member is expected to respect these rules during the project's duration. Some areas to consider when establishing ground rules are provided in Table 7.1.

Decision-Making Method

"Consensus decision making is the process recommended for major project issues. The approach is more time-consuming, and demanding, but provides the most thorough and cohesive results. It allows the team to work out different points of view on an issue and come to a conclusion that is acceptable. Consensus means that the decision is acceptable enough that all team members will support it. To achieve consensus, all team members must be allowed to express their views and must also objectively listen to the views of others. The decision and alternatives should be discussed until consensus is reached."[11]

Team Development

Although not in the Body of Knowledge for the ASQ CQPA certification, it's worth noting the stages that a team goes through as the members work together over time and the team grows and matures. Understanding these development stages is valuable for effective management of the team process.

- *Stage 1: Forming.* The team usually clarifies its purpose, identifies roles for each member, and defines rules of acceptable behavior (often called *norms*).

- *Stage 2: Storming.* Team members are still thinking primarily as individuals and so basing decisions on their individual experiences rather than pooling information with other members. Collaboration is not yet the norm, leading to behavior that may involve confrontations.

- *Stage 3: Norming.* The team's focus has shifted to meeting the team-related challenges. Team members are more agreeable to working out their differences for the sake of the team, resulting in more cooperation and dialogue.

- *Stage 4: Performing.* The team has a better appreciation of their processes. The team's members are supportive of each other, allowing the team to make significant progress toward achieving its goals.

If needed, team members should have training on how to interact with one another in a positive manner, deal with difficult people or situations, contribute to accomplishing the team's goals, and give and receive constructive feedback. With a basic understanding of these techniques, team members will be more productive, enhancing the development and performance of their team.

3. ROLES AND RESPONSIBILITIES

> Explain the various team roles and responsibilities, such as sponsor, champion, facilitator, team leader, and team member, and responsibilities, with regard to various group dynamics, such as recognizing hidden agendas, handling distractions and disruptive behavior, keeping on task, etc. (Understand)
>
> **Body of Knowledge I.G.3**

Of the seven team roles described in Table 7.2, those of timekeeper and scribe are the only ones that are optional, depending on the mission of the team. While the remaining five roles are essential, they may, as discussed above, be combined in a variety of ways. However, the most crucial roles for the success of the team, once it is formed, are those of the team leader and the facilitator. The team leader is responsible for the content, the work done by the team. The facilitator is responsible for ensuring that the work process of the team is the best for the stage and situation the team is in.[12]

Table 7.2 Team roles, responsibilities, and performance attributes.

Role name	Responsibility	Definition	Attributes of good role performance
Champion	Advocate	The person initiating a concept or idea for change/improvement	• Is dedicated to seeing it implemented • Holds absolute belief it is the right thing to do • Has perseverance and stamina
Sponsor	Backer; risk taker	The person who supports a team's plans, activities, and outcomes	• Believes in the concept/idea • Has sound business acumen • Is willing to take risk and responsibility for outcomes • Has authority to approve needed resources • Will be listened to by upper management
Team leader	Change agent; chair; head	A person who: • Staffs the team or provides input for staffing requirements • Strives to bring about change/improvement through the team's outcomes • Is entrusted by followers to lead them • Has the authority for and directs the efforts of the team • Participates as a team member • Coaches team members in developing or enhancing necessary competencies • Communicates with management about the team's progress and needs • Handles the logistics of team meetings • Takes responsibility for team records	• Is committed to the team's mission and objectives • Has experience in planning, organizing, staffing, controlling, and directing • Is capable of creating and maintaining channels that enable members to do their work • Is capable of gaining the respect of team members; serves as a role model • Is firm, fair, and factual in dealing with a team of diverse individuals • Facilitates discussion without dominating • Actively listens • Empowers team members to the extent possible within the organization's culture • Supports all team members equally • Respects each team member's individuality
Facilitator	Helper; trainer; adviser; coach	A person who: • Observes the team's processes and team members' interactions and suggests process changes to facilitate positive movement toward the team's goals and objectives • Intervenes if discussion develops into multiple conversations • Intervenes to skillfully prevent an individual from dominating the discussion or to engage an overlooked individual in the discussion • Assists the team leader in bringing discussions to a close • May provide training in team building, conflict management, and so forth	• Is trained in facilitating skills • Is respected by team members • Is tactful • Knows when and when not to intervene • Deals with the team's process, not content • Respects the team leader and does not override his or her responsibility • Respects confidential information shared by individuals or the team as a whole • Will not accept facilitator role if expected to report to management information that is proprietary to the team • Will abide by the ASQ Code of Ethics

Continued

Part I.G.3

Continued

Role name	Responsibility	Definition	Attributes of good role performance
Timekeeper	Gatekeeper; monitor	A person designated by the team to watch the use of allocated time and remind the team members when their time objective may be in jeopardy	• Is capable of assisting the team leader in keeping the team meeting within the predetermined time limitations • Is sufficiently assertive to intervene in discussions when the time allocation is in jeopardy • Is capable of participating as a member while still serving as a timekeeper
Scribe	Recorder; note taker	A person designated by the team to record critical data from team meetings. Formal "minutes" of the meetings may be published and distributed to interested parties.	• Is capable of capturing on paper or electronically the main points and decisions made in a team meeting and providing a complete, accurate, and legible document (or formal minutes) for the team's records • Is sufficiently assertive to intervene in discussions to clarify a point or decision in order to record it accurately • Is capable of participating as a member while still serving as a scribe
Team members	Participants; subject matter experts	The persons selected to work together to bring about a change/improvement, achieving this in a created environment of mutual respect, sharing of expertise, cooperation, and support	• Are willing to commit to the purpose of the team • Are able to express ideas, opinions, and suggestions in a nonthreatening manner • Are capable of listening attentively to other team members • Are receptive to new ideas and suggestions • Are even-tempered and able to handle stress and cope with problems openly • Are competent in one or more fields of expertise needed by the team • Have favorable performance records • Are willing to function as team members and forfeit "star" status

Source: R. T. Westcott, *The Certified Manager of Quality/Organizational Excellence Handbook*, 3rd ed. (Milwaukee: ASQ Quality Press, 2006): 75–76.

The need for a trained facilitator should be considered when:

- The team has been meeting for some time and is incapable of resolving conflicting issues

- A new member has been added, thus upsetting established relationships

- A key contributor has been lost to the group

- There are other factors, such as lack of sufficient resources, the threat of project cancellation, or a major change in requirements, with the potential of disrupting the smooth functioning of the team

Supplementing the team with on-call experts can often compensate for a shortfall in either the number of members or members' competencies. Selected members must willingly share their expertise, listen attentively, abide by the team's ground rules, and support all team decisions.

It's helpful to select a team member to serve as a timekeeper until the team has become more adept at self-monitoring its use of time. The role of timekeeper is often rotated, giving consideration to whether the selected member has a full role to play in the deliberations at a particular meeting.

Some team missions require very formal documentation, and a scribe or note taker may be needed. This role can be distracting for a member whose full attention is needed on the topics under discussion. For this reason, an assistant, not a regular member of the team, is sometimes assigned to take the minutes and publish them. Care should be taken not to select a team member for this role solely on the basis of that team member's gender or position in the organization.

Another function that should be assigned early in the development of a team is the responsibility of booking a meeting room, sending the invites, and ensuring that supplies are at hand in the meeting room.

All team members must adhere to expected standards of quality, responsibility, ethics, and confidentiality. (See "ASQ Code of Ethics," Chapter 1.) It is imperative that the most competent individuals available are selected for each role. See Table 7.2 for the attributes of good team role performance.

Very frequently a team must function in parallel with day-to-day assigned work and with the members not relieved of responsibility for their regularly assigned work. This, of course, places a burden and stress on the team members. The day-to-day work and the work of the team must both be conducted effectively. The inability to be in two places at once calls for innovative time management, conflict resolution, and negotiation skills.[13]

Dealing with Team Process Problems

Team members are most productive in an environment in which others are responsive and friendly, encourage contributions, and promote a sense of worth. Peter Scholtes spelled out 10 problems that frequently occur within teams and are typical of the types of situations for which team leaders and facilitators must be prepared. Following is the list along with recommended actions:[14]

Problem 1. Floundering or difficulty in starting or ending an activity. *Solution:* Redirect team to the project plan and written statement of purpose.

Problem 2. Team members attempt to influence the team process based on their position of authority in the organization. *Solution:* Talk to the members outside of the meeting; clarify the impact of their organizational role and the need for consensus and ask for cooperation and patience.

Problem 3. Participants who talk too much. *Solution:* Structure meetings so that everyone is encouraged to participate (for example, have members write down their opinions, then discuss them in the meeting one person at a time).

Problem 4. Participants who rarely speak. *Solution:* Practice gatekeeping by using phrases such as, "John, what's your view on this?" or divide tasks into individual assignments and have all members report.

Problem 5. Unquestioned acceptance of opinions as facts, or participants making opinions sound like facts. *Solution:* Do not be afraid to ask whether this is an opinion or a fact. Ask for supporting data.

Problem 6. Rushing to get to a solution before the problem-solving process is worked through. *Solution:* Remind the group of the cost of jumping to conclusions.

Problem 7. Attempting to explain other members' motives. *Solution:* Ask the other person to clarify.

Problem 8. Ignoring or ridiculing another's values or statements made. *Solution:* Emphasize the need for listening and understanding. Support the discounted person.

Problem 9. Digression/tangents creating unfocused conversations. *Solution:* Remind members of the written agenda and time estimates. Continually direct the conversation back on track. Remind team of its mission and the norms established at the first meeting.

Problem 10. Conflict involving personal matters. *Solution:* Request that these types of conflict be taken outside of the meeting. Reinforce ground rules.

Solutions to conflicts should be in the best interest of the team. Team members should be nonjudgmental, listening to team discussions and new ideas. Group feelings should be verbalized by periodically surfacing undercurrents or by giving feedback.

One important skill needed in working with teams is the ability to provide constructive feedback during and/or at the end of a meeting. Feedback is an important vehicle to help the team mature. This feedback can be provided by the facilitator or by team members.

There are two types of feedback: motivational and coaching. Motivational feedback must be constructive, that is, specific, sincere, clear, timely, and descriptive of what actually occurred in the meeting. Coaching or corrective feedback specifically states the improvements that need to be made. Scholtes provides the following guidelines for providing constructive feedback:[15]

Part I.G.3

- Be specific
- Make observations, not conclusions
- Share ideas or information, not advice
- Speak for yourself
- Restrict feedback to known things
- Avoid using labels
- Do not exaggerate
- Phrase the issue as a statement, not a question

Having a team do a self-evaluation at the end of each meeting (or occasionally) can be useful for helping the team to further develop their team skills and to take more responsibility for team progress. Team members can be asked to write down how well the team is doing on each of the norms (for example, on a scale of 1 to 5) and to list any additional norms they believe need to be added. A group discussion of the information can then result in revised norms and specific actions the team will take in the future to improve.

Chapter 8

H. Training Components

> Define and describe methods that can be used to
> train individuals on new or improved procedures
> and processes, and use various tools to measure the
> effectiveness of that training, such as feedback from
> training sessions, end-of-course test results, on-the-job
> behavior or performance changes, department or area
> performance improvements, etc. (Understand)
>
> **Body of Knowledge I.H**

TRAINING METHODS

There are important considerations in selecting a training method. Once training needs have been identified and objectives are in place, it's time to translate those training concepts into practice. Decisions on where and how the training is to be delivered need to happen. When selecting a delivery method, the following must be taken into account:

- Organizational culture

- Financial issues

- Administrative issues

- Instructor competencies

- Safety issues

Figure 8.1 outlines a checklist of important questions to consider when determining an appropriate delivery system for training.

On-the-Job or Off-the-Job Training

This is an important consideration when determining the appropriate location. Table 8.1 summarizes selection criteria used when deciding whether training should be away from the work environment to develop employee knowledge and skills or if on-the-job training is more appropriate.

Part I.H

- Training content
  - Abstract or concrete?
  - Technical or nontechnical?
  - Didactic or experiential?
  - Mandated or required (for example, by a regulatory body, or management in support of a strategic objective)?
- Learning constraints
  - Short or long time for learning mastery?
  - Short or long time to apply content?
- Training audience
  - Size—individual learners or groups?
  - Geographic dispersion—one location or multiple locations?
  - Education level—basic or advanced?
  - Training experience on a topic—new or refresher?
  - Internal employees or external personnel (for example, customers and suppliers)?
  - Low or high experience?
  - Low or high motivation?
- Instructor competence/train-the-trainer requirements (if applicable)
  - Professional or nonprofessional training capability?
  - Internal employee or external consultant?
- Training facilities
  - Ad hoc or dedicated training facilities?
  - On site or external?
  - Media/equipment available or required?
- Organizational preferences and capabilities
  - Self-directed or instructor-led?
  - Traditional modes or technology-driven?
- Budget practices for training expenditures
  - Departmental funding or companywide coverage?
  - Individually funded or budgeted line items?
- Evaluation provisions
  - Outputs (for example, new/enhanced skills, improved productivity)?
  - Outcomes (for example, improved customer satisfaction, increased profits, return on training investment)?

Figure 8.1 Considerations for selecting a training system.

Source: ASQ's Foundations in Quality Self-Directed Learning Series, Certified Quality Manager, Module 7 (Milwaukee: ASQ Quality Press, 2001): 7-26.

Table 8.1 Job training selection criteria.

Selection factors	Consider on-the-job training when . . .	Consider off-the-job training when . . .
Task frequency	Key tasks occur on a regular basis	Key tasks do not occur on a regular basis
Skill mastery	Skills can be mastered only over time with practice	Skills can be acquired rapidly with little practice
Subject matter/content	Content does not change frequently	Content requires frequent revisions
Numbers of prospective trainees	Few trainees are to be trained at the same time	Many employees have the same training needs and there are cost savings in training a group
Training facility	Required equipment and/or other resources are not available on site	Required equipment and/or other resources can be brought into a classroom
Work environment	The work environment is conducive to learning	The work environment is not appropriate or comfortable for learning
Instructor availability	Qualified instructors are not available on site but master performers are	Qualified instructors are available
Accuracy requirements	Actual job requirements are best experienced firsthand	Actual job requirements can be accurately simulated in a classroom environment
Target audience level	Potential trainees have diverse backgrounds and experience levels	Potential trainees have similar backgrounds and experience levels
Target audience motivation	Learners are well motivated and can work on their own with limited supervision	Learners are not sufficiently motivated to work on their own
Time requirements	Time to complete training is not a critical factor	Time to complete training is critical

Source: Adapted from "Job Training" by J. J. Phillips, in M. Nolan, *The ASTD Training and Development Handbook*, 1st ed. (New York: McGraw-Hill, 1996).

On-the-job training is training usually conducted at the workstation, typically done one-on-one,[1] and is best for learning motor tasks, such as:

- Procedures for installing software

- Conducting specific part or product inspections

- Procedures to perform routine troubleshooting

Off-the-job training is training that takes place away from the actual work site,[2] and may be preferable for knowledge-based tasks that require some degree of problem solving and decision making, such as:

- Continuous improvement tools
- Communication skills
- Data analysis

There certainly are many cases where training plans would utilize both for location selection. Training can start off-the-job learning concepts in a classroom setting, then continue by practicing or doing what was learned in the on-the-job setting. Another example would be learning general concepts as a group in a classroom setting and then training on-the-job for the job-specific skills by a more experienced person with the same job.

Training Delivery Methods

These can be traditional or technology-driven. Traditional methods include:

- Instructors
- Printed handouts
- Media aids (flipcharts, whiteboards, computerized projection, transparencies, and so on)
- Lectures

Technology-driven methods include:

- Video training (with teleconferencing)
- Computer-based (CD-ROM for self-paced training)
- Web-based (virtual classrooms, online programs, and 'Webinars')

Learner-Controlled Instruction

With the evolvement of new technology, training delivery has more options. By using learner-controlled instruction (LCI), also called self-directed learning (SDL), adult learners are training at their own pace, without an instructor. Delivery of this type of training is most popular using interactive CD-ROM-based training as it combines both printed text and interactive questioning from the CD. *ASQ's Foundations in Quality Self-Directed Learning Series* is a good example of this type of training method. LCI/SDL can be appropriate for the following situations:

- Individual training
- Learners that are geographically distant
- Training simultaneously at multiple sites

- Necessary consistency of skill execution

- Refresher training

Other common training delivery methods are presented in Table 8.2.

Table 8.2   Common training delivery methods.

Method	Description	Advantages	Disadvantages
Workbooks or self-study	Material-centered; self-paced	• Provides a common baseline of knowledge • Trainees control their progress • Cost-effective to deliver • Can be used in a variety of settings	• Requires self-motivated trainees • Well designed, may be expensive to develop • Poorly designed, materials may be boring • Trainees may get stuck and flounder
Lectures	One-way format; an instructor orally teaches information to a group	• Economical to develop and present • Easy to revise	• Difficult to master knowledge and skills • Low retention rate
Job aids	A repository of information, processes, or perspectives	• Cost-effective • Individuals can use them when and where they need them • Diminish the need for individuals to rely on their memory	• Can not provide a substitute for required expertise, fluid performance, or high production • Can not replace good judgment or prescribe intangible behavior
Video training	Best integrated with other methods; involves both auditory and visual senses	• Fast distribution • Good control over content and quality • Repeatability/reuse • Familiarity with the format	• Expensive production costs • Special equipment requirements (video teleconferencing) • Limited interaction • Tends to provide entertainment rather than instruction
Computer-based training	Interactive learning through a computer; also refers to computer-aided instruction, multimedia (for example, CD-ROMs), and computer-supported learning resources	• Supports wide geographic dispersion • Allows for self-pacing • Provides feedback • Adapts to individual learner needs and schedules • Can be retained as a reference • Provides testing administration	• High development costs • Expensive to change • Not designed for group learning (except through some Internet/intranet scenarios)

Source: ASQ's Foundations in Quality Self-Directed Learning Series, Certified Quality Manager, Module 7 (Milwaukee: ASQ Quality Press, 2001): 7-39.

There are many other training delivery techniques, including:

- Training and qualifying in-house trainers to deliver an externally developed program

- Augmenting in-house training with subject matter experts from outside the organization

- Establishing book reading discussion groups or forums on specific material

- Utilizing 'webinars' (training prepared for delivery via the Web)

- Sending individual to manufacturer's training course

- Partnering with local college or university to bring programs to the organization

- Rotating jobs and cross-training workers

Training Effectiveness

Evaluation is essential in determining whether the training program meets the objectives of the training plan. Evaluation includes:

1. Verification of the design (for example, subject content and delivery methods)

2. Applicability of facilities, equipment and tools, and media to be used

3. Qualifications of the trainer

4. Selection of the participants

5. Measures to be used to assess training effectiveness (validation)

6. Measures to be used to assess outcomes resulting from training (validation)

Technical training effectiveness can be greatly enhanced by four key considerations:

1. Appropriate preparation of the training facilities

2. Appropriate timing of the training relative to when performance is required

3. Appropriate sequence of training relative to skills mastery

4. Appropriate feedback indicating performance difficulties and progress

Table 8.3 describes five levels for evaluating training. Kirkpatrick is credited with defining the first four levels (reaction, learning, behavior, and results) for evaluating training.[3] Robinson and Robinson expanded level three to distinguish between type A learning (Are participants applying the skills and behavior as taught?) and type B learning (Are participants applying nonobservable skills, for

Table 8.3 Levels of training evaluation.

No.	Level	Questions	Comments
1	Reaction	Were the participants pleased with the training?	The typical "smile sheets" collected at the end of a program, session, or module
2	Learning	Did the participants learn what was intended for them to learn?	Were the training objectives met? Typically determined from some kind of test.
3	Behavior	Did participants change their behavior on the basis of what was taught?	Typically determined from an assessment or how learned behaviors are applied on the job
4	Results	Was there a positive affect on the organization resulting from participants' changed behavior?	Typically measured from pretraining versus post-training analysis of the organization's outcomes. Usually a "quantity" type measure (increased production, lower rejects, number of time periods without a customer complaint, and so on).
5	ROTI	Was there a measurable net dollar payback for the organization resulting from participants' application of learned behavior?	Return on training investment (ROTI) is based on the dollar valued added by the investment in the training

Source: R. T. Westcott, *The Certified Manager of Quality/Organizational Excellence Handbook,* 3rd ed. (Milwaukee: ASQ Quality Press, 2006): 564.

example, mental skills or learning, to the job?).[4] Level 5 in Table 8.3 is a perspective addressed by Westcott and Phillips.[5]

In Figure 8.2, Phillips presents a matrix comparing each level of evaluation to value of information, power to show results, frequency of use, and difficulty of assessment. The five levels represent a hierarchy with evaluation progressing from lowest to highest.[6] For example:

1. The *reaction* level provides the lowest-value information and results and is frequently and easily used.

2. The *learning* level is used to obtain a more in-depth assessment of whether specific training objectives were met.

3. The *behavior* (job applications) level assesses whether or not the training is applied back on the job, usually after some time has passed.

4. The *results* level measures the quantitative difference between the outcomes of the work units affected prior to their people receiving training (the baseline) and the outcomes some time after the training ended. The objectives for training measured could be decrease in scrap rate, increase

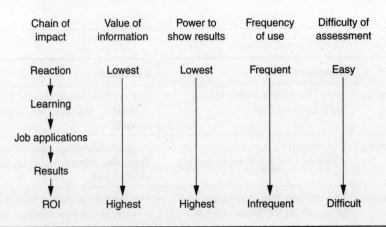

Chain of impact	Value of information	Power to show results	Frequency of use	Difficulty of assessment
Reaction	Lowest	Lowest	Frequent	Easy
Learning				
Job applications				
Results				
ROI	Highest	Highest	Infrequent	Difficult

Figure 8.2   Characteristics of evaluation levels.

Source: R. T. Westcott, *The Certified Manager of Quality/Organizational Excellence Handbook*, 3rd ed. (Milwaukee: ASQ Quality Press, 2006): 564.

in number of savings accounts opened, increased classroom attendance, elimination of medication errors in a hospital, and so on.

5. The *ROI* level offers the highest-value information and results, is not used as frequently, and is more difficult to assess. Level 5 is based on placing dollar values on the baseline, the results, and the improvement, then determining the value received for the dollars spent. When expressed as a ratio, a $3 return for every $1 spent is a useful minimum ROI. Quality training programs that score high at all levels are likely to be very successful.

Two formulas may be used to evaluate at level 5: the *benefits-to-cost ratio* (BCR) and the *return on training investment* (ROTI).

$$BCR = \frac{\text{Program benefits}}{\text{Program costs}}$$

$$ROTI = \frac{\text{Net program benefits (total benefits − costs)}}{\text{Program costs}}$$

A rule of thumb is that if ROTI is less than $3.00 (or a ratio of 3:1), meaning $3 of benefits obtained from every $1 spent, then training should not be considered (unless mandated).

Training program development costs are usually included in the first year (when training may extend over one year) unless the costs substantially exceed the first-year benefits. In this case, costs may be prorated over the period in which the training is delivered.

In this world of global competition between world-class quality organizations, a competent and empowered workforce could be the differentiating factor for success. It's important to use the results of training effectiveness as feedback. The feedback may show failures or ineffectiveness. As in the PDCA cycle, (explained in Chapter 10), use the feedback for continuous improvement to improve the quality training program. The ineffectiveness or failure is likely to be linked to one or more of the following causes:

- Cultural resistance by line managers
- Doubt as to the usefulness of the training occurring when:
  - No linkage exists with the strategic plans of the organization
  - No apparent connection is evident between the behaviors to be learned and their application on the job
  - No projected goals are established for post-training improvement
  - The question of "What's in it for me?" is inadequately addressed for participants and their supervisors
- Lack of participation (involvement) by line managers
- Technique versus problem orientation
- Inadequacies of leaders
- Mixing of levels of participants
- Lack of training application during the course
- Language too complex
- Lack of participation by the training function
- Operational and logistical deficiencies[7]

A successful quality training program will allow the workforce the skills necessary to carry out additional responsibilities, access to information for better decision making, and ultimately empowerment. An empowered workforce results in their being more responsive to customers, increased business performance, and a better quality of work life.

Part II
Problem Solving and Improvement

Part II

Chapter 9

A. Basic Quality Tools

Select, apply, and interpret these tools: flowcharts, Pareto charts, cause-and-effect diagrams, check sheets, scatter diagrams, and histograms. (Analyze)

[Note: The application of control charts is covered in Chapter 20.]

Body of Knowledge II.A

The seven basic quality tools are scientific tools used in analyzing and improving process performance. These classic quality tools are also referred to as the *seven basic tools.* They are primarily a graphic means for process problem analysis through the examination of data.

Figure 9.1 shows that some of the basic quality tools are:

- *Quantitative:* used for organizing and communicating numerical data

- *Nonquantitative:* used to process useful information for decision making

Quantitative	Nonquantitative
• Pareto charts	• Cause-and-effect diagrams
• Control charts	• Flowcharts
• Check sheets	
• Scatter diagrams	
• Histograms	

Figure 9.1   The seven basic quality tools.

With a team in place ready to collect useful data, the next step is for the team to know how to select and apply the tools designed to manage the data. In some instances, just one tool can be used for problem identification, problem solving, process improvement, or process management. In many cases, it takes various combinations of the seven basic quality tools depending on the nature of the problem or objective.

CHECK SHEETS

The check sheet is a quality tool that is used for processes where there is a significant human element. A check sheet (sometimes called a tally sheet) is a structured form that is custom-designed by the user to enable data to be quickly and easily recorded for analysis. Generally, check sheets are used to gather data on the frequency or patterns of particular events, problems, defects, and so on. Check sheets must be designed to gather the specific information needed. For instance, depending on the problem-solving effort, organizing the data by machines versus shifts or versus suppliers can show different trends.

Figure 9.2 shows defects observed by a quality inspector or technician who inspected a batch of finished products. The check sheet in this case shows:

- The fact (not opinion) that incomplete lenses are not a big problem

- Groupings to be investigated:

    a. Why are most cracked covers black?

    b. Why are all the fogged lenses green?

- Rows and columns can be easily totaled to obtain useful subtotals

- Requires definitions for the various defects like "fogged lens"

Line 12 Dec. 4, 2007	Lens color		
	Red	Green	Black
Cracked cover	II	I	IIIIII
Fogged		IIII	
Pitted	IIII	III	IIIII
Incomplete	I		I

Figure 9.2   Defects table.
Source: D. W. Benbow, A. K. Elshennawy, and H. F. Walker, *The Certified Quality Technician Handbook* (Milwaukee: ASQ Quality Press, 2003): 12.

Deficiencies noted July 5–11, 2007	5	6	7	8	9	10	11
Towels incorrectly stacked				IIIII			
Soap or shampoo missing	IIIIIIII	IIIIIII		IIIII			
TV Guide missing		II	III	IIIII	IIIIIII	IIIIIIIII	I
Mint missing from pillow	I		I	IIIII		I	
Toilet paper not folded into V			I	IIIII			

Figure 9.3 Time–related check sheet.

Source: D. W. Benbow, A. K. Elshennawy, and H. F. Walker, *The Certified Quality Technician Handbook* (Milwaukee: ASQ Quality Press, 2003): 12.

As depicted in Figure 9.3, time-related check sheets show multiple examples of the variety of trends that can be investigated. Two observations worth noting:

- "TV Guide missing" category shows a trend worth investigating
- Something amiss appears to have happened on July 8 and should be investigated

In some cases, recording the location of defects is necessary to provide a more visually oriented data collection device. This type of check sheet is called a *measles chart* or a *defect concentration diagram.* For example, if inspecting a vehicle's paint finish, a diagram of the vehicle's body would be useful for marking where each of any defects are located. Clusters will then indicate the biggest problem areas.

PARETO CHARTS

Pareto charts are based on the *Pareto principle,* which suggests that most effects are the result of relatively few causes. Vilfredo Pareto, a 19th-century Italian economist, noted that 80 percent of the wealth in Italy was held by 20 percent of the population. This observation was later defined by Joseph Juran in 1950 as the Pareto principle.

Pareto principle—The suggestion that most effects are the result of relatively few causes, that is, 80 percent of effects come from 20 percent of the possible causes (for example, machines, raw materials, operators).

Pareto chart—A basic tool used to graphically rank causes from most significant to least significant. It utilizes a vertical bar graph in which bar height reflects the relative frequency of causes.

Because the Pareto chart displays the category bars in descending order of length, from longest to shortest, it gives a clear visual representation of their relative contribution to the total effect.

Part II.A

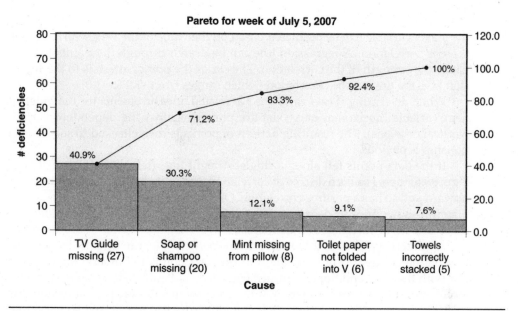

Figure 9.4   Pareto diagram.

Using the data from Figure 9.3, the Pareto diagram in Figure 9.4 shows that the problem-solving team working on reducing deficiencies should put their main efforts into preventing missing TV Guides. If the team were to focus their attention on finding a resolution to the towels being incorrectly stacked, it would only solve about 7.6 percent of occurrences. The right side of Figure 9.4 includes another feature, cumulative percent.

Although in Figure 9.4 "TV Guide missing" appears to be the biggest contributor to the overall problem, the most frequent is not always the most important. It's also important to do a similar analysis based on costs, (or severity for impacts to the goals of the business or customer satisfaction) as not all problems have equal financial impact. In this case, one would assume "TV Guide missing" does have a high financial impact compared to the other causes.

Sometimes when collecting data a team may find that there are a large number of causes but only a few occurrences of each. In those cases, it may be necessary to do some creative grouping to obtain a single cause that can amount to a 'big hitter.'

SCATTER DIAGRAM

A scatter diagram is a basic tool that graphically portrays the possible relationship between two variables or process characteristics. Two sets of data are plotted on a graph. The scatter diagram can help analyze data when a problem-solving team is working to determine the root cause of a problem but has several causes proposed.

If the diagram shows there is a correlation between two variables, it does not necessarily mean a direct cause-and-effect relationship exists. If it appears that the *y*-axis variable can be predicted based on the value of the *x*-axis variable, then there is correlation. A regression line can be drawn through the points to better gauge the strength of the correlation. The closer the point pattern is to the regression line, the greater the correlation (which ranges from –1.0 to +1.0).

When determing if two variables are related, measurements for the independent variable (horizontal or *x*-axis) are plotted against the dependent variable (vertical or *y*-axis). The resulting pattern of points is then checked to see if a relationship is obvious.

If the data points fall along a single, straight line, (left side of Figure 9.5 and Figure 9.6) there is a high degree of correlation. In Figure 9.5, the slope of the plot is upward and so has a positive correlation. On the right side of Figure 9.5, the scatter diagram shows a lower degree of positive correlation, indicating the existence of other sources of variation in the process.

A numeric value called the correlation coefficient (*r*) can be calculated to further define the degree of correlation. If the data points fall exactly on a straight line with the right end tipped upward (right side of Figure 9.5), $r = +1$ and so has a high degree of positive correlation. If the data points fall exactly on a straight line that tips down on the right end, (left side of Figure 9.6), $r = -1$ and so has a high

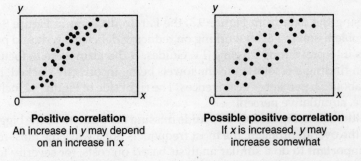

Positive correlation
An increase in *y* may depend on an increase in *x*

Possible positive correlation
If *x* is increased, *y* may increase somewhat

Figure 9.5 Scatter diagrams depicting high degree and low degree of positive correlation.

Source: Adapted from D. W. Benbow, R. W. Berger, A. K. Elshennawy, and H. F. Walker, *The Certified Quality Engineer Handbook* (Milwaukee: ASQ Quality Press, 2002): 281.

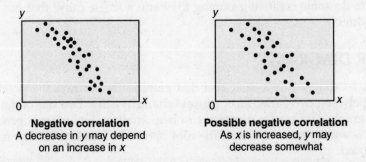

Negative correlation
A decrease in *y* may depend on an increase in *x*

Possible negative correlation
As *x* is increased, *y* may decrease somewhat

Figure 9.6 Scatter diagrams depicting high degree and low degree of negative correlation.

Source: Adapted from D. W. Benbow, R. W. Berger, A. K. Elshennawy, and H. F. Walker, *The Certified Quality Engineer Handbook* (Milwaukee: ASQ Quality Press, 2002): 281.

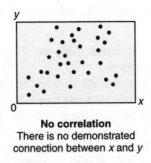

No correlation
There is no demonstrated
connection between *x* and *y*

Figure 9.7   Scatter diagram depicting no correlation.
Source: Adapted from D. W. Benbow, R. W. Berger, A. K. Elshennawy, and H. F. Walker, *The Certified Quality Engineer Handbook* (Milwaukee: ASQ Quality Press, 2002): 281.

degree of negative correlation. The value of *r* is always between –1 and +1 inclusive: $-1 \leq r \leq +1$.

If the correlation coefficient (*r*) has a value near zero, it indicates no correlation. See Figure 9.7.

Three considerations:

- There may be some relationships that are nonlinear

- The scales of graphs may be arbitrarily expanded or compressed on either axis and therefore a visual reading of the slope will not indicate the strength of the correlation

- A direct and strong correlation does not specify a cause-and-effect relationship

HISTOGRAMS

A histogram is the tool most commonly used to provide a graphical picture of the frequency distribution of large amounts of data. It also is used to show how often each different value occurs in a set of data. The data is shown as a series of rectangles of equal width but varying height. With a minimum of 50 data points, histograms can be used when:

- The data are numerical.

- Wanting to see the shape of the data's distribution. This is important when determining if the output of a process has a normal distribution.

- Analyzing whether a process can meet a customer's requirements.

- Analyzing what the output from a supplier's process looks like.

- Determining if a process change has occurred from one time period to another.

- Determining whether the outputs of two or more processes are different.

Part II.A

- Required to communicate the distribution of data quickly and easily to others.

Utilizing a minimum of 50 consecutive points, the worksheet in Figure 9.8 can help you construct a histogram.

The histogram worksheet helps determine the number of bars, the range of numbers that go into each bar, and the labels for the bar edges. After calculating W in step 2 of the worksheet, judgment is used to adjust it to a convenient number. For example, 0.9 can be rounded to an even 1.0. The value for W must not have more decimal places than the numbers being graphed.

Histogram Worksheet

Process: _____   Calculated by: _____

Data dates: _____   Date: _____

Step 1. Number of bars

Find how many bars there should be for the amount of data you have. This is a ballpark estimate. At the end, you may have one more or less.

# Data points	# of bars (B)	
50	7	
	8	
	9	
100	10	$B = $ _____
	11	
150	12	
	13	
200	14	

Step 2. Width of bars

Total range of the data = R = largest value − smallest value

$$= _____ - _____ = _____$$

Width of each bar $= W = \quad R \quad / \quad B$

$$= _____ / _____ = _____$$

Adjust for convenience. W must not have more decimal places than the data.

$$W = _____$$

Step 3. Find the edges of the bars

Choose a convenient number, L1, to be the lower edge of the first bar. This number can be lower than any of the data values. The lower edge of the second bar will be W more than L1. Keep adding W to find the lower edge of each bar.

L1	L2	L3	L4	L5	L6	L7	L8	L9	L10	L11	L12	L13	L14
—	—	—	—	—	—	—	—	—	—	—	—	—	—

Figure 9.8   Histogram worksheet.

Source: Adapted from http://www.asq.org/learn-about-quality/data-collection-analysis-tools/overview/histogram-worksheet.html.

As shown in Figure 9.9, for 100 data points and a range of scores from 300 to 800, the histogram will have 10 bars, each with a width of 50, the first bar starting with 300 and the last bar starting with 750. The resulting histogram is shown in Figure 9.10.

Histogram worksheet example using 100 students' Math SAT scores (data is fabricated) ranging from 300 to 800:

SAT score	Tally of 100 student scores	Frequency
300–350	‖‖‖ ‖	7
350–400	‖‖‖ ‖‖‖ ‖‖	13
400–450	‖‖‖ ‖‖‖ ‖‖‖ ‖	16
450–500	‖‖‖ ‖‖‖ ‖‖‖ ‖‖‖	19
500–550	‖‖‖ ‖‖‖ ‖‖‖ ‖‖‖	20
550–600	‖‖‖ ‖‖‖ ‖‖	13
600–650	‖‖‖ ‖‖	8
650–700	‖	2
700–750	‖	1
750–800	‖	1

Step 1: For 100 data points, # of bars $B = 10$

Step 2: The range $(R) = 800 - 300 = 500$

The width of each bar

$W = R/B$

$W = 500/10 = 50$

Step 3: L1 = 300

L2 = 350

.
.
.

L10 = 750

Figure 9.9 Tally and frequency distribution.

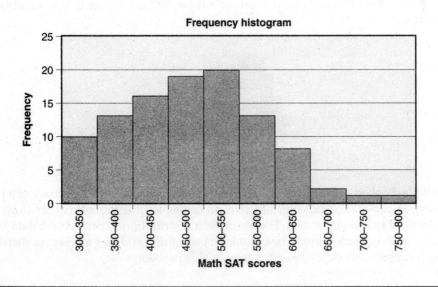

Figure 9.10 Histogram of student SAT scores.

Following are typical histogram shapes and what they mean. (Excerpted from Nancy R. Tague's *The Quality Toolbox*, Second Edition.)[1]

Normal. A common pattern is the bell-shaped curve known as the *normal distribution*. In a normal distribution, points are as likely to occur on one side of the average as on the other. Be aware, however, that other distributions look similar to the normal distribution.

Statistical calculations must be used to prove a normal distribution. Don't let the name "normal" confuse you. The outputs of many processes—perhaps even a majority of them—do not form normal distributions, but that does not mean anything is wrong with those processes. For example, many processes have a natural limit on one side and will produce skewed distributions. This is normal—meaning typical—for those processes, even if the distribution isn't called "normal"!

Normal distribution

Skewed. The skewed distribution is asymmetrical because a natural limit prevents outcomes on one side. The distribution's peak is off center toward the limit and a tail stretches away from it. For example, a distribution of analyses of a very pure product would be skewed, because the product cannot be more than 100 percent pure. Other examples of natural limits are holes that can not be smaller than the diameter of the drill bit or call-handling times that can not be less than zero. These distributions are called right- or left-skewed according to the direction of the tail.

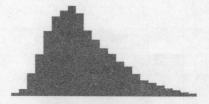

Right-skewed distribution

Double-Peaked or Bimodal. The bimodal distribution looks like the back of a two-humped camel. The outcomes of two processes with different distributions are combined in one set of data. For example, a distribution of production data from a two-shift operation might be bimodal if each shift produces a different distribution of results. Stratification often reveals this problem.

Bimodal (double-peaked) distribution

CAUSE-AND-EFFECT DIAGRAM

The cause-and-effect diagram is also known as the *Ishikawa diagram,* named after its developer, Kaoru Ishikawa. It's also known as the *fishbone diagram* and used to generate a list of possible root causes. Once a team has identified a problem, the diagram is used to divide root causes into broad categories. Results of further brainstorming of those root causes can be sorted into the identified categories.

Categories typical for the manufacturing environment are known as the four Ms:

- Manpower
- Machinery
- Methods
- Materials

Sometimes *measurement* or *environment* will be added (Figure 9.11). For service processes, the categories often used are the four Ps:

- People
- Policies
- Procedures
- Plant

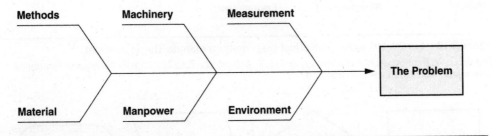

Figure 9.11   Cause-and–effect diagram for a manufacturing problem.

Source: D. W. Benbow, A. K. Elshennawy, and H. F. Walker. *The Certified Quality Technician Handbook* (Milwaukee: ASQ Quality Press, 2003): 9.

Part II.A

The cause-and-effect diagram provides a way of organizing the large number of subcategories generated from the brainstorming session. Figure 9.12 illustrates a completed cause-and-effect diagram.

The next step is to collect data to determine which of the causes is most likely to be acting on the problem being studied.

FLOWCHARTS

For a team to work on process improvement, they first need to make sure they have a good understanding of how the process works. A flowchart (also known as a process map) is a basic quality tool that provides graphical representation of the sequential steps in a process. The most frequently used flowchart symbols are shown in Figure 9.13. Effective flowcharts, as shown in Figure 9.14, include decisions, inputs, and outputs as well as process steps. Flowcharts can be used to document the as-is condition of a process or service and can point out missing, redundant, or incorrect steps.

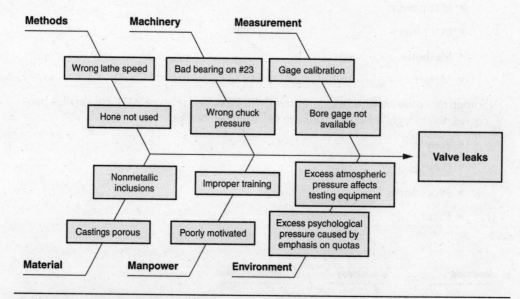

Figure 9.12   Completed cause-and-effect diagram for a manufacturing problem.

Source: D. W. Benbow, A. K. Elshennawy, and H. F. Walker, *The Certified Quality Technician Handbook* (Milwaukee: ASQ Quality Press, 2003): 9.

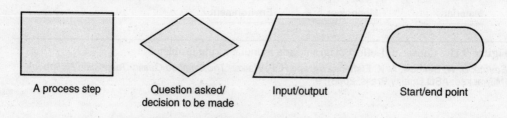

Figure 9.13   Some common flowchart symbols.

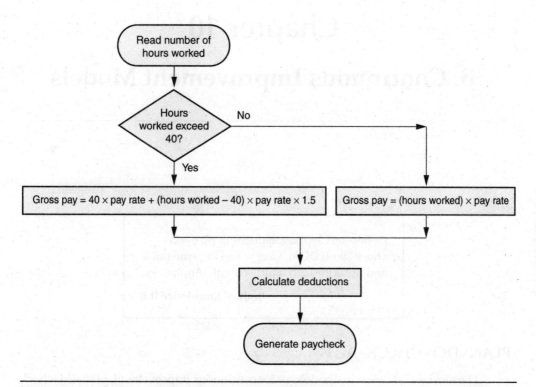

Figure 9.14   Flowchart for calculating weekly paycheck.

Source: D. W. Benbow, A. K. Elshennawy, and H. F. Walker, *The Certified Quality Technician Handbook* (Milwaukee: ASQ Quality Press, 2003): 11.

Chapter 10

B. Continuous Improvement Models

> Define and explain elements of plan–do–
> check–act (PDCA), kaizen, and incremental
> and breakthrough improvement. (Apply)
>
> **Body of Knowledge II.B**

PLAN–DO–CHECK–ACT

The most well-known methodology for continuous improvement, plan–do–check–act (PDCA), was developed by Walter Shewhart. W. Edwards Deming adapted the cycle to plan–do–study–act (PDSA), also known as the Deming wheel, emphasizing the role of learning in improvement.

PDCA is a model for continual process improvement. By developing a measurable action plan, decision making based on facts can happen, putting the plan into action. As Figure 10.1 shows, the cycle is endless. The cycle must be continuous to achieve significant improvements.

As depicted in Figure 10.1, the four steps of the PDCA/PDSA cycle are:

1. *Plan.* Recognize the opportunity for improvement and gather the data or do the research to answer, "What are the key targets the team expects this project to achieve?" When a problem is identified, the situation needs to be observed and data gathered to give a better definition of the problem. Once the problem is defined, root cause analysis (discussed further in Chapter 30) takes place. A popular and very effective tool for root cause analysis is the *five whys* tool. By asking the question "Why?" at least five times, a series of causes and effects results. This domino effect should end with the root cause. After the root cause has been identified, a countermeasure needs to be created and planned out. Know what the criteria are so progress can be measured—how will the team know if it was worth it? Or how will the team know if it is working? Data need to be gathered to determine whether the plan will work.

2. *Do.* Communicate the plan and implement according to the plan. As Paul Palmes says in his mp3 audiocast on the PDCA cycle, "If you change horses in midstream . . . people drown."[1] Communication is important. Everyone affected by the implementation of the countermeasure needs to be involved. If possible,

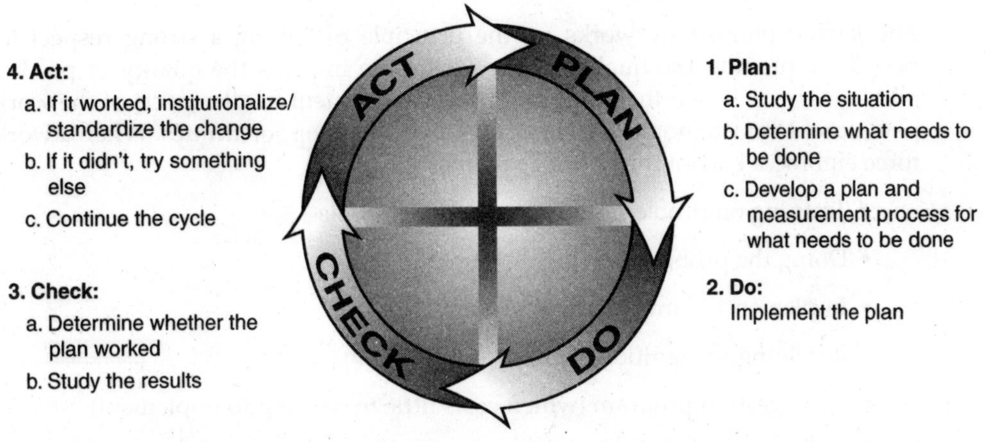

4. Act:

  a. If it worked, institutionalize/ standardize the change

  b. If it didn't, try something else

  c. Continue the cycle

3. Check:

  a. Determine whether the plan worked

  b. Study the results

1. Plan:

  a. Study the situation

  b. Determine what needs to be done

  c. Develop a plan and measurement process for what needs to be done

2. Do:

  Implement the plan

Figure 10.1   The PDCA/PDSA cycle.

Source: R. T. Westcott, *The Certified Manager of Quality/Organizational Excellence Handbook,* 3rd ed. (Milwaukee: ASQ Quality Press, 2006): 347.

the proposed solution should be deployed on a small scale first with as much data being collected as possible noting the conditions that occur as the change is being implemented.

    3. *Check.* Check according to the plan just as the team implements according to plan (in the *do* cycle). The team analyzes the information collected from the *do* phase. Did the action produce the desired results? Were new problems created?

    4. *Act.* A decision is made on whether it is feasible to adopt the change. If the change is adopted, the team uses the PDCA cycle as a continuous improvement cycle and looks for more opportunities for improvement. Sometimes as a result of the *check* phase it's decided that the plan should be altered and the PDCA cycle starts over again. Other possible actions include: abandoning the idea, increasing or reducing the scope, and then, always, beginning the PDCA cycle over again.

KAIZEN

Kaizen is a combination of two Japanese words (*kai* + *zen*) meaning incremental and orderly continuous improvement. Kaizen is a process-oriented system encompassing the whole organization, with the key to success being management's demonstration of complete support. Some of the goals of kaizen include:

- Continuous improvement
- Elimination of waste
- Just-in-time (JIT) manufacturing
- Standardized work
- Total quality control

The kaizen philosophy works on the principle of having a strong respect for people. Improving the quality of the workforce impacts the quality of product and/or service. When there is a focus on improvement in all aspects of the workplace, problems are not looked at as mistakes but as opportunities. When a workforce embraces kaizen they participate in:

- Training on problem solving and quality tools

- Doing the problem solving

- Implementing improvements

- Receiving recognition for their successes

- A suggestion program (which costs little to nothing to implement)

In Masaaki Imai's book *Kaizen,* he relates the kaizen approach to the PDCA cycle as shown in Figure 10.2.

The Toyota Production System is well known for its kaizen approach to productivity improvement, with the main goal of eliminating waste.

Kaizen events, also called *kaizen blitzes,* are typically five-day highly intensive events with a structured approach. These events focus on process improvement that may also result in the development of a new or revised work standard. Kaizen blitzes involve a team of workers that may be targeting eliminating waste, improving the work environment, reducing cycle time, reducing costs, or setup time reduction (for example, single minute exchange of dies [SMED], explained more in Chapter 14).

The goal of kaizen events is to provide a sense of urgency and generate enthusiasm for rapid results. Because of the immediacy, resources need to be in place and readily available for use.

Plan	What?	Definition of problem
		Analysis of problem
	Why?	Identification of causes
	How?	Planning countermeasures
Do		Implementation
Check		Confirmation of the result
Act		Standardization

Figure 10.2   Kaizen and the PDCA cycle.

Source: ASQ's Foundations in Quality Self-Directed Learning Series, Certified Quality Engineer, Module 5 (Milwaukee: ASQ Quality Press, 2000).

Part II.B

INCREMENTAL AND BREAKTHROUGH IMPROVEMENT

Quality programs that focus on continuous improvement are vital in providing incremental improvement of processes, products, and services in an existing technological paradigm. These programs view quality as being measurable in some quantitative way. Breakthrough improvement that leads to development of new paradigms requires thinking of quality in a different way. . . . [2]

The plan–do–check/study–act cycle and kaizen strategy typically focus on *incremental improvement* on a continuous basis. By taking one step at a time, process improvement is achieved on a gradual basis as opposed to 100 percent improvement upon completion of the first process improvement event. Although this strategy is being implemented on processes and subprocesses throughout the organization, the output of one improvement event often is the input to another new process improvement event, and businesses can lose patience because of the slow pace of progress.

Incremental improvement initiatives alone do not allow a business to keep up with the rapid pace of change in the technological arena as well as keep up with customer demands and competition. The Six Sigma philosophy and reengineering are two approaches to achieving breakthrough improvement. *Breakthrough improvement* is a method of solving chronic problems that results from the effective execution of a strategy designed to reach the next level of quality. Such change often requires a paradigm shift within the organization.

When initiating a method of improvement, the primary goal is to control variation. After reviewing the data, special cause problems and common cause problems are determined by using control charts (see Chapter 20). Once a state of controlled variation is reached, higher levels of quality are achieved by eliminating chronic problems. At this point, to achieve breakthrough improvement, organizations should focus on reducing variation in those areas most critical to meeting customers' requirements. The Six Sigma philosophy translates to the organizational belief that it is possible to produce totally defect-free products or services.

Sigma (σ) stands for the standard deviation of a process. The quantity of units processed divided into the number of defects actually occurring, multiplied by one million results in defects per million units:

$$\frac{\text{\# Defects}}{\text{\# Units produced}} \times 10^6$$

With the added statistical tolerance of a 1.5 sigma shift in the mean, a six sigma process is one that produces 3.4 defects per million opportunities (DPMO). See Table 10.1.

When an organization decreases the amount of variation occurring, the process sigma level will increase.

Table 10.1   Defects at various sigma levels.

Process sigma level		Defects per million opportunities
1 sigma	=	690,000 DPMO
2 sigma	=	308,000 DPMO
3 sigma	=	66,800 DPMO
4 sigma	=	6,210 DPMO
5 sigma	=	230 DPMO
6 sigma	=	3.4 DPMO

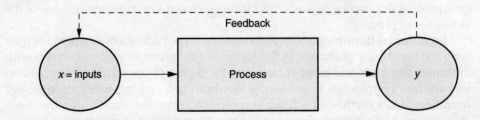

Figure 10.3   *y* is a function of *x*.

SIX SIGMA

This methodology provides businesses with the tools to improve the capability of their business processes. From Benbow and Kubiak's, *The Certified Six Sigma Black Belt Handbook*, "Six Sigma is a fact-based, data-driven philosophy of quality improvement that values defect prevention over defect detection. It drives customer satisfaction and bottom-line results by reducing variation and waste, thereby promoting a competitive advantage. It applies anywhere variation and waste exist, and every employee should be involved."[3] The handbook offers the definitions of Six Sigma as a philosophy, a set of tools, and a methodology.

Six Sigma Philosophy. Views all work as processes. Every process can be defined, measured, analyzed, improved, and controlled (DMAIC). Using the formula, $y = f(x)$, key process drivers can be identified. With a process input (x) being a causal factor, the process output (y) can be measured (y is a function of x).

x's are variables that will have the greatest impact on y. If y is customer satisfaction, then x will be key influences on customer satisfaction, be they inputs (that is, from customers or suppliers) or process variables (that is, materials, methods, environment, machine issues, and so on). If an organization controls the inputs (x) to the process, the outputs or process performance (y) can be controlled. It is important that all xs be identified in order to determine which of those inputs are controllable. Figure 10.3 shows a system where, with proper data collection and analysis, feedback is part of the process for improvement to the inputs to produce high quality outputs.

Six Sigma Tools. These are qualitative and quantitative techniques that drive process improvement. There are many tools that can be used in support of the Six Sigma methodology. The following are only a few:

- Affinity diagram

- Control charts

- Pareto chart

- Statistical process control

- Brainstorming

- Flowchart

- Mistake-proofing

- Process scorecards

- Cause-and-effect/fishbone diagram

- Histogram

- Failure mode and effects analysis

- Quality function deployment

Six Sigma Methodology. Six Sigma organizations actually utilize two methodologies: DMAIC (define, measure, analyze, improve, control) and DMADV (define, measure, analyze, design, verify). Both are measurement-based strategies that focus on process improvement and reducing variation.[4]

DMAIC is a rigorous and robust method that is the most widely adopted and recognized. A Six Sigma team member working on an improvement project will follow the five steps of DMAIC utilizing the tools necessary. See Table 10.2.

Where DMAIC is used on existing processes needing incremental improvement, the DMADV methodology is mostly applied to design for Six Sigma (DFSS) projects for a new design or, in cases where the process, product, or service has proven incapable of meeting customer requirements, a redesign.

For Six Sigma to be successful:

- Management must lead the improvement efforts and foster an environment that supports Six Sigma initiatives as a business strategy

- Every employee must focus on customer satisfaction

- Use teams that are assigned well-defined projects

- Provide Champions as leaders of the Six Sigma improvement process and to remove roadblocks for the Six Sigma improvement teams

- Provide experts (called Black Belts) for guidance and coaching

- Value and use the collected data

- View defects as opportunities

Part II.B

Table 10.2 DMAIC steps with some applicable tools.

DMAIC step	Some possible tools
Define the problem or improvement opportunity	Affinity diagram, charter, control charts, data collection, Pareto charts
Measure process performance	Control charts, data collection, flowchart, histogram, Pareto chart, process sigma, run charts
Analyze the process to determine the root causes of poor performance; determine whether the process can be improved or should be redesigned	Brainstorming, cause-and-effect diagram, design of experiments, histogram, interrelationship digraph, scatter diagram, tree diagram
Improve the process by attacking root causes	Activity network diagram, PDCA, prioritization matrix, brainstorming, control charts, failure mode and effects analysis (FMEA), histograms
Control the improved process to hold the gains	Communication plan, control charts, PDCA, run chart, Six Sigma storyboard

Source: Adapted from D. W. Benbow and T. M. Kubiak, *The Certified Six Sigma Black Belt Handbook* (Milwaukee: ASQ Quality Press, 2005).

- Implement statistical training at all levels, ensuring each employee a minimum of Six Sigma Green Belt status

Six Sigma can achieve improvements in quality and cost reduction quickly and on an incremental, continuous basis to achieve breakthrough levels of process quality.

REENGINEERING

This is an extreme action for breakthrough improvement, unlike the gradual improvement of processes resulting from incremental changes. When an organization undergoes reengineering, they are seeking drastic improvement results that often mean a paradigm shift has to happen within the organization.

In the case of total reengineering of an entire organization, its present structure and how the current work processes produce and deliver its products or services may not be a consideration. With a completely new perspective, they may ignore the current status and start with a fresh, new approach. Unfortunately, companies have used this tactic in desperation to cut costs by drastically reducing the number of employees. This, obviously, is a very unfavorable approach, and rather than reengineer an entire company all at once, process reengineering has become a more constructive method.

In process reengineering, a team will map out the structure and functions of a process to identify and purge the non-value-added activities, reducing waste and costs accordingly. This results in the development of a new and less costly process that performs the same functions.

Another aspect to reengineering is using creativity and innovation to help a business keep up with the rapid pace of change in the technological arena as well as keep up with customer demands and competition. Breakthrough improvement needs insight. To develop radical or paradigm-shifting products and services, incremental improvement must be combined with creativity and innovation. Some useful tools that can help promote creativity include:

- *Lateral thinking.* A process that includes recognizing patterns, becoming imaginative with the old ideas, and creating new ones

- *Imaginary brainstorming.* Breaks traditional thinking patterns

- *Knowledge mapping.* A process of visual thinking

- *Picture association.* Using pictures to generate new perspectives

- *Morphological box.* Used to develop the anatomy of a solution

Chapter 11

C. Basic Quality Management Tools

> Select and apply affinity diagrams, tree
> diagrams, process decision program charts,
> matrix diagrams, interrelationship digraphs,
> prioritization matrices, and activity network
> diagrams. (Apply)
>
> **Body of Knowledge II.C**

The quality management tools were developed to promote innovation, communicate information, and successfully plan major projects, and are used mostly by managers and knowledge workers. These tools analyze problems of a complex nature and are especially valuable in situations where there is little or no data available for decision making, that is, exploring problems, organizing ideas, and converting concepts into action plans.

The seven quality management tools, listed in an order that moves from abstract analysis to detailed planning, are:

1. *Affinity diagram.* Organizes a large number of ideas into their natural groupings.

2. *Interrelationship digraph (relations diagram).* Shows cause-and-effect relationships and helps you analyze the natural links between different aspects of a complex situation.

3. *Tree diagram.* Breaks down broad categories into finer and finer levels of detail, helping you move your thinking step by step from generalities to specifics.

4. *Matrix diagram.* Shows the relationship between two, three, or four groups of information and can give information about the relationship, such as its strength, the roles played by various individuals, or measurements.

5. *Matrix data analysis.* A complex mathematical technique for analyzing matrices, often replaced in this list by the similar *prioritization matrix.* One of the most rigorous, careful, and time-consuming of decision-

making tools, a prioritization matrix is an L-shaped matrix that uses pair-wise comparisons of a list of options to a set of criteria in order to choose the best option(s).

6. *Activity diagram (arrow diagram).* Shows the required order of tasks in a project or process, the best schedule for the entire project, and potential scheduling and resource problems and their solutions.

7. *Process decision program chart (PDPC).* Systematically identifies what might go wrong in a plan under development.[1]

AFFINITY DIAGRAM

This tool is used to organize a large group of facts or thoughts that are in chaos into smaller chunks that are easier to understand and work with. After conducting a brainstorming session, an affinity diagram helps to organize the resulting ideas. Affinity diagrams help break down large complicated issues into easy-to-understand categories, allowing teams to take ideas that seem unconnected and identify an inherent organization and pattern. This facilitates group consensus when necessary.

Each group member writes their ideas on sticky notes (one idea per note). Then, without talking, the group members sort the ideas simultaneously into related groupings or categories. Upon completion of grouping the ideas, discussion resumes and a heading is selected for each group. The resulting pattern is the affinity diagram, which at this point should promote discussion of the situation being analyzed. An example of the use of an affinity diagram is shown in Figure 11.1.

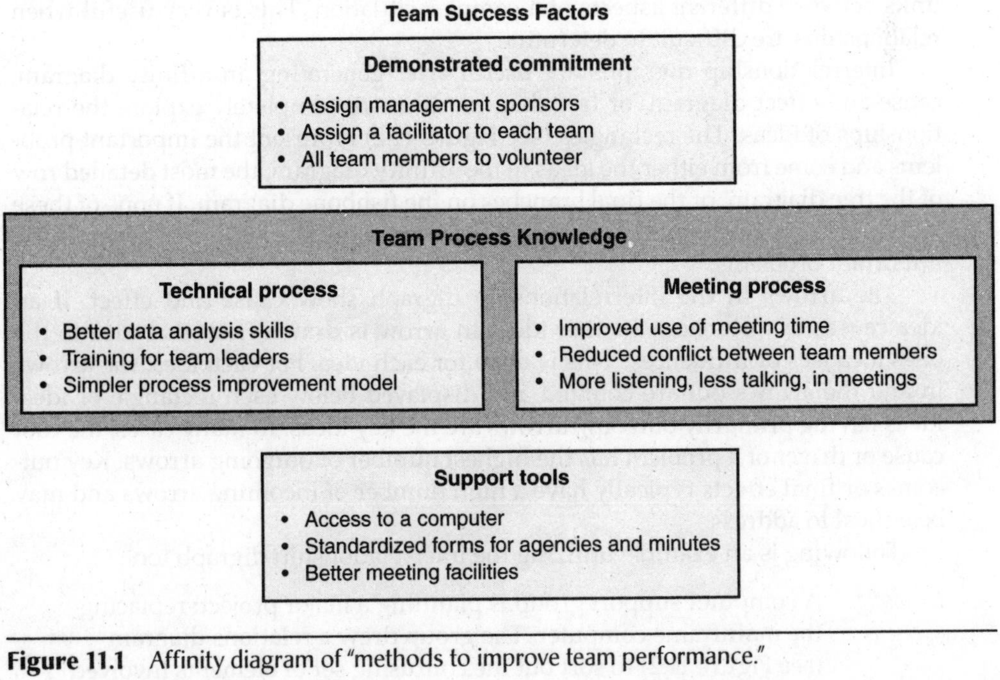

Team Success Factors

Demonstrated commitment

- Assign management sponsors
- Assign a facilitator to each team
- All team members to volunteer

Team Process Knowledge

Technical process

- Better data analysis skills
- Training for team leaders
- Simpler process improvement model

Meeting process

- Improved use of meeting time
- Reduced conflict between team members
- More listening, less talking, in meetings

Support tools

- Access to a computer
- Standardized forms for agencies and minutes
- Better meeting facilities

Figure 11.1   Affinity diagram of "methods to improve team performance."

Issues involved in computer replacement project

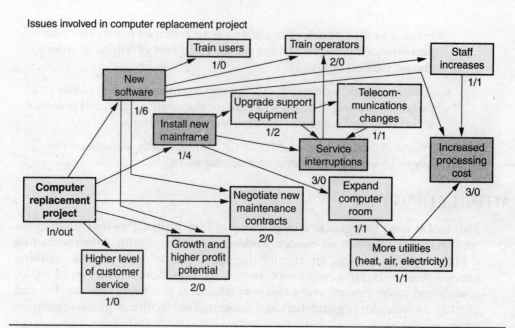

Figure 11.2   Interrelationship digraph (relations diagram) example.

INTERRELATIONSHIP DIGRAPH

This tool displays cause-and-effect relationships between factors, problems, and ideas in a complex situation using a "node and arrow" pictorial format. Also called *relationship diagrams,* interrelationship digraphs help a group analyze the natural links between different aspects of a complex situation. This is very useful when relationships are difficult to determine.

Interrelationship digraphs are useful after generating an affinity diagram, cause-and-effect diagram, or tree diagram to more completely explore the relationships of ideas. The rectangles (see Figure 11.2) represent the important problems and come from either the ideas in the affinity diagram, the most detailed row of the tree diagram, or the final branches on the fishbone diagram. If none of these tools were previously used, brainstorming ideas about the issue will generate the important problems.

The arrows in the interrelationship digraph show cause and effect. If an idea causes or influences another idea, an arrow is drawn from that idea to the ones it causes or influences. This is done for each idea. For each idea, the arrows in and the arrows out are counted and displayed below each rectangle or idea. Ideas having primarily outgoing arrows are the key ideas. In many cases, the root cause or driver of a problem has the highest number of outgoing arrows. Key outcomes or final effects typically have a high number of incoming arrows and may be critical to address.

Following is an example utilizing the interrelationship digraph tool:

Example: A computer support group is planning a major project: replacing the mainframe computer. The group drew a relations diagram (see Figure 11.2) to sort out the confusing set of elements involved in this project.

"Computer replacement project" is the card identifying the issue. The ideas that were brainstormed were a mixture of action steps, problems, desired results, and less-desirable effects to be handled. All these ideas went into the diagram together. As the questions were asked about relationships and causes, the mixture of ideas began to sort itself out.

After all the arrows were drawn, key issues became clear. They are in darker shaded boxes.

- "New software" has one arrow in and six arrows out. "Install new mainframe" has one arrow in and four out. Both ideas are basic causes.

- "Service interruptions" and "increased processing cost" both have three arrows in, and the group identified them as key effects to avoid.[2]

TREE DIAGRAM

This is a method used to separate a broad goal into increasing levels of detailed actions identified to achieve the stated goal. This analysis helps identify actions or subtasks required to solve a problem, implement a solution, or achieve an objective.

The tree diagram can supplement the affinity diagram and interrelationship digraph by identifying items not previously detected. This process allows teams to go from thinking of broad goals to more specific details. The tree diagram is also useful as a communication tool when many details need to be explained to others. See an example of a tree diagram in Figure 11.3.

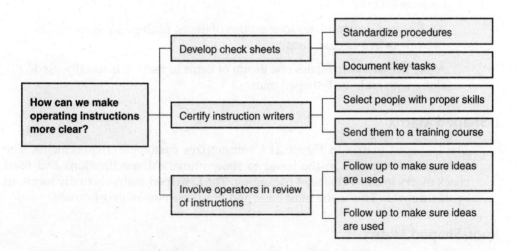

Figure 11.3   Tree diagram example.

Source: ASQ's Foundations in Quality Self-Directed Learning Series, Certified Quality Manager, Module 3 (Milwaukee: ASQ Quality Press, 2001): 3-26.

Part II.C

Table 11.1   When to use differently-shaped matrices.

Matrix type	# of groups	Relationships
L-shaped	2 groups	A ↔ B (or A ↔ A)
T-shaped	3 groups	B ↔ A ↔ C but not B ↔ C
Y-shaped	3 groups	A ↔ B ↔ C ↔ A
C-shaped	3 groups	All three simultaneously (3-D)
X-shaped	4 groups	A ↔ B ↔ C ↔ D ↔ A but not A ↔ C ↔ or B ↔ D
Roof-shaped	1 group	A ↔ A when also A ↔ B in L or T

MATRIX DIAGRAMS

This tool presents information in a table format of rows and columns to graphically reveal the strength of relationships between two or more sets of information. Patterns of project responsibilities become apparent so that tasks may be assigned appropriately. Depending on how many groups are being compared, there are six differently shaped matrices: L, T, Y, X, C, and roof-shaped. Table 11.1 shows how to decide when to use each type of matrix.

An *L-shaped matrix* relates two groups of items to each other (or one group to itself).

A *T-shaped matrix* relates three groups of items: Groups B and C are each related to A. Groups B and C are not related to each other.

A *Y-shaped matrix* relates three groups of items. Each group is related to the other two in a circular fashion.

A *C-shaped matrix* relates three groups of items all together simultaneously, in 3-D.

An *X-shaped matrix* relates four groups of items. Each group is related to two others in a circular fashion.

A *roof-shaped matrix* relates one group of items to itself. It is usually used along with an L- or T-shaped matrix.[3]

L-Shaped Matrix

The L-shaped matrix in Figure 11.4 summarizes customers' requirements. The team placed numbers in the boxes to show numerical specifications and used check marks to show choice of packaging. The L-shaped matrix actually forms an upside-down L. This is the most basic and most common matrix format.

Roof-Shaped Matrix

The roof-shaped matrix is used with an L- or T-shaped matrix to show one group of items relating to itself. It is most commonly used with a house of quality (refer to Chapter 26), where it forms the 'roof' of the 'house.' In Figure 11.5,

Customer Requirements

	Customer D	Customer M	Customer R	Customer T
Purity %	> 99.2	> 99.2	> 99.4	> 99.0
Trace metals (ppm)	< 5	—	< 10	< 25
Water (ppm)	< 10	< 5	< 10	—
Viscosity (cp)	20–35	20–30	10–50	15–35
Color	< 10	< 10	< 15	< 10
Drum		✓		
Truck	✓			✓
Railcar			✓	

Figure 11.4   L-shaped matrix example.

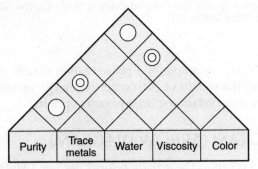

| Purity | Trace metals | Water | Viscosity | Color |

Frequently Used Symbols

◎ Strong relationship ○ Moderate relationship △ Weak or potential relationship No relationship	+ Positive relationship ○ Neutral relationship – Negative relationship
S Supplier C Customer D Doer O Owner	↑ Item on left influences item at top ← Item at top influences item on left The arrows usually are placed next to another symbol indicating the strength of the relationship.

Figure 11.5   Roof-shaped matrix example.

customer requirements are related to one another. For example, a strong relationship links color and trace metals, while viscosity is unrelated to any of the other requirements.

PRIORITIZATION MATRIX

This tool is an L-shaped matrix that shows the highest priority options or alternatives relative to accomplishing an objective. The prioritization matrix can portray comparisons of items like different product and market characteristics (see

Question: What are the effects of various interventions on the costs of the operation?			

Criteria Variables	Severity (5)	Detectability (2)	Occurrence (3)
Wrong color	3 15	2 4	2 6
Folding wrong	4 20	3 6	5 15
Color outside line	1 5	2 4	2 6
Missing paper clip	4 20	1 2	3 9

Figure 11.6   Prioritization matrix example.

Source: ASQ's Foundations in Quality Self-Directed Learning Series, Certified Quality Manager, Module 3 (Milwaukee: ASQ Quality Press, 2001): 3-23.

Figure 11.4). Another example of a prioritization matrix (see Figure 11.6) shows that multiplying the weight of each criterion by the variable's value will allow a team the ability to ascertain the high-priority options.

ACTIVITY DIAGRAM (ARROW DIAGRAM)

This tool, also known as the *activity network diagram* (AND), is an umbrella term that includes the *critical path method* (CPM) and the *program evaluation and review technique* (PERT). The AND is used when a task is complex and/or lengthy and, because of schedule constraints, it is desirable to determine which activities can be done in parallel. The AND is often used as a planning aid for construction projects and for large manufacturing contracts that are best handled as projects. The AND shows who is going to do what and when, and arrows are drawn to show what must be done in series and what can be done in parallel so that action can be taken if the overall time estimate is too high to meet goals.

In the AND shown in Figure 11.7, each numbered node represents one task that must be completed as a part of a project. The numbers shown between nodes are estimates of the number of weeks to complete the work in a given activity. The critical path can then be computed. The critical path is the longest path in the series of activities—those activities that must be completed on time to meet the planned project completion date.[4]

PROCESS DECISION PROGRAM CHART

No matter how well planned an activity is, things do not always turn out as expected. The process decision program chart, similarly to fault tree analysis, can be used to identify contingency plans for several situations that might contribute to failure of a particular project. The tasks to be completed for the project are listed, followed by a list of problems that could occur. For each problem a decision can be made in advance as to what will be done if each problem does occur. The contingency plan should also include ensuring availability of the resources needed to take the action. Figure 11.8 shows an example process decision program chart, a tool that is especially useful when a process is new, meaning that there are no previous data on types and frequencies of failure.

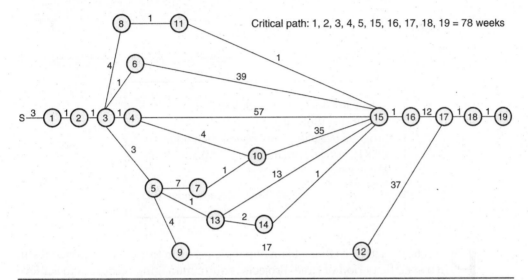

Figure 11.7   Activity network diagram example (activity on node).

Source: R. T. Westcott, *The Certified Manager of Quality/Organizational Excellence Handbook,* 3rd ed. (Milwaukee: ASQ Quality Press, 2006): 338.

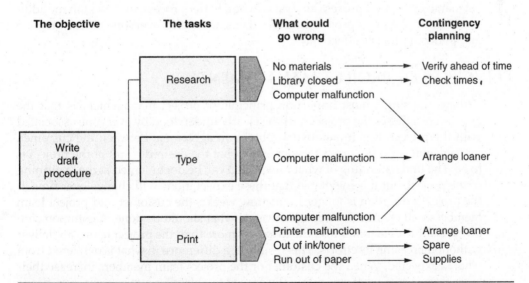

Figure 11.8   Process decision program chart example.

Source: R. T. Westcott, *The Certified Manager of Quality/Organizational Excellence Handbook,* 3rd ed. (Milwaukee: ASQ Quality Press, 2006): 343.

Part II.C

Chapter 12

D. Project Management Tools

> Select and interpret scheduling and
> monitoring tools such as Gantt charts,
> program evaluation and review technique
> (PERT), critical path method (CPM), etc.
> (Analyze)
>
> **Body of Knowledge II.D**

Projects are activities that a team of people perform to achieve a unique result in a specific period of time. Projects require managing resources to achieve a specific objective on a schedule.

Project management is a supplementary skill that augments an individual's primary skill. A good engineer will be a better engineer if he or she possesses, in addition to his or her primary skills as an engineer, the secondary skills of a project manager. A good process analyst will be a better process analyst with the additional skills of a good project manager. Thus, it benefits everyone to learn to plan and perform projects effectively.

PRINCIPLES OF PROJECT MANAGEMENT

Perhaps the single most important principle of project management is that the goal or objective of the project must be clearly understood by everyone associated with the project. It is imperative to obtain an agreement between the customer, management, and the project leader as to what the desired result of the project is to be. The understanding of what is expected can be documented at the beginning of the project, but it is inevitable that these expectations will change over time. If the project's duration is longer than a few weeks, the customer and project team members will change their perspectives of what should be done. A common condition is for the customer to come to expect more than the project team can deliver without increasing its efforts. This results in a difference in what is delivered from what is expected. When the customer or the project team members increase their expectations over what was originally agreed to, this is known as *scope creep*. Scope creep is a natural condition and must be managed by clear, written agreements at the beginning of the project and frequent reviews that remind the customers and project team members of what is expected as well as what is being done.

An essential principle of project management is that work must be planned in advance and that the plan must be followed. Many of the tools that have been developed for project management assist teams and individuals in planning the project and then monitoring progress of the project to ensure that the plan is followed. A project plan consists of a clear statement of what is to be accomplished (the desired outcome or result) and the detailed steps to reach that goal. Project plans also identify the necessary resources (people, equipment, facilities, tools, and materials) and describe how the progress and ultimate success of the project will be measured (when the deliverables will be provided, how much it will cost, what quantities will be delivered, and what quality is required).

The project plan must also describe the steps or activities required to achieve the project's result. Projects require that a job be broken down into doable tasks. A task is a job an individual or a small group of people can perform in a short time. Projects are comprised of tasks. It is the project manager's job to break the larger job (that is, the project) down into component tasks which, if all accomplished in the planned order, will result in the successful completion of the project.

Projects are constrained by three factors: time, resources, and quality. The project's schedule is a plan for managing the time, the project's budget is a plan for managing the project's resources, and the project's scope is the plan for managing the quality of the project's result. Schedule, budget, and scope are nearly always in conflict with each other—if you change one of them, you affect the others. For example, reducing the schedule in order to deliver a project result earlier nearly always requires additional resources, reduces the quality that can be delivered, or both. This triple constraint of time, resources, and quality is often depicted as a triangle to illustrate the interrelationship between the three constraints. See Figure 12.1.

An important responsibility of the project manager is to ensure that the progress of the project is monitored and that all the project stakeholders are kept aware of the status of the project. To this end, project managers and teams hold frequent progress reviews. The progress of the project is determined by comparing how well the current actual performance of the team compares with what was planned.

Figure 12.1   The triple constraints of project management.

Part II.D

Actual expenditures are compared with the planned budget, actual delivery and performance activities are compared to the scheduled times these were expected, and the deliverables' quality is compared to their planned quality. When there is a difference between what has been accomplished and what was planned, project managers refer to this difference as a *project variance.* Project managers must understand what causes project variances and must be prepared to explain when deliverables are late, costs are higher than budgeted, or the number and quality of the results are less than was expected.

To aid project managers in planning and monitoring the progress of projects, many project management tools have been developed. The use of a standard set of tools is an important practice for an organization to adopt. The use of standard tools improves communication between customers, management, project stakeholders, and project team members, and often increases the efficiency of the project team.

Project management practices, tools, and principles have been defined by the international organization Project Management Institute (PMI). PMI has created a Project Management Body of Knowledge (PMBOK)[1] that serves as an excellent reference for individuals and organizations who perform projects.

SOME STANDARD PROJECT MANAGEMENT TOOLS

Project Charter

The first "tool" of project management is the project charter. This is a document that authorizes the project to be conducted and gives the project manager the authority to manage the project. Companies often develop a standard form for chartering projects. The charter may identify constraints such as the budget that is available and the required timelines. Since one of the most important principles of project management is a clear understanding of what results are expected, the project charter is an extremely important tool.

Work Breakdown Structure

Once the objectives of the project have been defined, it is necessary to break the project down into its component tasks. To this end, a work breakdown structure (WBS) is useful. The WBS does two things: it guides a logical development of tasks to be performed and it provides a structure that summarizes project costs.

Work breakdown structures can be used to break projects down into many levels. Each project activity is broken down into smaller activities until the project planner can easily and convincingly estimate the time it will take to do an individual task and how much it will cost to perform it. The WBS then combines the individual task costs into a total project estimated cost. See Figure 12.2.

When the tasks have been identified, they can be listed. The project manager should add to this task list milestones, which are times on the schedule when the status of a task, an activity (a combination of tasks), or the project needs to be determined. Examples of milestones are "end of the project" and "decision point: should the next task be funded?"

Gantt Chart

When the tasks have been identified, they may be organized into a schedule. There are two types of schedules commonly used in project management: Gantt charts and network diagrams. A Gantt chart is a graphic depiction of a schedule using bars on a calendar to indicate when the work is planned to be performed. See the sample Gantt chart in Figure 12.3.

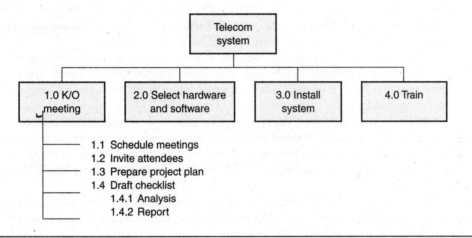

Figure 12.2 An example WBS.

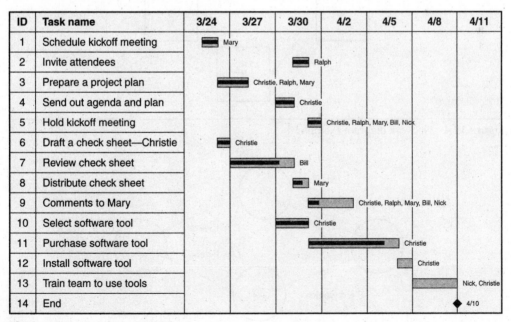

Figure 12.3 A sample Gantt chart.

Gantt charts are the most commonly used graphic presentation of project schedules because they are easy to understand and can show exactly when an individual is needed for a specific task on a project. Managers of the individuals assigned to projects can schedule the use of manpower more easily with Gantt charts to guide him or her.

Network Diagram

The second method for showing project schedule information is the network diagram. The network diagram in Figure 12.4 shows the relationship between tasks: task C cannot start until task A and task B have been completed.

Two methods for depicting network diagrams of projects are *program evaluation and review technique* (PERT) and *critical path method* (CPM). Today, the difference between these two methods is not important since they are seldom used in their original forms.

PERT is a method to analyze the tasks involved in completing a given project, especially the time needed to complete each task, and identifying the minimum time needed to complete the total project. The Project Management Institute no longer defines nor describes PERT.[2] See the example simple PERT chart in Figure 12.5.

CPM also models the activities and events of a project as a network, but events that signify the beginning or end activities are depicted as arrows between nodes. See Figure 12.6 for a CPM chart example.

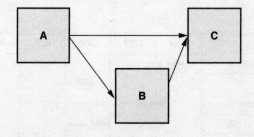

Figure 12.4   Network diagram example.

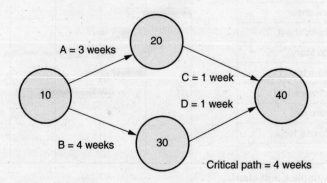

Figure 12.5   PERT chart example.

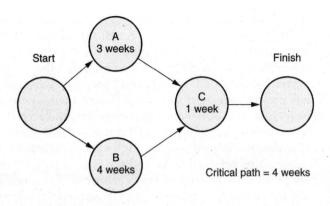

Figure 12.6   CPM chart example.

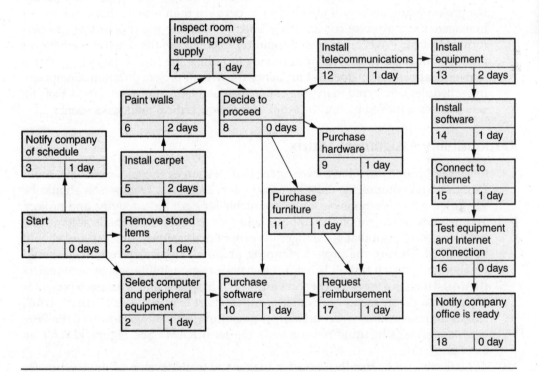

Figure 12.7   Network diagram example created by MS Project.

The network diagram helps project managers in two ways: it graphically shows relationships between tasks and it depicts the critical path. Software tools such as MS Project, PS8, and Primavera automatically create network diagrams at the same time they create a Gantt chart. Figure 12.7 is a network diagram example created by MS Project.

A primary reason for using network diagrams is that they show the critical path. The critical path refers to the longest path through the project. The duration of the entire project is defined by the critical path. Accordingly, it is useful to know the critical path so the duration of the entire project can be estimated.

When changes occur and some tasks take longer than was anticipated or when some tasks are completed earlier than expected, the duration of the entire path will change only if the affected tasks are on the critical path. It is conceivable that tasks not on the critical path might become so long that they change the critical path, and this often happens on large and complex projects. Accordingly, it is recommended that project managers employ software tools to create the network diagram. Software tools will automatically change the critical path when the duration of some tasks changes significantly, so it is useful to use software tools rather than to try to calculate the longest path through a project manually. Software tools often show relationships between tasks and the critical path on the Gantt chart, but for complex projects that becomes very confusing. It is much easier and less confusing to use the network diagram to display and modify the critical path.

Original estimates of how long it will take to perform projects and tasks are often too long to meet the time constraints imposed on the project. Also, during the project many tasks will take longer to perform than was expected, so project managers must exercise constant vigilance in reducing the time it takes to perform each task. Project managers should try to compress the schedules whenever possible but expending effort and money to reduce the time it takes to perform non–critical path tasks does not influence the overall project duration. Compressing schedules is referred to in project management language as *crashing*. Crashing schedules is achieved by finding ways to complete critical path tasks sooner.

Responsibility Assignment Matrix

Once the project tasks have been identified, resources (people, materials, tools, facilities, and equipment) must be assigned to each task. One person should be identified who is responsible and accountable for each task, activity, and project. Today many organizations attempt to hold two or more individuals accountable for a single task, but that is a dangerous project management practice and should be avoided. During the project planning phase all necessary resources should be allocated to each task. It is helpful to have a responsibility assignment matrix (RAM), which is a spreadsheet showing who is assigned and what each person is expected to do on each task. Sometimes this chart is called a RACI chart; RACI stands for responsibility, accountability, consulting, informing. However, the Project Management Institute recommends the term RAM.[3] See Figure 12.8 for an example RAM.

It is mandatory that the project manager and team develop accurate estimates of how many resources will be required to perform each task and the entire project. *Resource leveling* is a term used to describe moving scheduled tasks and milestones to accommodate limited resources—the project manager and team must ensure that resources are sufficient to perform the assigned tasks or the project schedules will be impossible to achieve.

Project Budget

When resources have been assigned to the project tasks, the project budget can be developed. There are two ways that budgets are created: top down and bottom up. Top-down budgeting refers to considering available monies or the cost

Responsibility Assignment Matrix						
Person/organization assigned →	SE 01	SE 02	HE 01	HE 02	TT 01	PM 01
Task AB726	A			P	R	
Task AB727		P	P	P	A	R
Task AB728			A	P		
Task AB729	P		A	P	R	R

Where: SE, HE, TT, and PM refer to job classifications

A = Accountable

P = Participant

R = Reviewer

Figure 12.8   Responsibility assignment matrix (RAM) example.

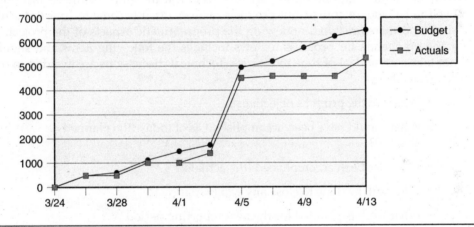

Figure 12.9   Line graph budget example.

of performing entire projects (perhaps the new project is analogous to one that was performed recently and the cost of doing the new one should be similar to the old one). Bottom-up budgeting refers to estimating the cost of each task and allowing the WBS to summarize the overall costs for the entire project.

The most common method for displaying project cost data, including budgets, is with a spreadsheet. Unfortunately, this method does not facilitate understanding and is often confusing to individuals not trained in reading financial reports. A graphic method for displaying budget information known as the *line graph budget chart* is an ideal way to show the relationship between cumulative actual expenses and estimated costs.

Line graph budgets like the one shown in Figure 12.9 are simple to create using modern software tools such as Excel chart wizard, and because they are easy to understand, they facilitate communication among interested project stakeholders.

Part II.D

Project Variance Analysis

During the execution of the project, it is mandatory that the progress of the project be monitored and tracked. When a difference between what has actually happened and what was planned occurs, this is known as a variance. Some sophisticated tools and techniques exist to analyze variances (see the Project Management Institute's Guide to the Project Management Body of Knowledge, PMBOK). Even when these sophisticated techniques are not employed, the project manager must be prepared to explain why there is a variance and what will be done to eliminate or mitigate it.

Project Progress Review

Routine progress review meetings become the primary instrument for monitoring and controlling project performance. The routine progress review should focus on the programmatic aspects of the project (schedule, budgets, and administration) rather than becoming another technical meeting. Technical meetings need to be held, of course, but progress reviews often become technical discussions and therefore avoid reviewing the programmatic aspects of the project. A standard agenda for progress reviews includes the following items and should not take a great deal of time to discuss. Technical discussions are avoided in the progress review:

- What are the project's objectives?
- What should have been accomplished as of today (the planned activity)?
- What has been accomplished (the actuals)?
- What will be done to close any gap (variances)?
- What work is planned for the next reporting period?
- What problems are anticipated?

Chapter 13

E. Taguchi Concepts

> Identify and describe Taguchi concepts and
> techniques such as signal-to-noise ratio,
> controllable and uncontrollable factors, and
> robustness. (Understand)
>
> **Body of Knowledge II.E**

D r. Genichi Taguchi developed a means of dramatically improving processes in the 1980s. Since then his methods have been widely used in achieving "robustness of design."

"Robust" as used here means strong, healthy, enduring, not easily broken, and not fragile. When we refer to robust design, we are describing a product design that is sturdy and intrinsically avoids defects. The concept of "designing quality into the product or process" is another expression of robust design.

Taguchi's approach for making product designs robust involved the application of five practices:

1. Classify the variables associated with the product into signal (input), response (result), control, and noise factors.

2. Mathematically specify the ideal form of the signal-to-response process that will produce a product that will make the system in which it is used work perfectly.

3. Compute the loss experienced when the output response deviates from the ideal (this practice is also known as the *Taguchi loss function*).

4. Predict the field quality through laboratory experiments, given a measured signal-to-noise ratio.

5. Employ design of experiments (DOE) to determine in a small number of experiments the control factors that create the best performance for the process.

To illustrate how this Taguchi method reduces variation and produces higher quality products, let us turn to one of the examples Taguchi provided.[1]

AN EXAMPLE OF ROBUST DESIGN VIA THE TAGUCHI METHOD

In 1953 the Ina Seito tile manufacturing company experienced nonuniform heat distribution in its brick baking kiln causing the tiles nearest the walls of the kiln to warp and break. Replacing the kiln would be expensive so Taguchi proposed the less costly option of reformulating the clay recipe so the tiles were less sensitive to temperature variation. Taguchi was able to show via laboratory experiments that if five percent more lime were added to the clay mix they would be more robust, that is, less likely to break and warp. The Ina Seito plant did not need to purchase a new, expensive kiln but was rather able to solve its problem in a much less expensive fashion. This experiment has become the showcase example of how intelligent use of DOE can produce products that are robust to process noise.

CLASSIFY THE VARIABLES

Figure 13.1 shows the relationships between the four classes of variables.

Signal factors are the process inputs and the responses are the results of the process.

Control factors are those factors that can be controlled. In the tile example they include limestone content in the clay mixture, fineness of the additives, amalgamate content, type of amalgamate, raw material quantity, waste return content, and type of feldspar.

Noise factors are those factors that can not be controlled. In Dr. Taguchi's application of robustness experiments, he treated temperature as a noise factor since it is difficult to control and its variation is a necessary evil. He investigated varying other factors that were much less expensive to control than temperature in order to make more robust products. Robust design is also called *minimum sensitivity design*.

THE TAGUCHI LOSS FUNCTION

The Taguchi loss function is a conception of quality that is more demanding than most. Genichi Taguchi contended that as product characteristics deviate from the

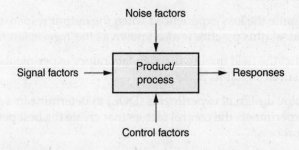

Figure 13.1   The four classes of variables in the Taguchi method.

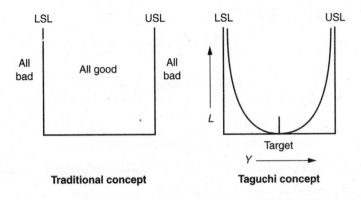

Figure 13.2   Comparison between the traditional and the Taguchi concept for assessing the impact of product quality.

design aim, the losses increase parabolically. The diagram on the left in Figure 13.2 illustrates the traditional concept, in which the product is considered good if its characteristics fall within the lower to upper specification limits, and is considered bad if it falls outside the specification limits. The diagram on the right in Figure 13.2 shows the Taguchi concept where the loss that society suffers increases parabolically as the product's characteristics deviate from the design target.

The formula expressing the Taguchi loss function is:

$$L = K(Y - T)^2$$

where

L = Loss in dollars

K = Cost coefficient

T = Design target or aim

Y = Actual quality value

The point of this equation is that merely attempting to produce a product within specifications doesn't prevent loss. Loss occurs even if the product passes final inspection by falling within the specification limits but doesn't exactly meet the design target. Consider an automobile driveshaft. Suppose its target diameter is 3.3 ±0.1 cm. If the housing that contains the driveshaft is built extremely close to its tolerances (that is, to accommodate a driveshaft exactly 3.3 cm), and the drive-shaft is slightly larger or smaller, even though it is within the specification limits, over time it will wear out faster than it would if it were exactly 3.3 cm. If it is 3.26 cm, it will pass final inspection and will power the automobile for a long time. But after hundreds of thousands of revolutions, it will wear out faster. The automobile owner will have to have her car repaired earlier than she would if the driveshaft were exactly 3.3 cm. The loss function for this drive shaft would be:

$$L = K(3.3 - 3.26)^2 = K(0.04)^2 = 0.0016K$$

SIGNAL-TO-NOISE RATIO

The signal-to-noise (S/N) ratio is used to evaluate system performance.

$$\eta = \frac{S}{N} = +10\log_{10}\frac{1}{r}\left[\frac{S_\beta - V_e}{V_N}\right]$$

where:

r is a measure of the magnitude of the input signal

S_β is the sum of squares of the ideal function (useful part)

V_e is the mean square of nonlinearity

V_N is an error term of nonlinearity and linearity

In assessing the results of experiments, the S/N ratio is calculated at each design point. The combinations of the design variables that maximize the S/N ratio are selected for consideration as product or process parameter settings. There are three cases of S/N ratios:

Case 1: S/N ratio for "smaller is better":

where S/N ratio = $-10\log_{10}$(mean-squared response).

This value would be used for minimizing the wear, shrinkage, deterioration, and so on, of a product or process.

Case 2: S/N ratio for "bigger is better":

$$\eta = \frac{S}{N} = -10\log_{10}\frac{1}{r}\left[\frac{\Sigma\frac{1}{y_i^2}}{n}\right]$$

where S/N ratio = $-10\log_{10}$(mean-squared of the reciprocal response).

This value would be used for maximizing values like strength, life, fuel efficiency, and so on, of a product or process.

Case 3: S/N ratio for "normal is best":

$$\eta = \frac{S}{N} = +10\log_{10}\frac{\text{mean}^2}{\text{variance}} = +10\log_{10}\frac{y^2}{S^2}$$

This S/N ratio is applicable for dimensions, clearances, weights, viscosities, and so on.

Chapter 14

F. Lean

> Identify and apply lean tools and processes, including
> setup reduction (SUR), pull (including just-in-time [JIT]
> and kanban), 5S, continuous flow manufacturing (CFM),
> value stream, poka-yoke, and total preventive/predictive
> maintenance (TPM) to reduce waste in areas of cost,
> inventory, labor, and distance. (Apply)
>
> Body of Knowledge II.F

The *lean* approach, or lean thinking, is "A focus on reducing cycle time and waste using a number of different techniques and tools, for example, value stream mapping, and identifying and eliminating monuments and non-value-added steps. Lean and agile are often used interchangeably." This definition is taken from the ASQ *CQM/OE Handbook*.[1] Lean practices use both incremental and breakthrough improvement approaches to eliminate waste and variation, making organizations more competitive, agile, and responsive to markets. As noted in George Alukal's article, "Create a Lean, Mean Machine," "Waste of resources has direct impact on cost, quality, and delivery . . . the elimination of waste results in higher customer satisfaction, profitability, throughput, and efficiency." George Alukal goes on to say that cutting out the following eight types of waste (called *muda* in Japanese) is the major objective of lean implementation:

1. *Overproduction.* Making more, earlier, or faster than required by the next process.

2. *Inventory waste.* Any supply in excess of a one-piece flow (make one batch and move one batch) through the manufacturing process, whether it is raw materials, work in process, or finished goods.

3. *Defective product.* Product requiring inspection, sorting, scrapping, downgrading, replacement, or repair.

4. *Overprocessing.* Extra effort that adds no value to the product (or service) from the customer's point of view.

5. *Waiting.* Idle time waiting for such things as manpower, materials, machinery, measurement, or information.

6. *Underutilized people.* Not fully using people's mental and creative skills and experience.*

7. *Motion.* Any movement of people, tooling, and equipment that does not add value to the product or service.

8. *Transportation waste.* Transporting parts or materials around the plant.[2]

There are a number of lean tools and techniques and the selection of which to use greatly depends on the root causes of variation and waste.

SETUP REDUCTION (SUR)

The goal of *setup reduction* is to provide a rapid and efficient way of converting a manufacturing process from running the current product to running the next product. This method can also be referred to as *quick changeover, single minute exchange of die (SMED),* or *rapid exchange of tooling and dies* (RETAD). The time that is lost in production for machine setups can be greatly reduced by applying the SUR method. The concept is that all changeovers (and startups) can and should take less than 10 minutes.

There are seven basic steps to reducing changeover time using the SMED system:

1. *Observe* the current methodology.

2. Separate the internal and external activities. *Internal* activities are those that can only be performed while the process is stopped, where *external* activities can be done while the last batch is being produced or once the next batch has started, for example, gathering the required tools for the job *before* the machine stops.

3. Convert (where possible) internal activities into external ones (preheating of tools is a good example of this).

4 Streamline the remaining internal activities by simplifying them. Focus on fixings—Shigeo Shingo rightly observed that it's only the last turn of a bolt that tightens it—the rest is just movement.

5. Streamline the external activities so that they are of a similar scale to the internal ones.

6. Document the new procedure and actions that are yet to be completed.

7. Do it all again: for each iteration of the above process, a 45 percent improvement in setup times should be expected, so it may take several iterations to cross the 10-minute line.[3]

---

* The traditional definition for the major objectives of lean implementation is of the seven types of waste. George Alukal describes an additional waste as people who are underutilized.

The SMED concept is credited to Shigeo Shingo, one of the main contributors to the consolidation of the Toyota Production System, along with Taiichi Ohno.[4]

KANBAN

This tool takes its name from the Japanese word meaning *sign* or *card*. Although kanban is related to just-in-time (JIT), it is in fact the means through which JIT is managed to form a 'pull' system. The pull system (Figure 14.1) ensures that product is provided to each customer on time in the amount needed. This is done with kanban cards. Kanban cards are visible signs that tell what is needed, when it's needed, and how much is needed. A kanban card signals when to refill goods removed from a stocking location or when a subassembly has been used. After a certain quantity is used, the kanban card is returned upstream to the supplying process, thereby signaling replenishment.

You need kanban cards to have a pull system. Visible cards let you control inventory so you have the correct amount of product (that is, no more, no less) ready to ship when your customer needs product (not before or after). There are many types of kanban systems and sundry complicated pull systems, but the concept is straightforward.

Think of kanban as a tool for managers to control the flow of product in a manufacturing operation. It is only as good as those who design it, so you need to be sure the people who design your kanban system use and understand it.[5]

JUST-IN-TIME

This lean methodology is an inventory strategy that provides for the delivery of material or product at the exact time and place where the material or product

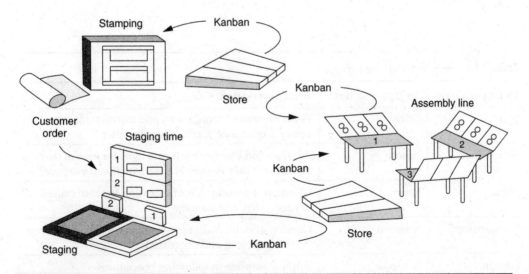

Figure 14.1   Pull system.

Source: Adapted from T. J. Shuker, "The Leap to Lean," in *54th Annual Quality Congress Proceedings* (Indianapolis, IN: Lean Concepts, 2000): 109.

will be used. When this material requirements planning system is implemented, there is a reduction of in-process inventory and its related costs (such as warehouse space), which in turn can dramatically increase the return on investment, quality, and efficiency of an organization.

By implementing JIT, buffer stock is eliminated or reduced and new stock is ordered when stock reaches the reorder level (facilitated by the use of kanban cards/signals).

5S

This tool is used with other lean concepts as the first step in clearing things out that are not necessary to create a clean and orderly work space. Maintaining cleanliness and efficiency allows waste and errors to be exposed. In Table 14.1, a brief description of each 5S term is provided along with the actual Japanese word from which 5S originated.

By implementing 5S, the workforce is empowered to control their work environment and consequently promote awareness of the 5S concept and principles of improvement.

CONTINUOUS FLOW MANUFACTURING

This lean methodology is also known as *one-piece flow,* meaning product moves through the manufacturing process one unit at a time. This is in contrast to batch processing, which produces batches of multiples of the same item and then sends the product through the process batch by batch.

As each processing step is completed, the unit is then pulled by the next process downstream. See Figure 14.2. The worker does not have to deal with waiting (as you would with batch processing), transporting products, and storing inventory.

Table 14.1   5S terms and descriptions.

5S English term	5S Japanese term	Brief description
Sort	Seiri	Separate needed tools, parts, and instructions from unneeded materials, removing the latter
Set in order	Seiton	Arrange and identify parts and tools for ease of use. Where will they best meet their functional purpose?
Shine	Seiso	Clean up the workplace, eliminating the root causes of waste, dirt, and damage
Standardize	Seiketsu	Develop and maintain agreed-upon conditions to keep everything clean and organized
Sustain	Shitsuke	Apply discipline in following procedures

Please note: these are adopted descriptions and English terms adapted from and not literally translated from the original 5S Japanese terms.

Continuous-Flow Processing

Figure 14.2 Continuous flow.
Source: Adapted from T. J. Shuker, "The Leap to Lean," in *54th Annual Quality Congress Proceedings* (Indianapolis, IN: Lean Concepts, 2000): 108.

For continuous flow to make sense, you need to understand takt time, standardized work, and pull system. Henry Ford used continuous flow when he 'pulled' cars along an assembly line as workers performed one operation or did one job at a time. Since Ford's Model-T assembly line, the idea of continuous flow has embraced increasingly sophisticated and novel manufacturing concepts. Now, products with dozens of colors and hundreds of options must flow continuously, just in time, without waste or stagnation (muda).

Think of takt time as a metronome that fixes the tempo at which work is done.

Takt time measures the pace of production, like an orchestra conductor's baton making sure everyone is on the beat and the process is smooth and continuous. If the pace is disrupted, continuous flow is lost and corrections must be done.[6]

Takt time is the available production time divided by the rate of customer demand. Operating to takt time sets the production pace to customer demand.[7]

$$\frac{\text{Daily available work time}}{\text{Parts required per day}} = \frac{\text{Time}}{\text{Volume}} = \text{Takt time}$$

Standardized work is documented and agreed-upon work instructions that express the best known methods and work sequence for each manufacturing or assembly process.[8]

VALUE STREAM

This describes the primary actions required to bring a product from concept to placing the product in the hands of the end user.[9] Value stream mapping (VSM) is charting the sequence of movements of information, materials, and production activities in the value stream (all activities involving the designing, ordering, producing, and delivering of products and services to the organization's customers). An advantage to this is that a "before action is taken" value stream map depicts the current state of the organization and enables identification of how value is created and where waste occurs. Plus, employees see the whole value stream rather than

Part II.F

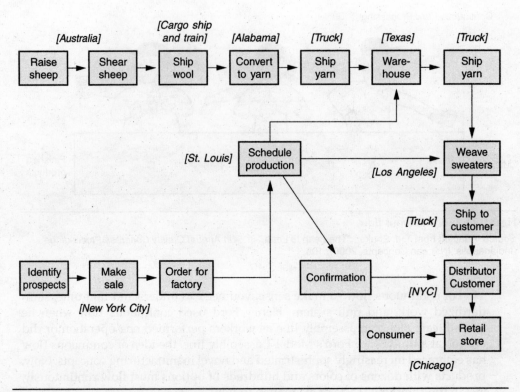

Figure 14.3 Value stream map example—macro level (partial).

Source: R. T. Westcott, *The Certified Manager of Quality/Organizational Excellence Handbook*, 3rd ed. (Milwaukee: ASQ Quality Press, 2006): 391.

just the one part in which they are involved. This improves understanding and communications and facilitates waste elimination. A value stream map (VSM) is used to identify areas for improvement. At the macro level, a VSM identifies waste along the value stream and helps strengthen supplier and customer partnerships and alliances. At the micro level, a VSM identifies waste—non-value-added activities—and identifies opportunities that can be addressed with a kaizen blitz. Figures 14.3 and 14.4 are value stream map examples (macro level and micro level).[10]

POKA-YOKE

This tool also takes its name from a Japanese term that means to mistake-proof a process by building safeguards into the system that avoid or immediately find errors. It comes from *poka*, which means *error*, and *yokeru*, which means *to avoid*.[11]

Poka-yoke devices can provide defect prevention for an entire process. These mechanisms can give warning or control, or can even shut down an operation:

- *When a defect is about to occur, a* warning *buzzer or light can be activated in order to alert workers. This is similar to the "lights on" warning buzzer that sounds in most automobiles when the driver leaves a vehicle with the engine off and lights on. Hopefully, the driver will notice the warning and correct the*

Part II.F

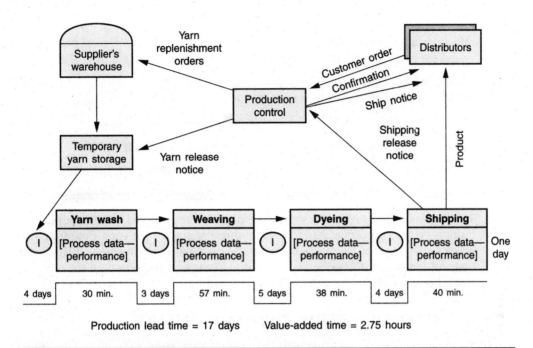

Figure 14.4   Value stream map example–plant level (partial).

Source: R. T. Westcott, *The Certified Manager of Quality/Organizational Excellence Handbook*, 3rd ed. (Milwaukee: ASQ Quality Press, 2006): 391.

mistake (turn off the lights) before a defect (dead battery) occurs. The driver may elect to ignore the warning, pointing out the fact that warnings, while good, are not always the most effective poka-yoke devices that can be deployed.

• Probably the most effective mistake-proofing is achieved through control of operations. By deploying poka-yoke devices throughout a process in order to prevent errors, defects will not occur. For example, lights that are automatically turned off by a timer in the car when the engine is off are controlled in order to prevent a defect (dead battery). Good poka-yoke devices can even prevent intentional errors. This factor is especially important in the area of computer and automatic teller machine security.

• Another function that can be performed by poka-yoke devices is shutdown. When an error is detected, an operation can be shut down, preventing defects from occurring. Missing screws in chassis assemblies are a common manufacturing headache.

   Workers can be told to make sure that all screws are installed. Banners and quality signs can be hung; however, missing screws will still be a problem. Poka-yoke can come to the rescue!

   One way to prevent the missing screw problem is to install a counter on the power screwdriver used to put in the chassis screws. The counter will register the number of installations performed on each unit. When the required number of screws has been installed, a switch in the conveyer *assembly can be closed, allowing the chassis to move to the next work position. The conveyor is "shut down" until all the screws have been put in.*[12]

In all cases, whether detecting or preventing defects, defects are caused by errors. Mistakes can be classified into four categories:

- *Information errors.* Misread or misinterpreted information

- *Misalignment.* Parts misaligned or machine is mistimed

- *Omission or commission.* Material or part is added or parts are omitted

- *Selection errors.* Wrong part, wrong orientation, or wrong operation

To implement poka-yoke, the workforce needs to be involved to determine where human errors can occur. After determining the source for each potential error, a poka-yoke method or mechanism can be devised for prevention. If there is no method to prevent the error, then a method to lessen the potential for that error should be devised.

TOTAL PREVENTIVE/PREDICTIVE MAINTENANCE

Total preventive (predictive) maintenance (TPM) aims to reduce loss due to equipment-related wastes such as:

- Downtime (due to inefficient setup or work stoppages)

- Speed losses

- Defective product (needing rework or repair)

- Constant adjustments

- Machine breakdowns

TPM allows the worker to not only run the machine or equipment but to proactively maintain it. The machine operator can maintain the machinery with proper routine care, minor repairs, and routine maintenance. With the operators being responsible and more involved in the machinery's operation, they understand the equipment better and are motivated to take better care, leading to less abuse and accidental damage. Cooperation amongst the operators and maintenance workers to perform maintenance inspections and preventive maintenance activities is critical to fully achieving implementation of TPM. "The aim of TPM is to attain the 'four zeros,' that is, zero defects, zero downtime, zero accidents, and zero waste."[13]

Chapter 15

G. Benchmarking

> Define and describe this technique and how
> it can be used to support best practices.
> (Understand)
>
> **Body of Knowledge II.G**

A carpenter may make a mark on his workbench so he can easily and efficiently cut every table to the exact length. In business, benchmarks are standards against which performance can be measured. Benchmarks are commonly used by companies to achieve quality performance. In theory, benchmarks are those standards that the organization strives to achieve. They are identified from many possible sources:

- Customer specifications

- Customer expectations

- The competition's product performance

- The best in the company's own or a related industry

- The best in the world

Benchmarking can be informal (when an individual or a team observes and copies the work that an outstanding individual or team is doing) or formal. Formal benchmarking is a term normally referring to emulating the *best practices in the industry* or even the *best practices in the world*.

Formal benchmarking consists of following these steps:

1. Obtain a commitment to change the organization's business practices

2. Determine which processes the company needs to improve

3. Identify the best practices

4. Form a benchmarking partnership with a firm that displays the best practices

5. Study the benchmarking partner's operations and products for best measures and practices

6. Create benchmarks that are measures of outstanding performance

7. Compare current practices with the benchmark partner's practices

8. Develop a plan to change the firm's operational practices to achieve the benchmark standards

OBTAIN A COMMITMENT TO CHANGE THE ORGANIZATION'S BUSINESS PRACTICES

Upper management must endorse the policy to employ benchmarking to drive change in the organization. This commitment must not be sought or rendered lightly. Significant changes will often be necessary, and the resistance to change old practices is very strong in most firms.

Obtaining a commitment from upper management is a good start toward making benchmarking an effective driver of change in the organization, but it is not sufficient. Employees and leaders at all levels who participate in the processes that need to be improved must be invited to contribute to planning and implementing the necessary changes or the benchmarking initiative will almost certainly fail.

DETERMINE WHICH PROCESSES THE COMPANY NEEDS TO IMPROVE

Means for determining which processes need to be improved consist of identifying waste (*muda*) and determining which business processes created the waste. Techniques for identifying waste are described in Chapters 9, 10, and 11 of this handbook. Internal weaknesses in the company and threats as determined in a strengths–weaknesses–opportunities–threats (SWOT) analysis are good candidates for benchmarking.

W. Edwards Deming warned that unless the company possesses deep understanding of its own processes, benchmarking (copying) processes that have been successful in other companies will not produce significant improvement. It is unlikely that a company will select an appropriate benchmarking partner unless its own processes to be benchmarked are fully understood.

IDENTIFY THE BEST PRACTICES

Determining which companies and agencies possess the processes with the best practices requires some research. Professional societies and trade organizations frequently publish a list of those companies that practice best-in-class processes in a particular industry. The Malcolm Baldrige National Quality Award is granted each year to organizations in various categories that have demonstrated that they possess best quality practices, and the MBNQA award winners have agreed in advance to serve as benchmark partners for organizations seeking support for their own quality improvement programs. Winners of other national and international award programs are also nearly always proud of their accomplishments and accordingly available to provide information about their processes. The marketing department of most large firms includes some individuals or teams who are tasked with continuous scanning of the competitive marketplace to identify companies and agencies that are likely to pose a threat in the future to the parent organization. These internal competitive intelligence experts can be consulted or even tasked to locate competitors who posses processes that exhibit the best practices in a given industry.

It is not always essential to find benchmarking practices from within a company's own industry. One large aerospace company was having trouble implementing statistical process control in its space-qualified component manufacturing process. A team of manufacturing and industrial engineers from the aerospace firm partnered with a large cookie manufacturer in the same city and were able to model the cookie manufacturer's implementation of SPC for its own operation.

Benchmarking against best-in-class may not be possible since:[1]

- The best in the world is not known

- There is no related process available

- The best in class is inaccessible due to geography or expense

- The best in class is not willing to partner

FORM A BENCHMARKING PARTNERSHIP

There are two considerations in selecting a benchmarking partner. First, the benchmarking partner should be best-in-class with respect to the processes that need to be improved. There is little point in performing a benchmarking activity if the benchmarking partner is not significantly better at something than the company seeking to improve. Second, the benchmarking partner must want to provide information that can be used to develop measures of outstanding performance. Again, there is little value in performing a benchmarking activity if the necessary information cannot be obtained.

The best benchmarking partners are companies that believe it is in their own best interest to share information about their practices and business processes. Most companies are reluctant to share information about their operations with other firms, particularly competitors. Accordingly, in order to form a benchmarking partnership it may be necessary to seek companies and organizations in different industries. Regardless of which industry a company is in, the benchmarking partner should feel that there is some value to itself in sharing information with a benchmarking partner. Some companies believe that to share information about themselves is a method for improving their public image and increasing sales. There is great public image and marketing value to a company in winning the Malcolm Baldrige National Quality Award, and a condition of applying for that award is to agree to make information about your processes available to other companies.

Quid quo pro is a term that describes giving something of value to get something of value—offering a benchmarking partner something that they value is an excellent way to encourage them to share information about their practices. Some companies will accept monetary payments in exchange for information about their processes, but most will not. If one of a company's business processes is already recognized as best in class, this company would have no difficulty in forming a partnership with another company some of whose processes are also best in class in a *quid quo pro* exchange for information about each other's best-in-class processes.

STUDY THE BENCHMARKING PARTNER'S OPERATIONS AND PRODUCTS

This step consists of collecting and analyzing information and data on the performance of the benchmark leader. Visiting the benchmark leader is always desirable. Source information by be obtained from Internet searches, books and trade journals, professional associations, corporate annual reports and other financial documents, and consulting firms that specialize in collecting, collating, and presenting benchmark information for a fee. Caution should be exercised whenever benchmark information is collected and analyzed since the data any company presents about itself is bound to be biased and will always be presented in as favorable a light as possible.

CREATE BENCHMARK MEASURES

There are many sorts of comparisons that benchmarking can highlight. These include organizational, process, performance, and project benchmarks.

Organizational benchmarking compares the overall strategic activities of the benchmark leader with another company. How the benchmark leader sets and accomplishes goals, integrates all the aspects of its business, satisfies or delights its customers, and drives its financial successes are examples of questions one would ask to understand how this best-in-class firm operates.

Process benchmarking compares a specific business process performed by the leading firm to processes the other firm desires to improve. Enterprise resource planning (ERP), strategic project management, customer relationship management (CRM), and employee retention are examples of specific best-in-class processes that a company who is striving to improve its business might seek understand.

Performance benchmarking enables managers to compare their own company's products and services to those recognized as the best in class. Price, technical quality, customer service, packaging, bundling, speed, reliability, and attractiveness are examples of factors performance benchmarking compares.

Project benchmarking consists of comparing a best-in-class project with other projects. Of all the types of benchmarking, this is perhaps the easiest to execute since it allows comparison to be made with organizations that are not direct competitors. While the objectives and even the practices between two projects are different, they will share many of the same elements. Projects share the common constraints of time, cost, resources, and performance. Before a new project is begun, benchmarking can stimulate the selection and application of better techniques and tools for planning, scheduling, monitoring, and controlling the project. Many companies insist on a postmortem, a formal meeting following the completion of each major project, to surface lessons learned during the execution of the project. These lessons learned can lead to benchmarks that can be used to stimulate improvements in subsequent projects.

COMPARE CURRENT PRACTICES WITH THE PARTNER'S

One of the main reasons benchmarking projects do not work when they don't is that the company fails to establish specific plans for closing the gap between current practices and the benchmark leader's practices. To avoid this pitfall, it is mandatory that specific gaps be identified

DEVELOP A PLAN TO CHANGE OPERATIONAL PRACTICES

When gaps exist between the firm's current practices and the benchmark leader's practices, a plan to close those gaps must be developed and implemented. This plan must address overcoming the inevitable resistance to change that will surface. Employees and even managers who have performed a job in the same way for a long time—sometimes decades—always find it difficult to change what they do. The benchmarking activity will not have impact if it does not result in an action plan to improve current processes.

Part II.G

Part III
Data Analysis

Part III

Chapter 16

A. Terms and Definitions

1. BASIC STATISTICS

> Define, compute, and interpret mean, median, mode, standard deviation, range, and variance. (Apply)
>
> **Body of Knowledge III.A.1**

Measures of Central Tendency

The values that represent the center of a data set are called *measures of central tendency*. There are three measures of central tendency: mean, median, and mode.

Mean. *Mean* is the statistical term for the more common word "average." It is calculated by finding the sum of the values in the data set and dividing by the number of values. The symbol used for "sum" is the Greek capital sigma, Σ. The values of the data set are symbolized by x's and the number of values is usually referred to as n. The symbol for mean is an x with a bar above it ($\bar{X}$). This symbol is pronounced "x bar." The formula for mean is:

$$\bar{X} = \frac{\Sigma X}{n} = \frac{X_1 + X_2 + \ldots X_n}{n}$$

This formula translates to addition of all the x's and dividing the resulting sum by n, the number of values. In other words, what is commonly known as "finding the average."

Example: Find the mean of the data set: 4   7   8   2   9   2

There are six members in the data set, so the mean is the total divided by 6:

$$\bar{X} = \frac{\Sigma X}{n} = \frac{4+7+8+2+9+2}{6} \approx 5.3$$

110

The ≈ symbol is used to indicate "approximately equal" because the value has been rounded to 5.3. It is common practice to calculate the mean to one more digit of accuracy than that of the original data.

Median. The *median* is the value that has approximately 50 percent of the values above it and 50 percent below. To find the median, first sort the data in ascending order. If there are an odd number of values, the median is the middle value of the sorted list. If there are an even number of values, the median is the mean of the two middle values in the list.

Example: The list in the previous example is first sorted into ascending order: 2   2   4   7   8   9

Since there are six values, the median is the mean of the third and fourth values in the sorted set, in this case, the average of 4 and 7, which in this case is 5.5.

Example: Find the median of the seven-element set: 2   2   4   5   7   8   9

Since the list is already sorted and has an odd number of values, the median is the middle value in the list, in this case 5.

Mode. The *mode* is the value that occurs most often in a data set. If no value occurs more than once, the set has no mode. If there is a tie for the value that occurs most often, the set will have more than one mode.

Example: Find the mode: 2   2   4   5   7   8   9

Since 2 is the only value that occurs more than once, the mode of the data set is 2.

Example: Find the mode: 2   2   4   5   5   5   7   8   8   8   9

There are two modes is this data set, 5 and 8.

Modes appear as high points or peaks on histograms. If a histogram has two peaks, it is referred to as *bimodal* even if the peaks aren't exactly the same height. A bimodal histogram usually indicates that a process variable may have been influenced by two different sources of variation, such as two different raw materials or different shifts that are using two different methods, either to run the process or to measure. A bimodal histogram often presents an opportunity to reduce variation by using more consistent raw materials or standardizing measuring or processing methods.

Measures of Dispersion

The *dispersion*, also called the *spread* or the *variation*, may be measured using the *range*, which is defined as the largest value minus the smallest value.

$$\text{Range} = R = (\text{largest value}) - (\text{smallest value})$$

Example: For the data set: 2   2   4   5   5   7   8   8

the range = 8 − 2 = 6

The range is often plotted on control charts, as discussed in Chapter 20 on $\bar{X}$ and R charts.

What would be the disadvantage of using the range as a measure of dispersion? The drawback to using the range is that it uses only two of the values from the data set: the largest and the smallest. If the data set is large, the range does not make use of much of the information contained in the data. For this and other reasons, the *standard deviation* is frequently used to measure dispersion. The value of the standard deviation may be approximated using numbers from the control chart. This method is explained in Chapter 20 under Process Capability. The standard deviation can also be found by entering the values of the data set into a calculator that has a standard deviation key. Refer to the calculator manual for appropriate steps.

Although the standard deviation is seldom calculated by hand, the following discussion provides some insight into its meaning.

Suppose it is necessary to estimate the standard deviation of a very large data set. One approach would be to randomly select a sample from that set. Suppose the randomly selected sample consists of the values 2, 7, 9, and 2. Naturally, it would be better to use a larger sample, but this will illustrate the steps involved. It is customary to refer to the sample values as "x-values" and list them in a column headed by the letter x. The first step, as illustrated in Figure 16.1(a), is to calculate the sum of the column (Σx) and the mean of the column ($\bar{X}$). Recall that $\bar{X} = \Sigma x / n$, where n is the sample size, four in this case. The next step as illustrated in Figure 16.1(b) is to calculate the deviation of each of the sample values from the mean. This is done by subtracting $\bar{X}$ from each of the sample values. The values in this column are called the *deviations from the mean*. The total or sum of the $x - \bar{X}$ values will typically be zero, so this total is not of much use. In Figure 16.1(c), a third column, labeled $(x - \bar{X})^2$, has been added. The values in this column are obtained by squaring each of the four values in the previous column. Recall that the square of a negative number is a positive number. The values in this column are called the *squares of the deviations from the mean*, and the sum of this column is the *sum of*

x
2
7
9
2

Σ 20

(a) Step 1: Find Σx and $\bar{X}$

x	$x - \bar{X}$
2	$2 - 5 = -3$
7	$7 - 5 = 2$
9	$9 - 5 = 4$
2	$2 - 5 = -3$

Σ 20   0

(b) Step 2: Form $x - \bar{X}$ column by subtracting $\bar{X}$ from each x-value

x	$x - \bar{X}$	$(x - \bar{X})^2$
2	-3	9
7	2	4
9	4	16
2	-3	9

Σ 20   0   38

(c) Step 3: Form the $(x - \bar{X})^2$ column by squaring the values in the $(x - \bar{X})$ column

Figure 16.1   Standard deviation calculation.

Source: D. W. Benbow, A. K. Elshennawy, and H. F. Walker, *The Certified Quality Technician Handbook* (Milwaukee: ASQ Quality Press, 2003): 27.

the squares of the deviations from the mean. The next step is to divide this sum by $n - 1$. Recall that $n = 4$ in this example, so the result is $38/3 \approx 12.7$. This quantity is called the *variance* and its formula is:

$$\text{Sample variance} = s^2 = \frac{\Sigma(x - \bar{X})^2}{n-1}$$

One disadvantage of the variance is that, as the formula indicates, it is measured in units that are the square of the units of the original data set. That is, if the *x*-values are in inches, then the variance is in square inches. If the *x*-values are in degrees Fahrenheit, then the variance is in square degrees Fahrenheit, which does not physically mean anything. For many applications, quality professionals and engineers need to use a measure of dispersion that is in the same units as the original data. For this reason, the preferred measure of dispersion is the square root of the variance, which is called the *standard deviation*. Its formula is:

$$\text{Sample standard deviation} = \sqrt{\text{variance}} = \sqrt{s^2} = s = \sqrt{\frac{\Sigma(x - \bar{X})^2}{n-1}}$$

As indicated at the beginning of this example, the sample standard deviation is used to estimate the standard deviation of a data set by using a sample from that data set. In some situations, it may be possible to use the entire data set rather than a sample. Statisticians refer to the entire data set as the *population* and the standard deviation is then called the population standard deviation, symbolized by the small case Greek sigma, σ. It is common to use capital N to refer to the number of values in the population. The only difference in the formula is that the divisor in the fraction is N rather than $n - 1$.

$$\text{Population standard deviation} = \sigma = \sqrt{\frac{\Sigma(x - X)^2}{N}}$$

When using the standard deviation function on a calculator, care should be taken to use the appropriate key. Unfortunately, there is not a universal labeling practice among calculator manufacturers. Some label the sample standard deviation key σ_{n-1} and the population standard deviation σ_n, while others use S_x and σ_x. Consult the calculator manual for details. Try entering the values 2, 7, 9, 2 in a calculator and verify that the sample standard deviation rounds to 3.6 and the population standard deviation is 3.1.

Of just what use is the standard deviation? One application is the comparison of two data sets. Suppose two machines can produce a certain shaft diameter. Sample parts from the two machines are collected and the diameters are measured and their sample standard deviations are calculated. Suppose the sample standard deviation of the parts from machine A is much smaller than the sample standard deviation of the parts from machine B. In other words, the diameters from machine A are closer to themselves than are the diameters from machine B. This means that the diameters from machine A have less variation, or smaller dispersion, than those produced by machine B and, other things being equal, machine A would be the preferred machine for these parts. The sample standard deviation is used rather than the population standard deviation because the population would be all the parts of this type that the machine would ever produce. In most practical applications, the sample standard deviation is the appropriate choice. In fact, some calculators do not have a population standard deviation key.

Notice that as the sample size gets large, the difference between the values of the sample standard deviation and the population standard deviation becomes quite small. For instance, if the sample size is 300, the divisor in the sample standard deviation formula is

$$300 - 1 = 299$$

and in the population standard deviation formula, the divisor would be 300.

The standard deviation also has applications to statistical inference, control charts, and process capability that will be discussed later.

2. BASIC DISTRIBUTIONS

> Define and explain frequency distributions (normal, binomial, Poisson, and Weibull) and the characteristics of skewed and bimodal distributions. (Understand)
>
> **Body of Knowledge III.A.2**

Normal Distribution

The normal distribution results from "continuous" data, that is, data that come from measurement on a continuous scale such as length, weight, or temperature, for example. Between any two values on a continuous scale there are infinitely many other values. Consider that on an inch scale between 1.256 and 1.257 there are values such as 1.2562, 1.2568, 1.256999, and so on.

The next two distributions to be discussed in this section, the binomial distribution and the Poisson distribution, are called "discrete" distributions. In quality applications, these distributions typically are based on count or attributes data rather than data measured on a continuous scale. The items being counted are often defective products or products that have defects. The word *defective* is used when all products are divided into two categories such as good and bad. The word *defect* is used when a particular product may have several defects or flaws, none of which may cause the product to be defective. The word defective would be used for light bulbs that failed to light up. The word defect would be used to describe minor scratches, or paint runs, for example, that would not cause the product to be rejected. To add to the confusion, a defective product is sometimes referred to as a nonconforming product while a defect is sometimes called a nonconformity.

Binomial Distribution

The prefix "bi-" implies the number two, as in *bi*cycle (two wheels) and *bi*partisan (two political parties). The binomial distribution is used when every object fits into one of two categories. The most frequent application in the quality field is when every part is classified as either good or defective, such as in valve leak tests or circuit continuity tests. In these cases, there are two possibilities: the valve either leaks or it does not; the circuit either passes current or it does not. If data on the amount of leakage or the amount of resistance were collected, the binomial distribution would not be an appropriate or best choice for describing a process. The word defective is often used in binomial distribution applications. A typical example might consider the number of defectives in a random sample of size 10. Notice that the number of defectives is not a continous variable because, for instance, between two defectives and three defectives, there are not an infinite number of other values, that is, it does not make sense to have 2.8 defectives. That is why this distribution is called *discrete*.

Example: Suppose 20 percent of the parts in a batch of 100,000 are defective and a sample of 10 parts is randomly selected. What is the probability that exactly one of the 10 is defective? The correct answer, stated in the conventional probabilistic notation, is $P(X = 1) \approx .27$. This reads as: "The probability that the number of defectives equals (is exactly) one is approximately .27." The formula for finding this answer will be given later. It can be used to find the probability that exactly two of the 10 are defective, exactly three are defective, and so on, up to exactly 10 of the 10 are defective. The results of applying the formula 11 times for the 11 possible answers are shown in Figure 16.2. Figure 16.2 also displays a histogram of the 11 results.

As the legend of Figure 16.2 indicates, the histogram depicts the binomial distribution for sample size = 10 when the rate of defectives is 20 percent or $p = .20$.

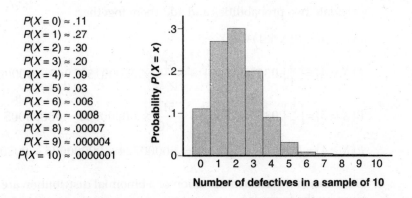

$P(X = 0) \approx .11$
$P(X = 1) \approx .27$
$P(X = 2) \approx .30$
$P(X = 3) \approx .20$
$P(X = 4) \approx .09$
$P(X = 5) \approx .03$
$P(X = 6) \approx .006$
$P(X = 7) \approx .0008$
$P(X = 8) \approx .00007$
$P(X = 9) \approx .000004$
$P(X = 10) \approx .0000001$

Number of defectives in a sample of 10

Figure 16.2 Binomial distribution with $n = 10$ and $p = .20$.

Source: D. W. Benbow, A. K. Elshennawy, and H. F. Walker, *The Certified Quality Technician Handbook* (Milwaukee: ASQ Quality Press, 2003): 40.

Statistics books sometimes refer to this as "ten trials with probability of success = .20 on each trial." *Success* in this example refers to selection of a defective part.

The formula for calculating the binomial probabilities is called the binomial formula:

$$P(X=x) = \binom{n}{x} p^x (1-p)^{n-x}$$

where

n = number of trials or sample size

x = number of successes (or defectives in this example)

p = probability of success in each trial (the probability of a part being defective)

and $\binom{n}{x}$ = number of combinations of x objects from a collection of n objects,

or stated another way, the number of combinations of x objects taken n at a time. (Refer to Section 3 of this chapter for a formula for combinations.)

To calculate $P(X = 3)$ using the binomial formula, where $n = 10$, $x = 3$, $p = .2$,

$$P(X=3) = \binom{10}{3} .2^3 \times (.8)^7 \approx \frac{10!}{3! \times 7!} \times .008 \times .21 \approx .2016$$

The probability of finding at most three defects, $P(X \le 3)$, can be found by calculating four probabilities and adding them:

$$P(X \le 3) = P(X=0) + P(X=1) + P(X=2) + P(X=3)$$

Example: A large batch of parts is 3.5 percent defective. A random sample of five is selected. What is the probability that this sample has at least four defective parts?

Calculate two probabilities and add them together:

$$P(X \ge 4) = P(X=4) + P(X=5)$$

$$P(X=4) = \binom{5}{4} .035^4 \times (.965)^1 \approx \frac{5!}{4! \times 1!} \times .0000015 \times .965 \approx .00000724$$

$$P(X=5) = \binom{5}{5} .035^5 \times (.965)^0 \approx \frac{5!}{5! \times 0!} \times .00000005 \approx .00000005$$

$$P(X \ge 4) = P(X=4) + P(X=5) \approx .00000724 + .00000005 \approx .00000729$$

The mean and standard deviation of a binomial distribution are given by the formulas

$$\mu = np \text{ and } \sigma = \sqrt{p(1-p)}$$

In this example,

$$\mu = 5 \times 0.35 = 1.75 \qquad \sigma = \sqrt{.035 \times 965} \approx .184$$

Poisson Distribution

When counting defects rather than defectives, the Poisson distribution is used rather than the binomial distribution. For example, a process for making sheet goods has an upper specification of eight bubbles per square foot. Every 30 minutes, the number of bubbles in a sample square foot are counted and recorded. If the average of these values is $\bar{c}$, the Poisson distribution is

$$P(X = x) = \frac{e^{-\bar{c}}\bar{c}^{x}}{x!}$$

where

x = number of defects

$\bar{c}$ = mean number of defects

e = natural log base (use the e^x key on the calculator)

In the example, if the average number of defects (bubbles) per square foot is 2.53, the probability of finding exactly four bubbles is

$$P(X = 4) = \frac{e^{-2.53}2.53^{4}}{4!} \approx .136$$

Consult the calculator instruction manual to evaluate $e^{-2.53}$.

The symbol in the position of the $\bar{c}$ in the above formula varies from book to book; however, many statistics texts use the Greek letter λ (lambda). The use of $\bar{c}$ in this formula is more consistent with the control limit formulas for the c-chart to be discussed later. The formulas for the mean and standard deviation of the Poisson distribution are

$$\mu = \bar{c} \text{ and } \sigma = \sqrt{\bar{c}}$$

Weibull Distribution

The Weibull distribution is often used in the field of life data analysis due to its flexibility—it can mimic the behavior of other statistical distributions such as the normal and the exponential. It requires the use of the natural logarithmic base for probability calculations. If the failure rate decreases over time, then $k < 1$. If the failure rate is constant over time, then $k = 1$. If the failure rate increases over time, then $k > 1$.

An understanding of the failure rate may provide insight as to what is causing the failures:

A decreasing failure rate would suggest "infant mortality." That is, defective items fail early and the failure rate decreases over time as they fall out of the population.

A constant failure rate suggests that items are failing from random events.

An increasing failure rate suggests "wear out"—parts are more likely to fail as time goes on.

When $k = 3.4$, then the Weibull distribution appears similar to the normal distribution. When $k = 1$, then the Weibull distribution reduces to the exponential distribution.[1]

3. PROBABILITY

Describe and apply basic terms and concepts (independence, mutual exclusivity, etc.) and perform basic probability calculations. (Apply)

Body of Knowledge III.A.3

The probability that a particular event occurs is a number between zero and one inclusive. For example, if a lot consisting of 100 parts has four defectives, we would say that the probability of randomly drawing a defective is .04 or four percent. Symbolically, this is written $P(\text{defective}) = .04$. The word "random" implies that each part has an equal chance of being drawn. If the lot had no defectives, the probability of drawing a defective would be 0 or zero percent. If the lot had 100 defectives, the probability of drawing a defective would be 1 or 100 percent.

Basic Probability Rules

Complementation rule: The probability that an event A will *not* occur is 1 − (the probability that event A does occur). Stated symbolically, $P(\text{not A}) = 1 - P(A)$. Some texts use symbols for "not A" such as −A, ~A, and sometimes A with a bar over it.

Special addition rule: Suppose a card is randomly selected from a standard 52-card deck. What is the probability that the card is a club? Since there are 13 clubs, $P(\clubsuit) = 13/52 = .25$. What is the probability that the card is either a club or a spade? Since there are 26 cards that are either clubs or spades,

$$P(\clubsuit \text{ or } \spadesuit) = P(\clubsuit) + P(\spadesuit) = 13/52 + 13/52 = 26/52 = .25 + .25 = .5$$

Therefore, it appears that $P(\clubsuit \text{ or } \spadesuit) = P(\clubsuit) + P(\spadesuit)$ which, generalized, becomes the special addition rule: $P(A \text{ or } B) = P(A) + P(B)$ with a word of warning: use only if A and B can not occur simultaneously. Note that the probability that event A or event B occurs can also be denoted by $P(A \cup B)$.

General addition rule: What is the probability of selecting either a king or a club? Using the special addition rule, if it were applicable (it will be demonstrated that it is not), $P(K \text{ or } \clubsuit) = P(K) + P(\clubsuit) = 4/52 + 13/52 = 17/52$. This is incorrect, because there are only 16 cards that are either kings or clubs or both (13 clubs, including one of which is a king of clubs, plus K$\diamondsuit$, K$\heartsuit$, and K$\spadesuit$). The reason that the special addition rule does not work here is that the two events (drawing a king and drawing a club) can occur simultaneously. The probability that events A and B both occur is denoted by $P(A \text{ and } B)$ or $P(A \cap B)$. This leads to the general addition rule:

$$P(A \text{ or } B) = P(A \cup B) = P(A) + P(B) - P(A \cap B)$$

The special addition rule has the advantage of being somewhat simpler but its disadvantage is that it is not valid when A and B can occur simultaneously. The general addition rule, although seemingly more complicated, is always valid! For the above example:

$$P(K \text{ and } \clubsuit) = 1/52$$

since only one card is both a king and a $\clubsuit$. To complete the example:

$$P(K \text{ or } \clubsuit) = P(K) + P(\clubsuit) - P(K \text{ and } \clubsuit) = 4/52 + 13/52 - 1/52 = 16/52$$

Two events that can not occur simultaneously are called *mutually exclusive*. So the warning for the special addition rule is sometimes stated as follows: "Use only if events A and B are mutually exclusive."

Contingency Tables

Suppose various parts in a lot or batch have two categories of attributes. The categories are color and size. Then further suppose that each part is one of four colors (red, yellow, green, blue) and one of three sizes (small, medium, and large). A useful tool that displays these attributes is the contingency table:

	Red	Yellow	Green	Blue
Small	16	21	14	19
Medium	12	11	19	15
Large	18	12	21	14

Each part belongs in exactly one column and exactly one row. So each part belongs in exactly one of the 12 cells in our example. When columns and rows are totaled, the table becomes:

	Red	Yellow	Green	Blue	Totals
Small	16	21	14	19	70
Medium	12	11	19	15	57
Large	18	12	21	14	65
Totals	46	44	54	48	192

Part III.A.3

Note that the total of all of the parts with the various attributes can be computed in two ways: by adding the subtotals for size vertically or by adding the subtotals for color horizontally.

Example: If one of the 192 parts is randomly selected, find the probability that the part is red.

Solution: $P(\text{red}) = P(\text{red and small}) + P(\text{red and medium}) + P(\text{red + large}) = 16/192 + 12/192 + 18/192 = 46/192 \approx .240$

Find the probability that the part is small.

Solution: $P(\text{small}) = P(\text{small} \cap \text{red}) + P(\text{small} \cap \text{yellow}) + P(\text{small} \cap \text{green}) + P(\text{small} \cap \text{blue}) = 16/192 + 21/192 + 148/192 + 19/192 = 70/192 \approx .365$

Find the probability that the part is red and small.

Solution: Since there are 16 parts that are both red and small, $P(\text{red} \cap \text{small}) = 16/192 \approx .083$

Find the probability that the part is red or small.

Solution: Since it is possible for a part to be both red and small simultaneously, the general addition rule must be used.

$P(\text{red or small}) = P(\text{red}) + P(\text{small}) - P(\text{red} \cap \text{small}) = 46/192 + 70/192 - 16/192 \approx .521$

Find the probability that the part is red or yellow.

Solution: Since no part can be both red and yellow simultaneously, the special addition rule may be used.

$P(\text{red or yellow}) = P(\text{red}) + P(\text{yellow}) = 46/192 + 44/192 \approx .469$

Notice that the general addition rule could also have been used:

$P(\text{red or yellow}) = P(\text{red}) + P(\text{yellow}) - P(\text{red} \cap \text{yellow}) = 46/192 + 44/192 - 0 \approx .469$

Conditional Probability

Example: Continuing with the above example, suppose that the selected part is known to be green. With this knowledge, what is the probability that the part is large?

Solution: Since the part is located in the green column of the table, it is one of the 54 green parts. So the lower number in the probability fraction is 54. Since 21 of the 54 parts are large, $P(\text{large, given that it is green}) = 21/54 \approx .389$

This is referred to as *conditional probability*. It is denoted $P(\text{large} \mid \text{green})$ and read as "The probability that the part is large given that it is green." It is useful to remember that the category to the right of the $\mid$ in the conditional probability symbol is the denominator in the probability fraction. Find the following probabilities:

P(small | red)   *Solution:* P(small | red) = 16/46 ≈ .348

P(red | small)   *Solution:* P(red | small) = 16/70 ≈ .229

P(red | green)   *Solution:* P(red | green) = 0/54 = 0

A formal definition for conditional probability is:

$$P(B \mid A) = P(A \cap B) / P(A)$$

Verifying that this formula is valid in each of the above examples will aid in understanding this concept.

General Multiplication Rule

Multiplying both sides of the conditional probability formula by $P(A)$:

$$P(A \cap B) = P(A) \times P(B \mid A)$$

This is called the *general multiplication rule.* It is useful to verify that this formula is valid using examples from the contingency table.

Independence and the Special Multiplication Rule

Consider the contingency table

	X	Y	Z	Totals
F	17	18	14	49
G	18	11	16	45
H	25	13	18	56
Totals	60	42	48	150

$$P(G \mid X) = 18/60 = .300 \text{ and } P(G) = 45/150 = .300$$
$$\text{so } P(G \mid X) = P(G)$$

The events G and X are called *statistically independent* or just independent events. Knowing that a part is of type X does not affect the probability that it is of type G. Intuitively, two events are called independent if the occurrence of one does not affect the probability that the other occurs. The formal definition of independence is: $P(B \mid A) = P(B)$. Making this substitution in the general multiplication rule produces the *special multiplication rule:*

$$P(A \cap B) = P(A) \times P(B)$$

So the warning for the special multiplication rule is sometimes stated as follows: "Use only if events A and B are independent."

Example: A box holds 129 parts, of which six are defective. A part is randomly drawn from the box and placed in a fixture. A second part is then drawn from the box. What is the probability that the second part is defective?

The probability can not be determined directly unless the outcome of the first draw is known. In other words, the probabilities associated with successive draws depend on the outcome of the previous draws. Using the symbol D_1 to denote the event that the first part is defective and G_1 to denote the event that the first part is good, and so on, here is one way to solve the problem:

There are two mutually exclusive events that result in a defective part for the second draw: good on the first draw and defective on the second draw, or else defective on the first draw and defective on the second draw. Symbolically, these two events are

$$(G_1 \text{ and } D_2) \text{ or else } (D_1 \text{ and } D_2)$$

The first step is to find the probability for each of these events.

By the general multiplication rule:

$$P(G_1 \cap D_2) = P(G_1) \times P(D_2 \mid G_1) = 123/129 \times 6/128 = .045$$

Also by the general multiplication rule:

$$P(D_1 \cap D_2) = P(D_1) \times P(D_2 \mid D_1) = 6/129 \times 5/128 = .002$$

Using the special addition rule:

$$P(D_2) = .045 + .002 = .047$$

When drawing two parts, what is the probability that one will be good and one defective? Drawing one good and one defective can occur in two mutually exclusive ways:

$$P(\text{one good and one defective}) = P((G_1 \text{ and } D_2) \text{ or}$$
$$(D_1 \text{ and } G_2)) = P((G_1 \cap D_2) \text{ or}$$
$$(D_1 \cap G_2)) = P(G_1 \cap D_2) + P(D_1 \cap G_2)$$
$$(\text{Mutually exclusive})$$

$$P(G_1 \cap D_2) = P(G_1) \times P(D_2 \mid G_2) = 123/129 \times 6/128 = .045$$

$$P(D_1 \cap G_2) = P(D_1) \times P(G_2 \mid D_2) = 6/129 \times 123/128 = .045$$

$$\text{So } P(\text{one good and one defective}) =$$
$$P(G_1 \cap D_2) + P(D_1 \cap G_2) = .045 + .045 = .090$$

Combinations

Example: A box of 20 parts has two defectives. The quality process analyst inspects the box by randomly selecting two parts. What is the probability that both parts selected are defective? The general formula for this type of problem is

$$P = \frac{\text{number of ways an event can occur}}{\text{number of possible outcomes}}$$

The "event" in this case is selecting two defectives so "number of ways an event can occur" refers to the number of ways two defective parts could be selected. There is only one way to do this since there are only two defective parts. Therefore the top number in the fraction is 1. The lower number in the fraction is the "number of possible outcomes." This refers to the number of different ways of selecting two parts from the box. This is also called the "number of combinations of two objects from a collection of 20 objects." The formula is

Number of combinations of n objects taken r at a time =

$$_nC_r = \frac{n!}{r!(n-r)!}$$

Note: Another way of writing this is $\binom{n}{r}$.

In this formula, the exclamation mark is pronounced "factorial," so $n!$ is pronounced "n factorial." The value of 6! is $6 \times 5 \times 4 \times 3 \times 2 \times 1 = 720$. The value of $n!$ is the result of multiplying the first n positive whole numbers. Most scientific calculators have a factorial key, typically labeled $x!$. To calculate 6! using this key, press 6 followed by the $x!$ key. Returning to the previous example, the lower number in the fraction is the number of possible combinations of 20 objects taken two at a time. Substituting into this formula:

$$\binom{20}{2} = \frac{20!}{2!(20-2)!} = \frac{20!}{2!18!}$$

This value would be calculated by using the following sequence of calculator keystrokes:

$$20x! \div (2x! \times 18x!) = 190$$

The answer to the example is $1/190 \approx .005$.

How might this be useful in the inspection process? Suppose that a supplier has shipped this box with the specification that it have no more than two defective parts. What is the probability that the supplier has met this specification? The answer $\approx .005$, which does not bode well for the supplier's quality system.

Example: A box of 20 parts has three defectives. The quality process analyst inspects the box by randomly selecting two parts. What is the probability that both parts selected are defective?

Part III.A.3

The bottom term of the fraction remains the same as in the previous example. The top term is the number of combinations of three defective objects taken two at a time:

$$\binom{3}{2} = \frac{3!}{2!(3-2)!} = \frac{3!}{2!1!} = \frac{3 \times 2 \times 1}{2 \times 1 \times 1} = 3$$

To see that this makes sense, name the three defectives A, B, and C. The number of different two-letter combinations of these three letters is AB, AC, BC. Note that AB is not a different combination from BA because it is the same two letters. If two defectives are selected, the order in which they are selected is not important. The answer to the problem is $3/190 \approx .016$.

An important thing to remember: *combinations are used when order is not important!*
   Note: Calculators have an upper limit to the value that can use the $x!$ key. If a problem requires a higher factorial, use the statistical function on a spreadsheet program. It is interesting to observe that a human can calculate the value of some factorial problems that a calculator can not.

Example: Find $\dfrac{1000!}{997!}$.

Most calculators cannot handle 1000!. But humans know that the terms of this fraction can be written as:

$$\frac{1000 \times 999 \times 998 \times 997 \times 996 \times 995 \times 994 \text{ and so on}}{997 \times 996 \times 995 \times 994 \text{ and so on}}$$

The factors in the bottom term cancel out all but the first three factors in the top term so that the answer is $1000 \times 999 \times 998$, which most of us prefer to calculate using a calculator.

Permutations

With combinations, the order of the objects does not matter. Permutations are very similar, except that the order does matter.

Example: A box has 20 parts labeled A through T. Two parts are randomly selected. What is the probability that the two parts are A and T in that order? The general formula applies:

$$P = \frac{\text{number of ways an event can occur}}{\text{number of possible outcomes}}$$

The bottom term of the fraction is the number of orderings or permutations of two objects from a collection of 20 objects. The general formula is:

Number of permutations of n objects taken r at a time =

$$_nP_r = \frac{n!}{(n-r)!}$$

In this case, $n = 20$ and $r = 2$:

$$_{20}P_2 = \frac{20!}{(20-2)!} = \frac{20!}{18!} = \frac{20 \times 19 \times 18 \times \ldots \times 1}{18 \times 17 \times \ldots \times 1} = 20 \times 19 = 380$$

This fraction can be calculated on a calculator using the following keystrokes: 20 $x! \div 18\ x! =$. The answer is 380. Of these 380 possible permutations, only one is AT, so the top term in the fraction is 1. The answer to the probability problem is 1/380 = .003.

Example: A team with seven members wants to select a task force of three people to collect data for the next team meeting. How many different three-person task forces could be formed? This is not a permutations problem because the order in which people are selected does not matter. In other words, the task force consisting of Barb, Bill, and Bob is the same task force as the one consisting of Bill, Barb, and Bob. Therefore the combinations formula will be used to calculate the number of combinations of three objects from a collection of seven objects, or seven objects taken three at a time:

$$_7C_3 = \frac{7!}{3!(7-3)!} = \frac{7!}{3!4!} = 35$$

Thirty-five different three-person task forces could be formed from a group of seven people.

Example: A team with seven members wants to select a cabinet consisting of a chairperson, facilitator, and scribe. How many different three-person cabinets could be formed? Here the order is important because the cabinet consisting of Barb, Bill, and Bob will have Barb as chairperson, Bill as facilitator, and Bob as scribe, while the cabinet consisting of Bill, Barb, and Bob has Bill as chairperson, Barb as facilitator, and Bob as scribe. This is a permutations problem because the order in which people are selected matters. Therefore the permutations formula will be used to calculate the number of combinations of three objects from a collection of seven objects, or seven objects taken three at a time:

$$_7P_3 = \frac{7!}{(7-3)!} = \frac{7!}{4!} = 210$$

Two hundred ten different three-person cabinets could be formed from a group of seven people.

Areas under the Normal Curve

There are several applications in the quality field for the area under a normal curve. This section discusses the basic concepts so that the individual applications will be simpler to understand when introduced later in the book.

Example: A 1.00000" gage block is measured 10 times with a micrometer and the readings are: 1.0008, 1.0000, 1.0000, .9996, 1.0005, 1.0001, .9990, 1.0003, .9999, and 1.0000.

The slightly different values were obtained due to variation in the micrometer, the technique used, and other factors contributing to measurement variation. The errors for these measurements can be found by subtracting 1.00000 from each observation. The errors are, respectively, .0008, 0, 0, −.0004, .0005, .0001, −.0010, .0003, −.0001, and 0. These error values have been plotted on a histogram in Figure 16.3(a). If 500 measurements had been made and their errors plotted on a histogram, it would look something like Figure 16.3(b).

The histogram in Figure 16.3(b) approximates the normal distribution. This distribution occurs frequently in various applications in the quality sciences. The normal curve is illustrated in Figure 16.4.

It has a fairly complex formula so locations on the normal curve are seldom calculated directly. Instead, a "standard normal table," such as the one in Appendix B, is used. Some properties of the "standard normal curve" are:

- The mean is zero and the curve is symmetric about zero.

- It has a standard deviation of 1. The units on the horizontal axis are standard deviations.

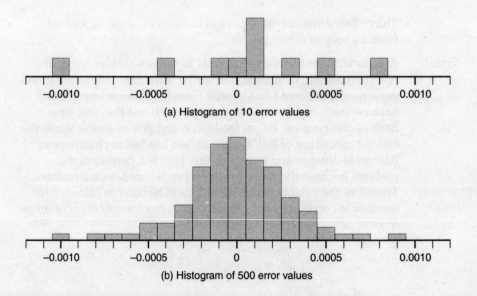

(a) Histogram of 10 error values

(b) Histogram of 500 error values

Figure 16.3 Histograms of error values.

Source: D. W. Benbow, A. K. Elshennawy, and H. F. Walker, *The Certified Quality Technician Handbook* (Milwaukee: ASQ Quality Press, 2003): 37.

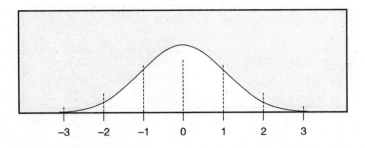

Figure 16.4   Standard normal curve.
Source: D. W. Benbow, A. K. Elshennawy, and H. F. Walker, *The Certified Quality Technician Handbook* (Milwaukee: ASQ Quality Press, 2003): 37.

- The total area under the curve is one square unit.

- The curve never touches the horizontal axis; it extends infinitely far in each direction to the right and the left.

Example: Use the standard normal table (Appendix B) to find the area under the standard normal curve to the right of 1. The values on the horizontal line are often referred to as z-values, so this problem is sometimes stated as: Find the area under the standard normal curve to the right of $z = 1$.

In Appendix B, find 1 in the z-column. The area associated with $z = 1$ is 0.1587, which is the correct answer to this problem.

Example: Find the area under the standard normal curve to the right of $z = 0$.

Intuitively, since the curve is symmetric about 0, we would suspect that the answer is 50 percent or .50. Verify this by finding zero in the z-column in Appendix B.

Example: Find the area under the standard normal curve to the right of $z = -1$. The area to the right of $z = 1$ is 0.1587, and because of the symmetry of the curve, the area to the *left* of $z = -1$ is also 0.1587. Since the total area under the curve is 1, the area to the right of $z = -1$ is $1 - 0.1587 = 0.8413$.

Example: Find the area under the standard normal curve between $z = 0$ and $z = 1$. This is the shaded area in Figure 16.5. The entire area to the right of $z = 0$ is 0.5, and the area to the right of $z = 1$ is 0.1587. The shaded area is $0.5 - 0.1587 = 0.3413$.

Example: Find the area under the standard normal curve between $z = -2$ and $z = 1$. This is the shaded area in Figure 16.6. From the table in Appendix B, the area to the right of $z = 2$ is 0.0228, so the area to the *left* of $z = -2$ is also 0.0228. Consequently, the area to the right of $z = -2$ is $1 - 0.0228 = 0.9772$, and the area to the right of $z = 1$ is .1587. The shaded area is $0.9772 - 0.1587 = 0.8185$.

Part III.A.3

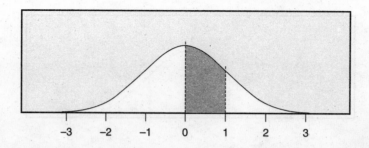

Figure 16.5   Area under the standard normal curve between $z = 0$ and $z = 1$.

Source: D. W. Benbow, A. K. Elshennawy, and H. F. Walker, *The Certified Quality Technician Handbook* (Milwaukee: ASQ Quality Press, 2003): 38.

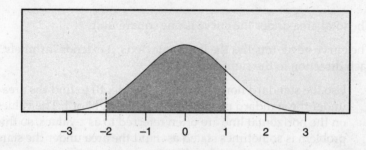

Figure 16.6   Area under the standard normal curve between $z = -2$ and $z = 1$.

Source: D. W. Benbow, A. K. Elshennawy, and H. F. Walker, *The Certified Quality Technician Handbook* (Milwaukee: ASQ Quality Press, 2003): 38.

Some normal distributions are not the standard normal distribution. The next example shows how the standard normal table in Appendix B can be used to find areas for these cases.

Example: An automatic bar machine produces parts whose diameters are normally distributed with a mean of 0.750" and standard deviation of 0.004". What percentage of the parts have diameters between 0.750" and 0.754"?

Figure 16.7 illustrates the problem. This is not the standard normal distribution, because the mean is not zero and the standard deviation is not 1. Figure 16.7 shows vertical dashed lines at standard deviations from –3 to 3. The horizontal axis has the mean scaled at the given value 0.750", and the distance between standard deviation markers is the given value of 0.004". The problem asks what percentage of the total area is shaded. From the diagram, this is the area between $z = 0$ and $z = 1$ on the standard normal curve, so the area is 0.5 – 0.1587 = 0.3413. Since the total area under the standard normal curve is 1, this represents 34.13 percent of the area and is the answer to the problem.

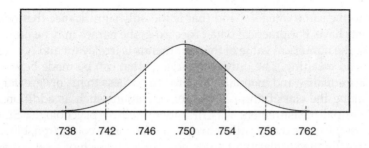

.738   .742   .746   .750   .754   .758   .762

Figure 16.7 Area under a normal curve between 0.750 and 0.754.

Source: D. W. Benbow, A. K. Elshennawy, and H. F. Walker, *The Certified Quality Technician Handbook* (Milwaukee: ASQ Quality Press, 2003): 39.

Further examples of this type are given in the process capability discussion (Chapter 20).

The standard normal curve can be related to probability in the previous example by asking the question, "If a part is selected at random from the output of the automatic bar machine, what is the probability that it will have a diameter between 0.750 and 0.754?" Since 34.13 percent of the parts have diameters in this range, the probability is .3413.

How accurate are the answers to the above examples? Suppose that 10,000 parts are produced. If the distribution is exactly normal and the population mean and standard deviation are exactly 0.750 and 0.004, respectively, then the number with diameters between 0.750 and 0.754 would be 3413. However, since exactly normal distributions exist only in theory and sample data are typically used to estimate the population mean and standard deviation, the actual number will vary somewhat.

4. MEASUREMENT SCALES

> Define and apply nominal, ordinal, interval, and ratio measurement scales. (Apply)
>
> **Body of Knowledge III.A.4**

Nominal Measurement

In this type of measurement, names are assigned to objects as labels. This assignment is performed by evaluating, by some procedure, the similarity of the to-be-measured instance to each of a set of named category definitions. The name of the category in the set is the "value" assigned by nominal measurement to the given instance. If two instances have the same name associated with them, they belong to the same category, and that is the only significance that nominal measurements have. For practical data processing, the names may be numerals, but in that case the numerical value of these numerals is irrelevant and hence their order also has no meaning. The only comparisons that can be made between variable values are equality and inequality. There are no "less than" or "greater than" relations among the classifying names, nor operations such as addition or subtraction. Nominal measurement was first identified by psychologist Stanley Smith Stevens in the context of a child learning to categorize colors (red, blue, and so on) by comparing the similarity of a perceived color to each of a set of named colors previously learned by ostensive definition. Other examples include: geographical location in a country represented by that country's international telephone access code, the marital status of a person, or the make or model of a car. The measure of central tendency of this type is the mode. Statistical dispersion may be measured with a variation ratio, index of qualitative variation, or via information entropy, but no notion of standard deviation exists. Variables that are measured only nominally are also called categorical variables. In social research, variables measured at a nominal level include gender, race, religious affiliation, political party affiliation, college major, and birthplace.

Ordinal Measurement

In this classification, the numbers assigned to objects represent the rank order (first, second, third, and so on) of the entities measured. The numbers are called ordinals. The variables are called ordinal variables or rank variables. Comparisons of greater than and less than can be made, in addition to equality and inequality. However, operations such as conventional addition and subtraction are still meaningless. Examples include the Mohs scale of mineral hardness, the results of a horse race that indicate only which horses arrived first, second, third, and so on, but with no time intervals, and most measurements in psychology and the other social sciences, for example, attitudes like preference, conservatism or prejudice, and social class. The central tendency of an ordinally measured variable can be represented by its mode or its median; the latter gives more information.

Interval Measurement

The numbers assigned to objects have all the features of ordinal measurements, and in addition, equal differences between measurements represent equivalent intervals. That is, differences between arbitrary pairs of measurements can be meaningfully compared. Operations such as addition and subtraction are therefore meaningful. The zero point on the scale is arbitrary; negative values can be used. Ratios between numbers on the scale are not meaningful, so operations such as multiplication and division can not be carried out directly. For instance, it can not be said that 30° Celsius is twice as hot as 15° Celsius. But ratios of differences can be expressed; for example, one difference can be twice another. The central tendency of a variable measured at the interval level can be represented by its mode, its median, or its arithmetic mean; the mean gives the most information. Variables measured at the interval level are called interval variables, or sometimes scaled variables, though the latter usage is not obvious and is not recommended.

Examples of interval measures are the year date on many calendars and temperature in the Celsius scale or Fahrenheit scale. About the only interval measures commonly used in social scientific research are constructed measures such as in standardized intelligence (IQ) tests.

Ratio Measurement

The numbers assigned to objects have all the features of interval measurement and also have meaningful ratios between arbitrary pairs of numbers. Operations such as multiplication and division are therefore meaningful. The zero value on a ratio scale is nonarbitrary. Variables measured at the ratio level are called ratio variables. Most physical quantities, such as mass, length, or energy are measured on ratio scales; so is temperature measured in kelvins, that is, relative to absolute zero. The central tendency of a variable measured at the ratio level can be represented by its mode, its median, its arithmetic mean, or its geometric mean; however, as with an interval scale, the arithmetic mean gives the most useful information. Social variables of ratio measure include age, length of residence in a given place, number of organizations belonged to, or number of church attendances in a particular time.

Interval and/or ratio measurement are sometimes called "true measurement," though it is often argued that this usage reflects a lack of understanding of the uses of ordinal measurement. Only ratio or interval scales can correctly be said to have units of measurement.

Debate on Classifications

There has been, and continues to be, debate about the merit of these classifications, particularly in the cases of the nominal and ordinal classifications.[2] Thus, while Stevens's classification is widely adopted, it is not universally accepted.[3]

Among those who accept the classification scheme, there is also some controversy in the behavioral sciences over whether the mean is meaningful for ordinal measurement. Mathematically it is not, but many behavioral scientists use it anyway. This is often justified on the basis that ordinal scales in behavioral science are really somewhere between true ordinal and interval scales—although the interval difference between two ordinal ranks is not constant, it is often of the same order of magnitude. Thus some argue that so long as the unknown interval difference between ordinal scale ranks is not too variable, interval scale statistics such as means can meaningfully be used on ordinal scale variables.

L. L. Thurstone made progress toward developing a justification for obtaining interval-level measurements based on the law of comparative judgment. Further progress was made by Georg Rasch, who developed the probabilistic Rasch model, which provides a theoretical basis and justification for obtaining interval-level measurements from counts of observations such as total scores on assessments.

Part III.A.4

Chapter 17

B. Data Types and Collection Methods

1. TYPES OF DATA

> Identify, define, and classify continuous
> (variables) data and discrete (attributes) data,
> and identify when it is appropriate to convert
> attributes data to variables measures. (Apply)
>
> **Body of Knowledge III.B.1**

Categorical, Qualitative, or Attributes Data

Answers to questions like "Which type of book do you prefer to read: mysteries, biographies, fiction, or nonfiction?" are examples of categorical, qualitative, or attributes data since the responses vary by type, and a numerical response would not make sense. Other types of qualitative variables would be gender, socio-economic status, or highest level of education.

Qualitative data is also called attributes data; that is, data that describes a characteristic, attribute, or quality. How many are good or bad? How many conform to specification? How many are wearing red, blue, green, or white? How many have a pleasant taste?

Quantitative Data

Quantitative data can be broken into two main subtypes: discrete and continuous. Discrete data take the form of countable measures such as:

1. Number of students enrolled in a statistics course

2. Number of pizzas ordered on a particular night

3. Number of games that a baseball team won last season

Continuous data can assume any one of the uncountable number of values on the number line. For example, the distance from home to work may be three miles, but in more exact terms, it may be 2.9849 miles. The time that it takes a plane to travel from Chicago to Los Angeles may be 4.5 hours, but it would be more accurate to say 4 hours, 33 minutes, 2.43 seconds for a particular flight, and another trip may be 4 hours, 33 minutes, 1.892 seconds long. And so on.

The need for understanding the difference between continuous and discrete data becomes apparent when discussing probability distributions in other chapters of this book.

2. METHODS FOR COLLECTING DATA

> Define and apply methods for collecting data such as using data coding, automatic gauging, etc. (Apply)
>
> **Body of Knowledge III.B.2**

Automatic Gauging

As industrial processes are automated, gauging must keep pace. Automated gauging is performed in two general ways. One is in-process or on-the-machine control by continuous gauging of the work. The second way is post-process or after-the-machine gauging control. Here, the parts coming off the machine are passed through an automatic gage. A control unit responds to the gage by sorting pieces by size and adjusts or stops the machine if parts are found beyond limits.

Data Coding

The computations of $\bar{X}$ and s can sometimes be made easier by using some important properties and methods of coding (transforming) data. These are used in changes of scale, for example. Consider the following measurements: 522.895, 522.992, 522.890, 522.900, 522.954.

What happens to the average $\bar{X}$ and standard deviation s of this data set if the values are "translated" or shifted by subtracting a constant value, say 522, from each of the measurements? The new translated data becomes 0.895, 0.992, 0.890, 0.900, 0.954. The average of this new set of numbers is the original average translated or shifted by an amount of 522, which equals 0.9262. So the original data's average is 522.9262.

The standard deviation of this new set of numbers is not changed by a translation. The standard deviation s of the new data set is the same as that of the original data set, which, in this case, is 0.044924.

What happens to the average $\bar{X}$ and standard deviation s if each value is multiplied by a constant, say, for example, 1/522? Each value from the example becomes

$$522.895 \times 1/522 = 1.001715$$
$$522.992 \times 1/522 = 1.0019$$
$$522.890 \times 1/522 = 1.001705$$
$$522.900 \times 1/522 = 1.001724$$
$$522.954 \times 1/522 = 1.001828$$

The new average is 1.001774, and to obtain the average of the original data set, multiply the new average by 522, obtaining 522.9262. Similarly, the standard deviation of the original data set can be obtained by multiplying the coding factor, in this case 522, by the new standard deviation, $0.0000861 \times 522 = .044924$.

Hence data coded in a form $y = a + bx$ where a and b are translation and factor, respectively, the new average is $\overline{Y} = a + b\overline{X}$ and the new standard deviation is $s_Y = bs_X$.

Chapter 18

C. Sampling

---

1. CHARACTERISTICS

> Identify and define sampling characteristics
> such as lot size, sample size, acceptance
> number, operating characteristic (OC) curve.
> (Understand)
>
> **Body of Knowledge III.C.1**

Sampling versus 100 Percent Inspection

Inspection can be done by sampling, that is, selecting a portion of a larger group of items, or inspection may be done by screening (also called sorting or 100 percent inspection), in which all units are inspected. *Acceptance sampling* is the process of inspecting a portion of the product in a lot for the purpose of making a decision regarding classification of the entire lot as either conforming or nonconforming to quality specifications. Sampling provides the economic advantage of lower inspection costs due to fewer units being inspected. In addition, the time required to inspect a sample is substantially less than that required for the entire lot and there is less risk of damage to the product due to reduced handling. Most inspectors find that selection and inspection of a random sample is less tedious and monotonous than inspection of a complete lot. Another advantage of sampling inspection is related to the supplier/customer relationship. By inspecting a small fraction of the lot, and forcing the supplier to screen 100 percent in case of lot rejection (which is the case in rectifying inspection), the customer emphasizes that the supplier must be more concerned about quality. On the other hand, variability is inherent in sampling, resulting in sampling errors: rejection of lots of conforming quality and acceptance of lots of nonconforming quality.

Acceptance sampling is most appropriate when inspection costs are high and when 100 percent inspection is monotonous and can cause inspector fatigue and boredom, resulting in degraded performance and increased error rates. Obviously, sampling is the only choice available for destructive inspection. *Rectifying*

135

sampling is a form of acceptance sampling. Sample units detected as nonconforming are discarded from the lot, replaced by conforming units, or repaired. Rejected lots are subject to 100 percent screening, which can involve discarding, replacing, or repairing units detected as nonconforming.

In certain situations, it is preferable to inspect 100 percent of the product. This would be the case for critical or complex products, where the cost of making the wrong decision would be too high. Screening is appropriate when the fraction nonconforming is extremely high. In this case, most of the lots would be rejected under acceptance sampling and those accepted would be so as a result of statistical variations rather than better quality. Screening is also appropriate when the fraction nonconforming is not known and an estimate based on a large sample is needed.

It should be noted that the philosophy now being espoused in supplier relations is that the supplier is responsible for ensuring that the product shipped meets the user's requirements. Many larger customers are requiring evidence of product quality through the submission of process control charts showing that the product was produced by a process that was in control and capable of meeting the specifications.

Sampling Plans

Sampling may be performed according to the type of quality characteristics to be inspected. There are three major categories of sampling plans:

1. Sampling by attributes

2. Sampling by variables

3. Special sampling plans

It should be noted that acceptance sampling is not advised for processes in continuous production and in a state of statistical control. For these processes, Deming provides decision rules for selecting either 100 percent inspection or no inspection.[1]

Lot-by-Lot versus Average Quality Protection

Sampling plans based on average quality protection from continuing processes have their characteristics based on the binomial and/or Poisson distributions. Plans used for lot-by-lot protection, not considered to have been manufactured by a continuing process, have their characteristics based on the hypergeometric distribution, which takes the lot size into consideration for calculation purposes.

Sampling plans based on the Poisson and binomial distributions are more common than those based on the hypergeometric distribution. This is due to the complexity of calculating plans based on the hypergeometric distribution. New software on personal computers, however, may eliminate this drawback.

The Operating Characteristic (OC) Curve

No matter which type of attributes sampling plan is being considered, the most important evaluation tool is the operating characteristic (OC) curve. The OC curve

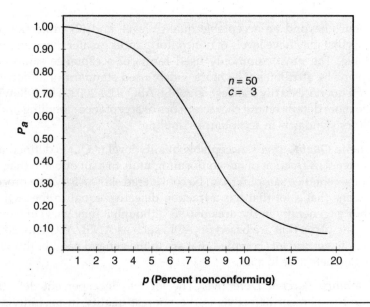

Figure 18.1 An operating characteristic (OC) curve.

Source: D. W. Benbow, A. K. Elshennawy, and H. F. Walker, *The Certified Quality Technician Handbook* (Milwaukee: ASQ Quality Press, 2003): 135.

allows a sampling plan to be almost completely evaluated at a glance, giving a pictorial view of the probabilities of accepting lots submitted at varying levels of percent defective. The OC curve illustrates the risks involved in acceptance sampling. Figure 18.1 shows an OC curve for a sample size n of 50 drawn from an infinite lot size, with an acceptance number c of 3.

As can be seen by the OC curve, if the lot were 100 percent to specifications, the probability of acceptance P_a would also be 100 percent. But if the lot were 13.4 percent defective, there would be a 10 percent probability of acceptance.

There are two types of OC curves to consider: type A OC curves and type B OC curves. Type A OC curves are used to calculate the probability of acceptance on a lot-by-lot basis when the lot is not a product of a continuous process. These OC curves are calculated using the hypergeometric distribution.

Type B OC curves are used to evaluate sampling plans for a continuous process. These curves are based on the binomial and/or Poisson distributions when the requirements for inspection and test usage are met. In general, the ANSI/ASQ Z1.4-2003 standard OC curves are based on the binomial distribution for sample sizes through 80 and the Poisson approximation to the binomial is used for sample sizes greater than 80.

Acceptance Sampling by Attributes

Acceptance sampling by attributes is generally used for two purposes: (1) protection against accepting lots from a continuing process whose average quality deteriorates beyond an acceptable quality level, and (2) protection against isolated lots that may have levels of nonconformances greater than can be considered acceptable. The most commonly used forms of acceptance sampling are sampling plans by attributes. The most widely used standard of all attributes plans, although not necessarily the best, is ANSI/ASQ Z1.4-2003. The following sections provide more details on the characteristics of acceptance sampling and discussion of military standards in acceptance sampling.

Acceptable Quality Level. Acceptable quality level (AQL) is defined as the maximum percent or fraction of nonconforming units in a lot or batch that, for the purposes of acceptance sampling, can be considered satisfactory as a process average. This means that a lot that has a fraction defective equal to the AQL has a high probability (generally in the area of 0.95, although it may vary) of being accepted. As a result, plans that are based on AQL, such as ANSI/ASQ Z1.4-2003, favor the producer in getting lots accepted that are in the general neighborhood of the AQL for fraction defective in a lot.

Lot Tolerance Percent Defective. The lot tolerance percent defective (LTPD), expressed in percent defective, is the poorest quality in an individual lot that should be accepted. The LTPD has a low probability of acceptance. In many sampling plans, the LTPD is the percent defective having a 10 percent probability of acceptance.

Producer's and Consumer's Risks. There are risks involved in using acceptance sampling plans. The risks involved in acceptance sampling are producer's risk and consumer's risk. These risks correspond with Type 1 and Type 2 errors in hypothesis testing. The definitions of producer's and consumer's risks are:

- *Producer's risk* (α)—The producer's risk for any given sampling plan is the probability of rejecting a lot that is within the acceptable quality level.[2] This means that the producer faces the possibility (at level of significance *a*) of having a lot rejected even though the lot has met the requirements stipulated by the AQL.

- *Consumer's risk* (β)—The consumer's risk for any given sampling plan is the probability of acceptance (usually 10 percent) for a designated numerical value of relatively poor submitted quality.[3] The consumer's risk, therefore, is the probability of accepting a lot that has a quality level equal to the LTPD.

Average Outgoing Quality. The average outgoing quality (AOQ) is the expected average quality of outgoing products, including all accepted lots plus all rejected lots that have been sorted 100 percent and have had all of the nonconforming units replaced by conforming units.

There is a given AOQ for specific fractions nonconforming of submitted lots sampled under a given sampling plan. When the fraction nonconforming is very low, a large majority of the lots will be accepted as submitted. The few lots that are rejected will be sorted 100 percent and have all nonconforming units replaced with conforming units. Thus, the AOQ will always be less than the submitted

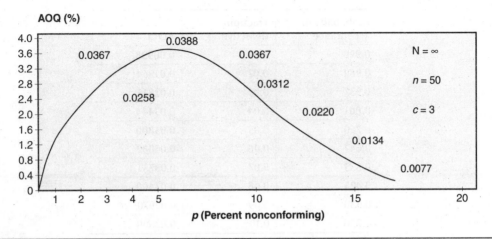

Figure 18.2   Average outgoing quality curve for $N = \infty$, $n = 50$, $c = 3$.

Source: D. W. Benbow, A. K. Elshennawy, and H. F. Walker, *The Certified Quality Technician Handbook* (Milwaukee: ASQ Quality Press, 2003): 137.

quality. As the quality of submitted lots becomes poor in relation to the AQL, the percent of lots rejected becomes larger in proportion to accepted lots. As these rejected lots are sorted and combined with accepted lots, an AOQ lower than the average fraction of nonconformances of submitted lots emerges. Therefore, when the level of quality of incoming lots is good, the AOQ is good; when the incoming quality is bad and most lots are rejected and sorted, the result is also good.

To calculate the AOQ for a specific fraction nonconforming and a sampling plan, the first step is to calculate the probability of accepting the lot at that level of fraction nonconforming. Then, multiply the probability of acceptance by the fraction nonconforming for the AOQ. Thus:

$$AOQ = P_a p[1 - \text{sample size/lot size}]$$

If the desired result is a percentage, multiply by 100.

The average outgoing quality limit is the maximum AOQ for all possible levels of incoming quality.

Average Outgoing Quality Limit. The average outgoing quality limit (AOQL) is a variable dependent on the quality level of incoming lots. When the AOQ is plotted for all possible levels of incoming quality, a curve as shown in Figure 18.2 results. The AOQL is the highest value on the AOQ curve.

Assuming an infinite lot size, the AOQ may be calculated as $AOQ = P_a p$. Probability of acceptance (P_a) may be obtained from tables as explained earlier and then multiplied by p (associated value of fraction nonconforming) to produce a value for AOQ as shown in the next example, using the previous equation.

Example: Given an OC with curve points (P_a and p) as shown, construct the AOQ curve. Note that P_a and p are calculated as explained in the previous example.

Probability of acceptance	Fraction defective	AOQ
0.998	0.01	0.00998
0.982	0.02	0.01964
0.937	0.03	0.02811
0.861	0.04	0.03444
0.760	0.05	0.03800
0.647	0.06	0.03882
0.533	0.07	0.03731
0.425	0.08	0.03400
0.330	0.09	0.02970
0.250	0.10	0.02500

As can be seen, the AOQ rises until the incoming quality level of 0.06 nonconforming is reached. The maximum AOQ point is 0.03882, which is called the AOQL. This is the AOQL for an infinite lot size, sample size = 50, accept on three or less nonconformances.

Lot Size, Sample Size, and Acceptance Number. For any single sampling plan, the plan is completely described by the lot size, sample size, and acceptance number. In this section, the effect of changing the sample size, acceptance number, and lot size on the behavior of the sampling plan will be explored along with the risks of constant percentage plans.

The effect on the OC curve caused by changing the sample size while holding all other parameters constant is shown in Figure 18.3. The probability of acceptance changes considerably as sample size changes. The P_a for the given sample sizes for a 10 percent nonconforming lot and an acceptance number of zero are shown below.

Sample size	Probability of acceptance (P_a%)
10	35
4	68
2	82
1	90

The effect of changing the acceptance number on a sampling plan while holding all other parameters constant is shown in Figure 18.4. Another point of interest is that for $c = 0$, the OC curve is concave in shape, while plans with larger accept numbers have a "reverse s" shape. Figure 18.4 and the following table show the effect of changing the acceptance number of a sampling plan on the indifference quality level (IQL: 50–50 chance of accepting a given percent defective).

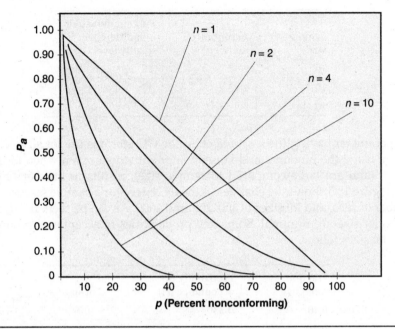

Figure 18.3   Effect on an OC curve of changing sample size *n* when accept number *c* is held constant.

Source: D. W. Benbow, A. K. Elshennawy, and H. F. Walker, *The Certified Quality Technician Handbook* (Milwaukee: ASQ Quality Press, 2003): 139.

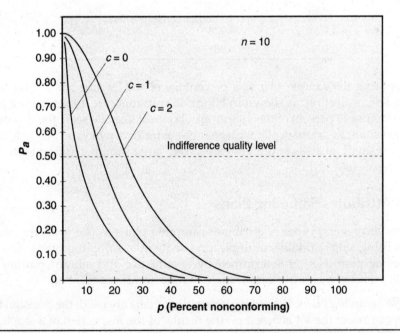

Figure 18.4   Effect of changing accept number *c* when sample size *n* is held constant.

Source: D. W. Benbow, A. K. Elshennawy, and H. F. Walker, *The Certified Quality Technician Handbook* (Milwaukee: ASQ Quality Press, 2003): 139.

Sample size	Acceptance number	Percent defective at indifference quality level (%)
10	2	27
10	1	17
10	0	7

The parameter having the least effect on the OC curve is the lot size N. For this reason, using the binomial and Poisson approximations, even when lot sizes are known (and are large compared to sample size), results in little error in accuracy. Figure 18.5 shows the changes in the OC curve for a sample size of 10, accept number of zero, and lot sizes of 100, 200, and 1000. As can be seen, the differences due to lot size are minimal. Some key probabilities of acceptance points for the three lot sizes follow.

Fraction defective	Probability of acceptance (P_a)	Lot size
0.10	0.330	100
0.30	0.023	100
0.50	0.001	100
0.10	0.340	200
0.30	0.026	200
0.50	0.001	200
0.10	0.347	1000
0.30	0.028	1000
0.50	0.001	1000

Computing the sample size as a percentage of the lot size has a large effect on risks and protection, as shown in Figure 18.6. In this case, plans having a sample size totaling 10 percent of the lot size are shown. As can be seen, the degree of protection changes dramatically with changes in lot size, which results in low protection for small lot sizes and gives excessively large sample requirements for large lot sizes.

Types of Attributes Sampling Plans

There are several types of attributes sampling plans in use, with the most common being single, double, multiple, and sequential sampling plans. The type of sampling plan used is determined by ease of use and administration, general quality level of incoming lots, average sample number, and so on.

Single Sampling Plans. When single sampling plans are used, the decision to either accept or reject the lot is based on the results of the inspection of a single sample of *n* items from a submitted lot. In the example shown earlier, the OC curve and AOQ curve were calculated for a single sampling plan where *n* = 50 and *c* = 3. Single sampling plans have the advantage of ease of administration, but due to the

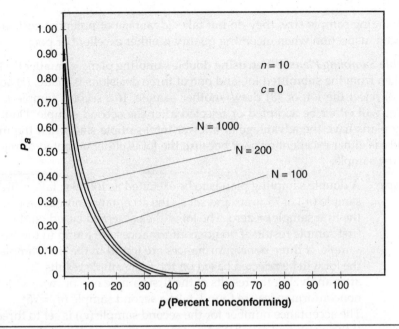

Figure 18.5   Effect of changing lot size N when accept number *c* and sample size *n* are held constant.

Source: D. W. Benbow, A. K. Elshennawy, and H. F. Walker, *The Certified Quality Technician Handbook* (Milwaukee: ASQ Quality Press, 2003): 140.

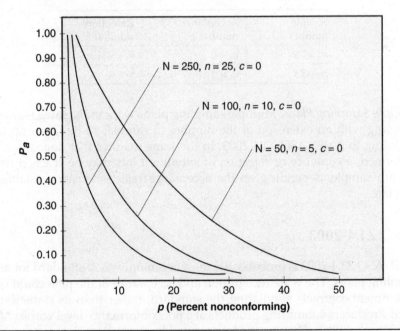

Figure 18.6   Operating characteristic curves for sampling plans having the sample size equal to 10 percent of the lot size.

Source: D. W. Benbow, A. K. Elshennawy, and H. F. Walker, *The Certified Quality Technician Handbook* (Milwaukee: ASQ Quality Press, 2003): 141.

unchanging sample size, they do not take advantage of potential cost savings of reduced inspection when incoming quality is either excellent or poor.

Double Sampling Plans. When using double sampling plans, a smaller first sample is taken from the submitted lot, and one of three decisions is made: (1) accept the lot, (2) reject the lot, or (3) draw another sample. If a second sample is drawn, the lot will either be accepted or rejected after the second sample. Double sampling plans have the advantage of a lower total sample size when the incoming quality is either excellent or poor because the lot is either accepted or rejected on the first sample.

Example: A double sampling plan is to be executed as follows: take a first sample (n_1) of 75 units and set c_1 (the acceptance number for the first sample) = zero. The lot will be accepted based on the first sample results if no nonconformances are found in the first sample. If three nonconformances are found in the first sample, the lot will be rejected based on the first sample results. If after analyzing the results of the first sample one or two nonconformances are found, take a second sample ($n_2 = 75$). The acceptance number for the second sample (c_2) is set to three. If the combined number of nonconformances in the first and second samples is three or less, the lot will be accepted, and if the combined number of nonconformances is four or more, the lot will be rejected. The plan is represented as follows:

Sample number	Acceptance number c	Rejection number r
$n_1 = 75$	$c_1 = 0$	$r_1 = 3$
$n_2 = 75$	$c_2 = 3$	$r_2 = 4$

Multiple Sampling Plans. Multiple sampling plans work in the same way as double sampling with an extension of the number of samples to be taken up to seven, according to ANSI/ASQ Z1.4-2003. In the same manner that double sampling is performed, acceptance or rejection of submitted lots may be reached before the seventh sample, depending on the acceptance/rejection criteria established for the plan.

ANSI/ASQ Z1.4-2003

ANSI/ASQ Z1.4-2003 is probably the most commonly used standard for attributes sampling plans. The wide recognition and acceptance of the plan could be due to government contracts stipulating the standard, rather than its statistical importance. Producers submitting products at a nonconformance level within AQL have a high probability of having the lot accepted by the customer.

When using ANSI/ASQ Z1.4-2003, the characteristics under consideration should be classified. The general classifications are critical, major, and minor defects:

- *Critical defect.* A critical defect is a defect that judgment and experience indicate is likely to result in hazardous or unsafe conditions for the individuals using, maintaining, or depending on the product or a defect that judgment and experience indicate is likely to prevent performance of the unit. In practice, critical characteristics are commonly inspected to an AQL level of 0.40 to 0.65 percent if not 100 percent inspected. One hundred percent inspection is recommended for critical characteristics if possible. Acceptance numbers are always zero for critical defects.

- *Major defect.* A major defect is a defect, other than critical, that is likely to result in failure or to reduce materially the usability of the unit of product for its intended purpose. In practice, AQL levels for major defects are generally about one percent.

- *Minor defect.* A minor defect is a defect that is not likely to reduce materially the usability of the unit of product for its intended purpose. In practice, AQL levels for minor defects generally range from 1.5 percent to 2.5 percent.

Levels of Inspection. There are seven levels of inspection used in ANSI/ASQ Z1.4-2003: reduced inspection, normal inspection, tightened inspection, and four levels of special inspection. The special inspection levels should only be used when small sample sizes are necessary and large risks can be tolerated. When using ANSI/ASQ Z1.4-2003, a set of switching rules must be followed as to the use of reduced, normal, and tightened inspection. The following guidelines are taken from ANSI/ASQ Z1.4-2003:

- *Initiation of inspection.* Normal inspection Level II will be used at the start of inspection unless otherwise directed by the responsible authority.

- *Continuation of inspection.* Normal, tightened, or reduced inspection shall continue unchanged for each class of defect or defectives on successive lots or batches except where the following switching procedures require change. The switching procedures shall be applied to each class of defects or defectives independently.

Switching Procedures. Switching rules are graphically shown in Figure 18.7.

- *Normal to tightened.* When normal inspection is in effect, tightened inspection shall be instituted when two out of five consecutive lots or batches have been rejected on original inspection (that is, ignoring resubmitted lots or batches for this procedure).

- *Tightened to normal.* When tightened inspection is in effect, normal inspection shall be instituted when five consecutive lots or batches have been considered acceptable on original inspection.

- *Normal to reduced.* When normal inspection is in effect, reduced inspection shall be instituted providing that all of the following conditions are satisfied:

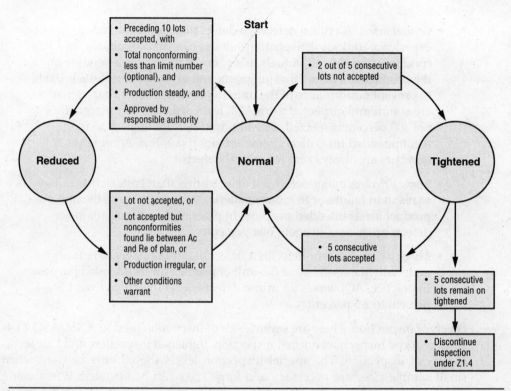

Figure 18.7 Switching rules for normal, tightened, and reduced inspection.

Source: D. W. Benbow, A. K. Elshennawy, and H. F. Walker, *The Certified Quality Technician Handbook* (Milwaukee: ASQ Quality Press, 2003): 143.

a. The preceding 10 lots or batches (or more), as indicated by the note on ANSI/ASQ Z1.4-2003 Table I, shown as Figure 18.10, page 155, have been on normal inspection and none has been rejected on original inspection.

b. The total number of defectives (or defects) in the sample from the preceding 10 lots or batches (or such other number as was used for condition [a] above) is equal to or less than the applicable number given in Table VIII of ANSI/ASQ Z1.4-2003 (shown as Figure 18.9, page 154). If double or multiple sampling is in use, all samples inspected should be included, not 'first' samples only.

c. Production is at a steady rate.

d. Reduced inspection is considered desirable by the responsible authority.

- *Reduced to normal.* When reduced inspection is in effect, normal inspection shall be instituted if any of the following occur on original inspection:

a. A lot or batch is rejected.

b. A lot or batch is considered acceptable under reduced inspection but the sampling procedures terminated without either acceptance or rejection criteria having been met. In these circumstances, the lot or batch will be considered acceptable, but normal inspection will be reinstated starting with the new lot or batch.

c. Production becomes irregular or delayed.

d. Other conditions warrant that normal inspection shall be instituted.

- *Discontinuation of inspection.* In the event that 10 consecutive lots or batches remain on tightened inspection (or such other number as may be designated by the responsible authority), inspection under the provisions of this document should be discontinued pending action to improve the quality of submitted material.

Types of Sampling. ANSI/ASQ Z1.4-2003 allows for three types of sampling:

1. Single sampling

2. Double sampling

3. Multiple sampling

The choice of the type of plan depends on many variables. Single sampling is the easiest to administer and perform but usually results in the largest average total inspection. Double sampling in ANSI/ASQ Z1.4-2003 results in a lower average total inspection than single sampling but requires more decisions to be made, such as:

- Accept the lot after first sample

- Reject the lot after first sample

- Take a second sample

- Accept the lot after second sample

- Reject the lot after second sample

Multiple sampling plans further reduce the average total inspection but also increase the number of decisions to be made. As many as seven samples may be required before a decision to accept or reject the lot can be made. This type of plan requires the most administration. A general procedure for selecting plans from ANSI/ASQ Z1.4-2003 is as follows:

1. Decide on an AQL.

2. Decide on the inspection level.

3. Determine the lot size.

4. Find the appropriate sample size code letter. See Table I from ANSI/ASQ Z1.4-2003, also shown as Figure 18.10 at the end of this section.

5. Determine the type of sampling plan to be used: single, double, or multiple.

6. Using the selected AQL and sample size code letter, enter the appropriate table to find the desired plan to be used.

7. Determine the normal, tightened, and reduced plans as required from the corresponding tables.

Example: A lot of 1750 parts has been received and is to be checked to an AQL level of 1.5 percent. Determine the appropriate single, double, and multiple sampling plans for general inspection Level II.

Steps to define the plans are as follows:

1. Table I of ANSI/ASQ Z1.4-2003, also shown as Figure 18.10 at the end of this section, stipulates code letter K.

2. Normal inspection is applied. For code letter K, using Table II-A of ANSI/ASQ Z1.4-2003, also shown as Figure 18.11 at the end of this section, a sample of 125 is specified.

3. For double sampling, two samples of 80 may be required. Refer to Table III-A of the standard, shown as Figure 18.12 at the end of this section.

4. For multiple sampling, at least two samples of 32 are required, and it may take up to seven samples of 32 before an acceptance or rejection decision is made. Refer to Table IV-A of the standard, shown as Figure 18.13 at the end of this section.

A breakdown of all three plans follows:

Sampling plan	Sample(s) size	Ac	Re
Single sampling	125	5	6
Double sampling			
First	80	5	
Second	80	7	
Multiple sampling			
First	32	*	4
Second	32	1	5
Third	32	2	6
Fourth	32	3	7
Fifth	32	5	8
Sixth	32	7	9
Seventh	32	9	10

Ac = Acceptance number
Re = Rejection number
* Acceptance not permitted at this sample size

Dodge-Romig Tables

Dodge-Romig tables were designed as sampling plans to minimize average total inspection (ATI). These plans require an accurate estimate of the process average nonconforming in selection of the sampling plan to be used. The Dodge-Romig tables use the AOQL and LTPD values for plan selection, rather than AQL as in ANSI/ASQ Z1.4-2003. When the process average nonconforming is controlled to requirements, Dodge-Romig tables result in lower average total inspection, but rejection of lots and sorting tend to minimize the gains if process quality deteriorates.

Note that if the process average nonconforming shows statistical control, acceptance sampling should not be used. The most economical course of action in this situation is either no inspection or 100 percent inspection.[4]

Variables Sampling Plans

Variables sampling plans use the actual measurements of sample products for decision making rather than classifying products as conforming or nonconforming, as in attributes sampling plans. Variables sampling plans are more complex in administration than attributes plans, thus they require more skill. They provide some benefits, however, over attributes plans. Two of these benefits are:

1. Equal protection to an attributes sampling plan but with a much smaller sample size. There are several types of variables sampling plans in use, including: (1) known, (2) unknown but can be estimated using sample standard deviation s, and (3) unknown and the range R is used as an estimator. If an attributes sampling plan sample size is determined, the variables plans previously listed can be compared as a percentage to the attributes plan.

2. Variables sampling plans allow the determination of how close to nominal or a specification limit the process is performing. Attributes plans either accept or reject a lot; variables plans give information on how well or poorly the process is performing.

Variables sampling plans, such as ANSI/ASQ Z1.9-2003, have some disadvantages and limitations:

1. The assumption of normality of the population from which the samples are being drawn.

2. Unlike attributes sampling plans, separate characteristics on the same parts will have different averages and dispersions, resulting in a separate sampling plan for each characteristic.

3. Variables plans are more complex in administration.

4. Variables gauging is generally more expensive than attributes gauging.

ANSI/ASQ Z1.9-2003

The most common standard for variables sampling plans is ANSI/ASQ Z1.9-2003, which has plans for: (1) variability known, (2) variability unknown—standard deviation method, and (3) variability unknown—range method. Using the afore-mentioned methods, this sampling plan can be used to test for a single specification limit, a double (or bilateral) specification limit, estimation of the process average, and estimation of the dispersion of the parent population.

As in ANSI/ASQ Z1.4-2003, several AQL levels are used and specific switching procedures for normal, reduced, and tightened inspection are followed. ANSI/ASQ Z1.9-2003 allows for the same AQL value for each specification limit of double specification limit plans or the use of different AQL values for each specification limit. The AQL values are designated M_L for the lower specification limit and M_U for the upper specification limit.

There are two forms used for every specification limit ANSI/ASQ Z1.9-2003 plan: form 1 and form 2. Form 1 provides only acceptance or rejection criteria, whereas form 2 estimates the percent below the lower specification and the percent above the upper specification limit. These percentages are compared to the AQL for acceptance/rejection criteria. Figure 18.8 summarizes the structure and organization of ANSI/ASQ Z1.9-2003.

There are 14 AQL levels used in ANSI/ASQ Z1.9-2003 that are consistent with the AQL levels used in ANSI/ASQ Z1.4-2003. Section A of ANSI/ASQ Z1.9-2003 contains both an AQL conversion table and a table for selecting the desired inspection level. Level IV is generally considered normal inspection, with Level V being

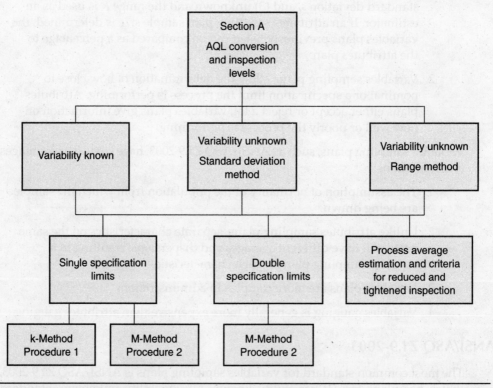

Figure 18.8 General structure and organization of ANSI/ASQ Z1.9-2003.

Source: D. W. Benbow, A. K. Elshennawy, and H. F. Walker, *The Certified Quality Technician Handbook* (Milwaukee: ASQ Quality Press, 2003): 148.

tightened inspection and Levels I, II, and III being reduced inspection. Table A-3 of ANSI/ASQ Z1.9-2003 contains the OC curves for the sampling plans in Sections B, C, and D.

Section B contains sampling plans used when the variability is unknown and the standard deviation method is used. Part I is used for a single specification limit, Part II is used for a double specification limit, and Part III is used for estimation of process average and criteria for reduced and tightened inspection.

Section C contains sampling plans used when the variability is unknown and the range method is used. Parts I, II, and III are the same as Parts I, II, and III in Section B.

Section D contains sampling plans used when variability is known. Parts I, II, and III are the same as Parts I, II, and III in Section B.

Variability Unknown—Range Method. An example from Section C will be used here to illustrate the use of the variability unknown—range method for a single specification limit. The quality indices for a single specification limit are:

$$\frac{U-\bar{X}}{\bar{R}} \text{ or } \frac{\bar{X}-L}{\bar{R}}$$

where:

U = upper specification limit

L = lower specification limit

$\bar{X}$ = sample average

$\bar{R}$ = average range of the sample

The acceptance criterion is a comparison of the quality $(U - \bar{X})/\bar{R}$ or $(\bar{X} - L)/\bar{R}$ to the acceptability constant k. If the calculated quantity is equal to or greater than k, the lot is accepted; if the calculated quantity is negative or less than k, the lot is rejected.

The following example illustrates the use of the variability unknown—range method, form I variables sampling plan and is similar to examples from Section C of ANSI/ASQ Z1.9-2003.

Example: The lower specification limit for electrical resistance of a certain electrical component is 620 ohms. A lot of 100 items is submitted for inspection. Inspection Level IV, normal inspection, with AQL = 0.4 percent, is to be used. From ANSI/ASQ Z1.9-2003 Table A-2 and Table C-1, shown on pages 160 and 161 as Figure 18.14 and Figure 18.15 respectively, it is seen that a sample of size 10 is required. Suppose that values of the sample resistances (in the order reading from left to right) are:

645, 651, 621, 625, 658 ($R = 658 - 621 = 37$)

670, 673, 641, 638, 650 ($R = 673 - 638 = 35$)

Determine compliance with the acceptability criterion.

Line	Information needed	Value	Explanation
1.	Sample size: n	10	
2.	Sum of measurement: ΣX	6472	
3.	Sample mean $\bar{X}$: $\Sigma X/n$	647.2	6472/10
4.	Average range $\bar{R}$: ΣR/no. of subgroups	36	(37 + 35)/2
5.	Specification limit (lower): L	620	
6.	The quantity: $(\bar{X} - L)/R$	.756	(647.2 − 620)/36
7.	Acceptability constant: k	.811	See Table C-1 (Figure 18.15 on page 161)
8.	Acceptability criterion: Compare $(X - L)/R$ with k	.756 ≤ .811	

The lot does not meet the acceptability criterion, since $(\bar{X} - L)/\bar{R}$ is less than k.

Note: If a single upper specification limit U is given, then compute the quantity $(U - \bar{X})/\bar{R}$ in line 6, and compare it with k. The lot meets the acceptability criterion if $(U - \bar{X})/\bar{R}$ is equal to or greater than k.

Variability Unknown—Standard Deviation Method. In this section, a sampling plan is shown for the situation where the variability is not known and the standard deviation is estimated from the sample data. The sampling plan will be that for a double specification limit and it is found in Section B of the standard with one AQL value for both upper and lower specification limits combined.

The acceptability criterion is based on comparing an estimated percent nonconforming to a maximum allowable percent nonconforming for the given AQL level. The estimated percent nonconforming is found in ANSI/ASQ Z1.9-2003 Table B-5, shown as Figure 18.16, page 162.

The quality indices for this sampling plan are:

$$\frac{U - \bar{X}}{s} \text{ or } \frac{\bar{X} - L}{s}$$

where

> U = upper specification limit
>
> L = lower specification limit
>
> $\bar{X}$ = sample average
>
> s = estimate of lot standard deviation

The quality level of the lot is in terms of the lot percent defective. Three values are calculated: P_U, P_L, and p. P_U is an estimate of conformance with the upper specification limit, P_L is an estimate of conformance with the lower specification limit, and p is the sum of P_U and P_L.

The value of p is then compared with the maximum allowable percent defective. If p is less than or equal to M (ANSI/ASQ Z1.9-2003 Table B-5, shown as Figure 18.16) or if either Q_U or Q_L is negative, the lot is rejected. The following example illustrates the above procedure.

Example: The minimum temperature of operation for a certain device is specified as 180°F. The maximum temperature is 209°F. A lot of 40 items is submitted for inspection. Inspection Level IV, normal inspection with AQL = 1 percent, is to be used. ANSI/ASQ Z1.9-2003 Table A-2, shown as Figure 18.14, page 160, gives code letter D, which results in a sample size of five from ANSI/ASQ Z1.9-2003 Table B-3, shown as Figure 18.17, page 163. The results of the five measurements in degrees Fahrenheit are as follows: 197, 188, 184, 205, 201. Determine if the lot meets acceptance criteria.

Line	Information needed	Value obtained	Explanation
1.	Sample size: n	5	
2.	Sum of measurements: ΣX	975	
3.	Sum of squared measurements: ΣX^2	190,435	
4.	Correction factor: $(\Sigma X)^2/n$	190,125	$975^2/5$
5.	Corrected sum of squares (SS): $\Sigma X^2 - CF$	310	$190,435 - 190,125$
6.	Variance V: $SS/n - 1$	77.5	310/4
7.	Standard deviation s: $\sqrt{V}$	8.81	$\sqrt{77.5}$
8.	Sample mean $\bar{X}$: $\Sigma X/n$	195	975/5
9.	Upper specification limit: U	209	
10.	Lower specification limit: L	180	
11.	Quality index: $Q_U = (U - \bar{X})/s$	1.59	$(209 - 195)/8.81$
12.	Quality index: $Q_L = (\bar{X} - L)/s$	1.7	$(195 - 180)/8.81$
13.	Estimate of lot percent defective above U: P_U	2.19%	$Q_U = 1.59$, sample size = 5. See Table B-5 (Figure 18.16 on page 162)
14.	Estimate of lot percent defective below L: P_L	0.66%	$Q_L = 1.7$, sample size = 5. See Table B-5 (Figure 18.16 on page 162)
15.	Total estimate of percent defective in lot: $p = P_U + P_L$	2.85%	2.19 + .66
16.	Maximum allowable percent defective: M	3.33%	See Table B-3 (Figure 18.17 on page 163)
17.	Acceptibility criterion: Compare $p = P_U + P_L$ with M	2.85% < 3.33%	

The lot meets the acceptability criterion, since $p = P_U + P_L$ is less than M.

ANSI/ASQ Z1.9-2003 provides a variety of other examples for variables sampling plans.

Part III.C.1

Part III.C.1

Acceptance Quality Limits

Number of sample units from last 10 lots or batches	0.010	0.015	0.025	0.040	0.065	0.10	0.15	0.25	0.40	0.65	1.0	1.5	2.5	4.0	6.5	10	15	25	40	65	100	150	250	400	650	1000
20–29	*	*	*	*	*	*	*	*	*	*	*	*	*	*	*	0	0	2	4	8	14	22	40	68	115	181
30–49	*	*	*	*	*	*	*	*	*	*	*	*	*	*	0	0	1	3	7	13	22	36	63	105	178	277
50–79	*	*	*	*	*	*	*	*	*	*	*	*	*	0	0	2	3	7	14	25	40	63	110	181	301	
80–129	*	*	*	*	*	*	*	*	*	*	*	*	0	0	2	4	7	14	24	42	68	105	181	297		
130–199	*	*	*	*	*	*	*	*	*	*	*	0	0	2	4	7	13	25	42	72	115	177	301	490		
200–319	*	*	*	*	*	*	*	*	*	*	0	0	2	4	8	14	22	40	68	115	181	277	471			
320–499	*	*	*	*	*	*	*	*	*	0	2	1	4	8	14	24	39	68	113	189						
500–799	*	*	*	*	*	*	*	*	0	0	4	3	7	14	25	40	63	110	181							
800–1249	*	*	*	*	*	*	*	0	0	2	7	7	14	24	42	68	105	181								
1250–1999	*	*	*	*	*	*	0	0	2	4	14	13	24	49	69	110	169									
2000–3149	*	*	*	*	*	0	0	2	4	8	24	22	40	68	115	181										
3150–4999	*	*	*	*	0	0	1	4	8	14	40	38	67	111	186											
5000–7999	*	*	*	0	0	2	3	7	14	25	68	63	110	181												
8000–12499	*	*	0	0	2	4	7	14	24	42	110	105	181													
12500–19999	*	0	0	2	4	7	13	24	40	69	181	169														
20000–31499	0	0	2	4	8	14	22	40	68	115																
31500 & Over	0	1	4	8	14	24	38	67	111	186																

* = Denotes that the number of sample units from the last ten lots or batches is not sufficient for reduced inspection for this AQL. Is this instance, more than ten lots or batches may be used for the calculation, provided that the lots or batches used are the most recent ones in sequence, that they have all been on normal inspection, and that none has been rejected while on original inspection.

Figure 18.9   ANSI/ASQ Z1.4–2003 Table VIII: Limit numbers for reduced inspection. Used by permission.

Lot or batch size	Special inspection levels				General inspection levels		
	S-1	S-2	S-3	S-4	I	II	III
2 to 8	A	A	A	A	A	A	B
9 to 15	A	A	A	A	A	B	C
16 to 25	A	A	B	B	B	C	D
26 to 50	A	B	B	C	C	D	E
51 to 90	B	B	C	C	C	E	F
91 to 150	B	B	C	D	D	F	G
151 to 280	B	C	D	E	E	G	H
281 to 500	B	C	D	E	F	H	J
501 to 1200	C	C	E	F	G	J	K
1201 to 3200	C	D	E	G	H	K	L
3201 to 10000	C	D	F	G	J	L	M
10001 to 35000	C	E	F	H	K	M	N
35001 to 150000	D	E	G	J	L	N	P
150001 to 500000	D	E	G	J	M	P	Q
500001 and over	D	E	H	K	N	Q	R

Figure 18.10   ANSI/ASQ Z1.4–2003 Table I: Sample size code letters per inspection levels. Used by permission.

Part III.C.1

Part III.C.1

Acceptance Quality Limits, AQLs, in Percent Nonconforming Items and Nonconformities per 100 Items (Normal Inspection)

Each cell gives Ac Re (Ac = Acceptance number, Re = Rejection number). ↓ = Use the first sampling plan below the arrow. If sample size equals, or exceeds, lot size, carry out 100 percent inspection. ↑ = Use the first sampling plan above the arrow.

Code	n	0.010	0.015	0.025	0.040	0.065	0.10	0.15	0.25	0.40	0.65	1.0	1.5	2.5	4.0	6.5	10	15	25	40	65	100	150	250	400	650	1000
A	2	↓	↓	↓	↓	↓	↓	↓	↓	↓	↓	↓	↓	↓	↓	↓	↓	0 1	1 2	2 3	3 4	5 6	7 8	10 11	14 15	21 22	30 31
B	3	↓	↓	↓	↓	↓	↓	↓	↓	↓	↓	↓	↓	↓	↓	↓	0 1	1 2	2 3	3 4	5 6	7 8	10 11	14 15	21 22	30 31	44 45
C	5	↓	↓	↓	↓	↓	↓	↓	↓	↓	↓	↓	↓	↓	↓	0 1	1 2	2 3	3 4	5 6	7 8	10 11	14 15	21 22	30 31	44 45	↑
D	8	↓	↓	↓	↓	↓	↓	↓	↓	↓	↓	↓	↓	↓	0 1	1 2	2 3	3 4	5 6	7 8	10 11	14 15	21 22	30 31	44 45	↑	↑
E	13	↓	↓	↓	↓	↓	↓	↓	↓	↓	↓	↓	↓	0 1	1 2	2 3	3 4	5 6	7 8	10 11	14 15	21 22	30 31	44 45	↑	↑	↑
F	20	↓	↓	↓	↓	↓	↓	↓	↓	↓	↓	↓	0 1	1 2	2 3	3 4	5 6	7 8	10 11	14 15	21 22	30 31	44 45	↑	↑	↑	↑
G	32	↓	↓	↓	↓	↓	↓	↓	↓	↓	↓	0 1	1 2	2 3	3 4	5 6	7 8	10 11	14 15	21 22	30 31	44 45	↑	↑	↑	↑	↑
H	50	↓	↓	↓	↓	↓	↓	↓	↓	↓	0 1	1 2	2 3	3 4	5 6	7 8	10 11	14 15	21 22	30 31	44 45	↑	↑	↑	↑	↑	↑
J	80	↓	↓	↓	↓	↓	↓	↓	↓	0 1	1 2	2 3	3 4	5 6	7 8	10 11	14 15	21 22	30 31	44 45	↑	↑	↑	↑	↑	↑	↑
K	125	↓	↓	↓	↓	↓	↓	↓	0 1	1 2	2 3	3 4	5 6	7 8	10 11	14 15	21 22	30 31	44 45	↑	↑	↑	↑	↑	↑	↑	↑
L	200	↓	↓	↓	↓	↓	↓	0 1	1 2	2 3	3 4	5 6	7 8	10 11	14 15	21 22	30 31	44 45	↑	↑	↑	↑	↑	↑	↑	↑	↑
M	315	↓	↓	↓	↓	↓	0 1	1 2	2 3	3 4	5 6	7 8	10 11	14 15	21 22	30 31	44 45	↑	↑	↑	↑	↑	↑	↑	↑	↑	↑
N	500	↓	↓	↓	↓	0 1	1 2	2 3	3 4	5 6	7 8	10 11	14 15	21 22	30 31	44 45	↑	↑	↑	↑	↑	↑	↑	↑	↑	↑	↑
P	800	↓	↓	↓	0 1	1 2	2 3	3 4	5 6	7 8	10 11	14 15	21 22	30 31	44 45	↑	↑	↑	↑	↑	↑	↑	↑	↑	↑	↑	↑
Q	1250	↓	↓	0 1	1 2	2 3	3 4	5 6	7 8	10 11	14 15	21 22	30 31	44 45	↑	↑	↑	↑	↑	↑	↑	↑	↑	↑	↑	↑	↑
R	2000	↓	0 1	1 2	2 3	3 4	5 6	7 8	10 11	14 15	21 22	30 31	44 45	↑	↑	↑	↑	↑	↑	↑	↑	↑	↑	↑	↑	↑	↑

↓ = Use the first sampling plan below the arrow. If sample size equals, or exceeds, lot size, carry out 100 percent inspection.

↑ = Use the first sampling plan above the arrow.

Ac = Acceptance number.

Re = Rejection number.

Figure 18.11   ANSI/ASQ Z1.4-2003 Table II-A: Single sampling plans for normal inspection. Used by permission.

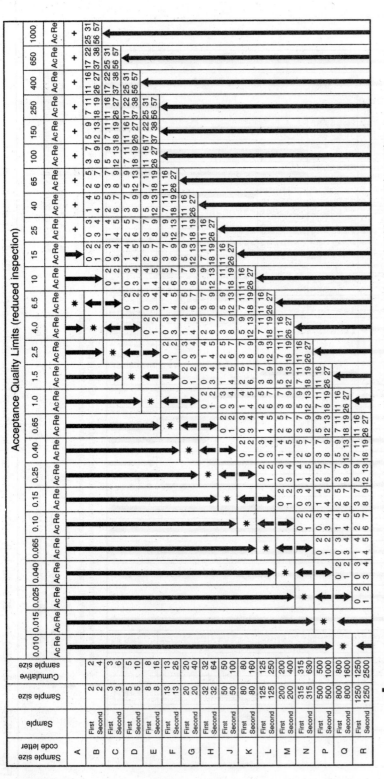

Part III.C.1

Figure 18.12 ANSI/ASQ Z1.4–2003 Table III–A: Double sampling plans for normal inspection. Used by permission.

Part III.C.1

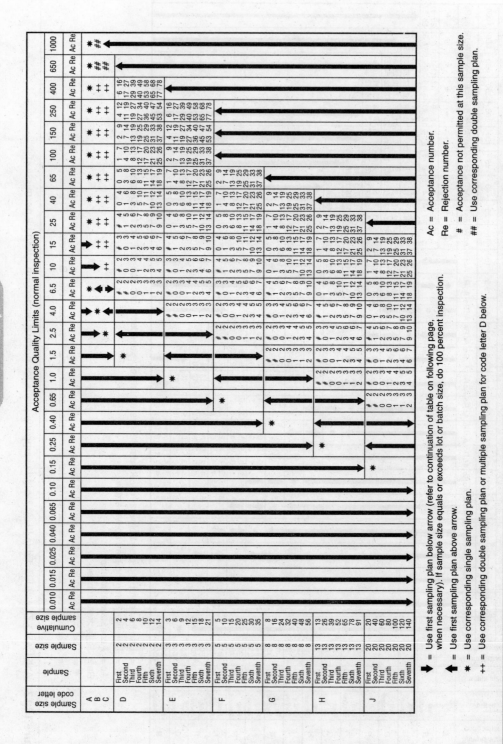

Figure 18.13   ANSI/ASQ Z1.4–2003 Table IV–A: Multiple sampling plans for normal inspection. Used by permission.

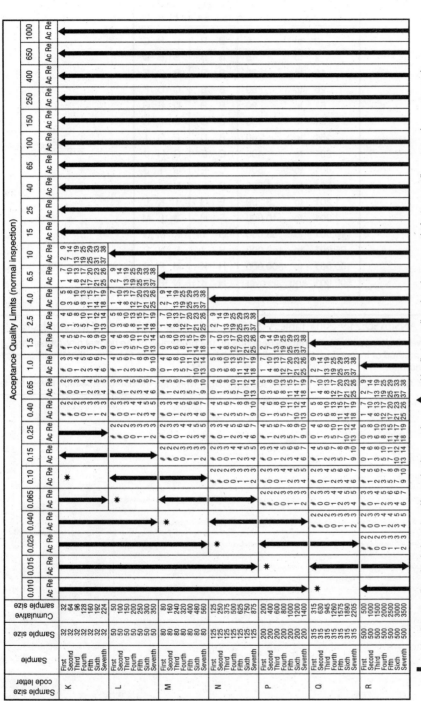

Figure 18.13 *Continued.*

Part III.C.1

Lot Size			Inspection Levels				
			Special S3 S4		General I II III		
2	to	8	B	B	B	B	C
9	to	15	B	B	B	B	D
16	to	25	B	B	B	C	E
26	to	50	B	B	C	D	F
51	to	90	B	B	D	E	G
91	to	150	B	C	E	F	H
151	to	280	B	D	F	G	I
281	to	400	C	E	G	H	J
401	to	500	C	E	G	I	J
501	to	1,200	D	F	H	J	K
1,201	to	3,200	E	G	I	K	L
3,201	to	10,000	F	H	J	L	M
10,001	to	35,000	G	I	K	M	N
35,001	to	150,000	H	J	L	N	P
150,001	to	500,000	H	K	M	P	P
500,001	and	over	H	K	N	P	P

Figure 18.14   ANSI/ASQ Z1.9-2003 Table A-2*: Sample size code letters.† Used by permission.

* The theory governing inspection by variables depends on the properties of the normal distribution and, therefore, this method of inspection is only applicable when there is reason to believe that the frequency distribution is normal.

† Sample size code letters given in body of table are applicable when the indicated inspection levels are to be used.

Sample Size Code Letter	Sample Size	Acceptance Quality Limits (normal inspection)											
		T	.10	.15	.25	.40	.65	1.00	1.50	2.50	4.00	6.50	10.00
		k	k	k	k	k	k	k	k	k	k	k	k
B	3							↓	↓	.587	.502	.401	.296
C	4						↓	.651	.598	.525	.450	.364	.276
D	5				↓	↓	.663	.614	.565	.498	.431	.352	.272
E	7			↓	.702	.659	.613	.569	.525	.465	.405	.336	.266
F	10	↓	↓	.916	.863	.811	.755	.703	.650	.579	.507	.424	.341
G	15	1.04	.999	.958	.903	.850	.792	.738	.684	.610	.536	.452	.368
H	25	1.10	1.05	1.01	.951	.896	.835	.779	.723	.647	.571	.484	.398
I	30	1.10	1.06	1.02	.959	.904	.843	.787	.730	.654	.577	.490	.403
J	40	1.13	1.08	1.04	.978	.921	.860	.803	.746	.668	.591	.503	.415
K	60	1.16	1.11	1.06	1.00	.948	.885	.826	.768	.689	.610	.521	.432
L	85	1.17	1.13	1.08	1.02	.962	.899	.839	.780	.701	.621	.530	.441
M	115	1.19	1.14	1.09	1.03	.975	.911	.851	.791	.711	.631	.539	.449
N	175	1.21	1.16	1.11	1.05	.994	.929	.868	.807	.726	.644	.552	.460
P	230	1.21	1.16	1.12	1.06	.996	.931	.870	.809	.728	.646	.553	.462
		.10	.15	.25	.40	.65	1.00	1.50	2.50	4.00	6.50	10.00	
		Acceptance Quality Limits (tightened inspection)											

All AQL values are in percent nonconforming. T denotes plan used exclusively on tightened inspection and provides symbol for identification of appropriate OC curve.

↓ Use first sampling plan below arrow; that is, both sample size as well as k value. When sample size equals or exceeds lot size, every item in the lot must be inspected.

Figure 18.15 ANSI/ASQ Z1.9-2003 Table C-1: Master table for normal and tightened inspection for plans based on variability unknown (single specification limit–form 1). Used by permission.

Part III.C.1

Q_U or Q_L	Sample Size														
	3	4	5	7	10	15	20	25	30	35	50	75	100	150	200
1.50	0.00	0.00	3.80	5.28	5.87	6.20	6.34	6.41	6.46	6.50	6.55	6.60	6.62	6.64	6.65
1.51	0.00	0.00	3.61	5.13	5.73	6.06	6.20	6.28	6.33	6.36	6.42	6.47	6.49	6.51	6.52
1.52	0.00	0.00	3.42	4.97	5.59	5.93	6.07	6.15	6.20	6.23	6.29	6.34	6.36	6.38	6.39
1.53	0.00	0.00	3.23	4.82	5.45	5.80	5.94	6.02	6.07	6.11	6.17	6.21	6.24	6.26	6.27
1.54	0.00	0.00	3.05	4.67	5.31	5.67	5.81	5.89	5.95	5.98	6.04	6.09	6.11	6.13	6.15
1.55	0.00	0.00	2.87	4.52	5.18	5.54	5.69	5.77	5.82	5.86	5.92	5.97	5.99	6.01	6.02
1.56	0.00	0.00	2.69	4.38	5.05	5.41	5.56	5.65	5.70	5.74	5.80	5.85	5.87	5.89	5.90
1.57	0.00	0.00	2.52	4.24	4.92	5.29	5.44	5.53	5.58	5.62	5.68	5.73	5.75	5.78	5.79
1.58	0.00	0.00	2.35	4.10	4.79	5.16	5.32	5.41	5.46	5.50	5.56	5.61	5.64	5.66	5.67
1.59	0.00	0.00	2.19	3.96	4.66	5.04	5.20	5.29	5.34	5.38	5.45	5.50	5.52	5.55	5.56
1.60	0.00	0.00	2.03	3.83	4.54	4.92	5.08	5.17	5.23	5.27	5.33	5.38	5.41	5.43	5.44
1.61	0.00	0.00	1.87	3.69	4.41	4.81	4.97	5.06	5.12	5.16	5.22	5.27	5.30	5.32	5.33
1.62	0.00	0.00	1.72	3.57	4.30	4.69	4.86	4.95	5.01	5.04	5.11	5.16	5.19	5.21	5.23
1.63	0.00	0.00	1.57	3.44	4.18	4.58	4.75	4.84	4.90	4.94	5.01	5.06	5.08	5.11	5.12
1.64	0.00	0.00	1.42	3.31	4.06	4.47	4.64	4.73	4.79	4.83	4.90	4.95	4.98	5.00	5.01
1.65	0.00	0.00	1.28	3.19	3.95	4.36	4.53	4.62	4.68	4.72	4.79	4.85	4.87	4.90	4.91
1.66	0.00	0.00	1.15	3.07	3.84	4.25	4.43	4.52	4.58	4.62	4.69	4.74	4.77	4.80	4.81
1.67	0.00	0.00	1.02	2.95	3.73	4.15	4.32	4.42	4.48	4.52	4.59	4.64	4.67	4.70	4.71
1.68	0.00	0.00	0.89	2.84	3.62	4.05	4.22	4.32	4.38	4.42	4.49	4.55	4.57	4.60	4.61
1.69	0.00	0.00	0.77	2.73	3.52	3.94	4.12	4.22	4.28	4.32	4.39	4.45	4.47	4.50	4.51
1.70	0.00	0.00	0.66	2.62	3.41	3.84	4.02	4.12	4.18	4.22	4.30	4.35	4.38	4.41	4.42
1.71	0.00	0.00	0.55	2.51	3.31	3.75	3.93	4.02	4.09	4.13	4.20	4.26	4.29	4.31	4.32
1.72	0.00	0.00	0.45	2.41	3.21	3.65	3.83	3.93	3.99	4.04	4.11	4.17	4.19	4.22	4.23
1.73	0.00	0.00	0.36	2.30	3.11	3.56	3.74	3.84	3.90	3.94	4.02	4.08	4.10	4.13	4.14
1.74	0.00	0.00	0.27	2.20	3.02	3.46	3.65	3.75	3.81	3.85	3.93	3.99	4.01	4.04	4.05
1.75	0.00	0.00	0.19	2.11	2.93	3.37	3.56	3.66	3.72	3.77	3.84	3.90	3.93	3.95	3.97
1.76	0.00	0.00	0.12	2.01	2.83	3.28	3.47	3.57	3.63	3.68	3.76	3.81	3.84	3.87	3.88
1.77	0.00	0.00	0.06	1.92	2.74	3.20	3.38	3.48	3.55	3.59	3.67	3.73	3.76	3.78	3.80
1.78	0.00	0.00	0.02	1.83	2.66	3.11	3.30	3.40	3.47	3.51	3.59	3.64	3.67	3.70	3.71
1.79	0.00	0.00	0.00	1.74	2.57	3.03	3.21	3.32	3.38	3.43	3.51	3.56	3.59	3.62	3.63
1.80	0.00	0.00	0.00	1.65	2.49	2.94	3.13	3.24	3.30	3.35	3.43	3.48	3.51	3.54	3.55
1.81	0.00	0.00	0.00	1.57	2.40	2.86	3.05	31.6	3.22	3.27	3.35	3.40	3.43	3.46	3.47
1.82	0.00	0.00	0.00	1.49	2.32	2.79	2.98	3.08	3.15	3.19	3.27	3.33	3.36	3.38	3.40
1.83	0.00	0.00	0.00	1.41	2.25	2.71	2.90	3.00	3.07	3.11	3.19	3.25	3.28	3.31	3.32
1.84	0.00	0.00	0.00	1.34	2.17	2.63	2.82	2.93	2.99	3.04	3.12	3.18	3.21	3.23	3.25
1.85	0.00	0.00	0.00	1.26	2.09	2.56	2.75	2.85	2.92	2.97	3.05	3.10	3.13	3.16	3.17
1.86	0.00	0.00	0.00	1.19	2.02	2.48	2.68	2.78	2.85	2.89	2.97	3.03	3.06	3.09	3.10
1.87	0.00	0.00	0.00	1.12	1.95	2.41	2.61	2.71	2.78	2.82	2.90	2.96	2.99	3.02	3.03
1.88	0.00	0.00	0.00	1.06	1.88	2.34	2.54	2.64	2.71	2.75	2.83	2.89	2.92	2.95	2.96
1.89	0.00	0.00	0.00	0.99	1.81	2.28	2.47	2.57	2.64	2.69	2.77	2.83	2.85	2.88	2.90

Figure 18.16   ANSI/ASQ Z1.9–2003 Table B-5: Table for estimating the lot percent nonconforming using standard deviation method.* Used by permission.

* Values tabulated are read in percent.

Sample Size Code Letter	Sample Size	Acceptance Quality Limits (normal inspection)											
		T	.10	.15	.25	.40	.65	1.00	1.50	2.50	4.00	6.50	10.00
		M	M	M	M	M	M	M	M	M	M	M	M
B	3	↓	↓	↓	↓	↓	↓	↓	↓	7.59	18.86	26.94	33.69
C	4					↓	↓	1.49	5.46	10.88	16.41	22.84	29.43
D	5		↓	↓	↓	0.041	1.34	3.33	5.82	9.80	14.37	20.19	26.55
E	7	↓	0.005	0.087	0.421	1.05	2.13	3.54	5.34	8.40	12.19	17.34	23.30
F	10	0.077	0.179	0.349	0.714	1.27	2.14	3.27	4.72	7.26	10.53	15.17	20.73
G	15	0.186	0.311	0.491	0.839	1.33	2.09	3.06	4.32	6.55	9.48	13.74	18.97
H	20	0.228	0.356	0.531	0.864	1.33	2.03	2.93	4.10	6.18	8.95	13.01	18.07
I	25	0.250	0.378	0.551	0.874	1.32	2.00	2.86	3.97	5.98	8.65	12.60	17.55
J	35	0.253	0.373	0.534	0.833	1.24	1.87	2.66	3.70	5.58	8.11	11.89	16.67
K	50	0.243	0.355	0.503	0.778	1.16	1.73	2.47	3.44	5.21	7.61	11.23	15.87
L	75	0.225	0.326	0.461	0.711	1.06	1.59	2.27	3.17	4.83	7.10	10.58	15.07
M	100	0.218	0.315	0.444	0.684	1.02	1.52	2.18	3.06	4.67	6.88	10.29	14.71
N	150	0.202	0.292	0.412	0.636	0.946	1.42	2.05	2.88	4.42	6.56	9.86	14.18
P	200	0.204	0.294	0.414	0.637	0.945	1.42	2.04	2.86	4.39	6.52	9.80	14.11
		.10	.15	.25	.40	.65	1.00	1.50	2.50	4.00	6.50	10.00	
		Acceptance Quality Limits (tightened inspection)											

All AQL values are in percent nonconforming. T denotes plan used exclusively on tightened inspection and provides symbol for identification of appropriate OC curve.

↓ Use first sampling plan below arrow; that is, both sample size as well as k value. When sample size equals or exceeds lot size, every item in the lot must be inspected.

Figure 18.17   ANSI/ASQ Z1.9-2003 Table B-3: Master table for normal and tightened inspection for plans based on variability unknown (double specification limit and form 2—single specification limit). Used by permission.

Part III.C.1

2. SAMPLING METHODS

> Define and distinguish between various sampling methods such as random, sequential, stratified, fixed sampling, attributes and variables sampling. (Understand)
>
> [NOTE: Reading sampling tables is not required.]
>
> **Body of Knowledge III.C.2**

Sampling is that part of statistical practice concerned with the selection of individual observations intended to yield some knowledge about a population of concern, especially for the purposes of statistical inference. In particular, results from probability theory and statistical theory are employed to guide practice.

The sampling process consists of five stages:

1. Definition of population of concern

2. Specification of a set of items or events that it is possible to measure

3. Specification of sampling method for selecting items or events from the set of items

4. Sampling and data collecting

5. Review of sampling process

Successful statistical practice is based on focused problem definition. Typically, we seek to take action on some population, for example when a batch of material from production must be released to the customer or dispositioned for scrap or rework.

Alternatively, we seek knowledge about the cause system of which the population is an outcome, for example when a researcher performs an experiment on rats with the intention of gaining insights into biochemistry that can be applied for the benefit of humans. In the latter case, the population of concern can be difficult to specify, as it is in the case of measuring some physical characteristic such as the electrical conductivity of copper.

However, in all cases, time spent in making the population of concern precise is usually well spent, often because it raises many issues, ambiguities, and questions that would otherwise have been overlooked at this stage.

Random Sampling

In random sampling, also known as probability sampling, every combination of items from the sample set, or stratum, has a known probability of occurring, but these probabilities are not necessarily equal. With any form of sampling there is a risk that the sample may not adequately represent the population, but with random sampling there is a large body of statistical theory that quantifies the risk and thus enables an appropriate sample size to be chosen. Furthermore, once the sample has been taken, the sampling error associated with the measured results can be computed. There are several forms of random sampling. For example, in simple random sampling, each element has an equal probability of occurring. Other examples of probability sampling include stratified sampling and multi-stage sampling.

Stratified Sampling

Where the population embraces a number of distinct categories, the sample set can be organized by these categories into separate strata or demographics. One of the sampling methods discussed below is then applied to each stratum separately. Major gains in efficiency (either lower sample sizes or higher precision) can be achieved by varying the sampling fraction from stratum to stratum. The sample size is usually proportional to the relative size of the strata. However, if variances differ significantly across strata, sample sizes should be made proportional to the stratum standard deviation. Disproportionate stratification can provide better precision than proportionate stratification. Typically, strata should be chosen to have:

- Means that differ substantially from one another
- Variances that are different from one another and lower than the overall variance

Sequential Sampling

It is often beneficial and possible to obtain and inspect samples over a period of time. Such sampling, if it is recorded on a control chart, is typically referred to as *time series analysis*. This type of sampling is of value when time may be a factor contributing to variation and may signal an assignable cause, such as tool wear or chemical depletion.

Systematic or Fixed Sampling

Selecting, for example, every 10th name from the telephone directory is called an "every 10th sampling scheme," which is an example of systematic sampling. It is a type of nonprobability sampling unless the directory itself is randomized. It is easy to implement and the stratification induced can make it efficient, but it is especially vulnerable to periodicities in the list. If periodicity is present and the period is a multiple of 10, then bias will result. Another type of systematic or fixed sampling scheme is to always sample, say, five percent of the batch or lot to make a decision to accept or reject the lot. There are several disadvantages to such a sampling scheme that should be resolved before undertaking such sampling:

1. Rules for accepting or rejecting the lot based on how many out of the five percent sample are found to be nonconforming

2. Estimation of consumer/producer risk and protection

Part III.C.2

3. Concurrence between consumer and producer to employ such a non–statistically based sampling scheme

Convenience Sampling

Sometimes called grab or opportunity sampling, this is the method of choosing items arbitrarily and in an unstructured manner from the sample set. Though almost impossible to treat rigorously, it is the method most commonly employed in many practical situations. In social science research, snowball sampling is a similar technique, where existing study subjects are used to recruit more subjects into the sample.[5]

Chapter 19

D. Measurement Terms

> Define and distinguish between accuracy,
> precision, repeatability, reproducibility, bias,
> and linearity. (Understand)
>
> **Body of Knowledge III.D**

MEASUREMENT

A *measurement* is a series of manipulations of physical objects or systems according to a defined protocol, which results in a number. The number is purported to uniquely represent the magnitude (or intensity) of a certain characteristic, which depends on the properties of the test object. This number is acquired to form the basis of a decision affecting some human goal or satisfying some human object need, the satisfaction of which depends on the properties of the test subject. These needs or goals can be usefully viewed as requiring three general classes of measurements:

1. *Technical.* This class includes those measurements made to assure dimensional compatibility, conformation to design specifications necessary for proper function, or, in general, all measurements made to ensure fitness for intended use of some object.

2. *Legal.* This class includes those measurements made to ensure compliance with a law or regulation. This class is the concern of weights and measures bodies, regulators, and those who must comply with those regulations. The measurements are identical in kind with those of technical metrology but are usually embedded in a much more formal structure. Legal metrology is more prevalent in Europe than in the United States, although this is changing.

3. *Scientific.* This class includes those measurements made to validate theories of the nature of the universe or to suggest new theories. These measurements, which can be called scientific metrology (properly the domain of experimental physics), present special problems.[1]

MEASUREMENT CONCEPTS

Measurement Error

Error in measurement is the difference between the indicated value and the true value of a measured quantity. The true value of a quantity to be measured is seldom known. Errors are classified as random and systematic. *Random errors* are accidental in nature. They fluctuate in a way that cannot be predicted from the detailed employment of the measuring system or from knowledge of its functioning. Sources of error such as hysteresis, ambient influences, or variations in the workpiece are typical but not completely all-inclusive in the category of random error. *Systematic errors* are those not usually detected by repetition of the measurement operations. An error resulting from either faulty calibration of a local standard or a defect in contact configuration of an internal measuring system is typical but not completely inclusive in the systematic class of errors.[2]

It is important to know all the sources of error in a measuring system, rather than merely to be aware of the details of their classification. Analysis of the causes of errors is helpful in attaining the necessary knowledge of achieved accuracy.[3]

There are many different sources of error that influence the precision of a measuring process in a variety of ways according to the individual situation in which such errors arise. The permutation of error sources and their effects, therefore, is quite considerable. In general, these errors can be classified under three main headings:[4]

1. Process environment

2. Measuring equipment

3. Operator fallibility

These factors constitute an interrelated three-element system for the measuring process as shown in Figure 19.1.

The areas in which operator fallibility arises can be grouped as follows:[5]

1. Identification of the measuring situation

2. Analysis of alternative methods

3. Selection of equipment

4. Application (or measurement)

The identification of measuring situations becomes increasingly complex in modern metrology. As parts become smaller and more precise, greater attention has to be paid to geometric qualities such as roundness, concentricity, straightness, parallelism, and squareness. Deficiencies in these qualities may consume all of the permitted design tolerance, so that a simple dimensional check becomes grossly insufficient.

Operators have to be knowledgable about what they have to measure and how satisfactorily the requirements of the situation will be met by the measuring instrument. Correct identification of the measuring situation will eliminate

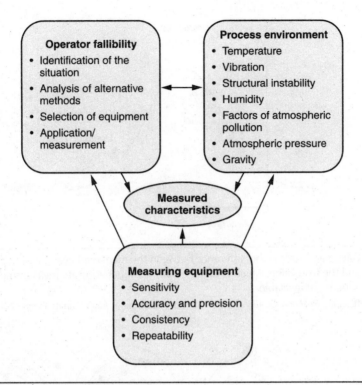

Figure 19.1 Factors affecting the measuring process.

Source: D. W. Benbow, A. K. Elshennawy, and H. F. Walker, *The Certified Quality Technician Handbook* (Milwaukee: ASQ Quality Press, 2003): 122.

those methods found unsuitable for the situation. The proper measuring equipment can therefore be selected from a smaller range of measuring process alternatives. Method analysis can then be applied to these alternatives to determine which best satisfies the situation. This usually involves examining each method for different characteristics and evaluating the relative accuracies between the different methods.

Accuracy

Accuracy is the degree of agreement of individual or average measurements with an accepted reference value, level, or standard.[6] See Figure 19.2.

Precision

Precision is the degree of mutual agreement among individual measurements made under prescribed like conditions, or simply, how well identically performed measurements agree with each other. This concept applies to a process or a set of measurements, not to a single measurement, because in any set of measurements, the individual results will scatter about the mean.[7] See Figure 19.3.

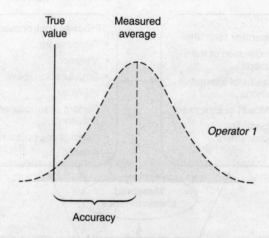

Figure 19.2 Gauge accuracy is the difference between the measured average of the gauge and the true value, which is defined with the most accurate measurement equipment available.

Source: E. R. Ott, et al., *Process Quality Control,* 4th ed. (Milwaukee: ASQ Quality Press, 2005): 527.

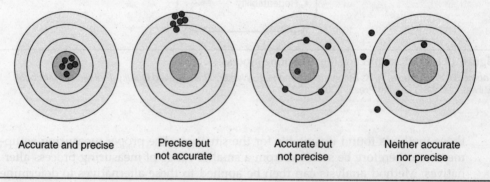

| Accurate and precise | Precise but not accurate | Accurate but not precise | Neither accurate nor precise |

Figure 19.3 Measurement data can be represented by one of four possible scenarios.
Source: E. R. Ott, et al., *Process Quality Control,* 4th ed. (Milwaukee: ASQ Quality Press, 2005): 527.

Repeatability and Reproducibility

Repeatability refers to how close the measurements of an instrument are to each other if such measurements were repeated on a part under the same measuring conditions. *Reproducibility* is a measure of the degree of agreement between two single test results made on the same object in two different, randomly selected measuring locations or laboratories. While repeatability is normally used to designate precision for measurements made within a restricted set of conditions (for example, individual operators), reproducibility is normally used to designate precision for measurements involving variation between certain sets (for example, laboratories) as well as within them.

Linearity

Linearity is the degree of accuracy of the measurements of an instrument throughout its operating range.[8] Linearity is related to hysteresis in that as an instrument measures an object, arriving at the measurement from below, it may provide a particular answer. As the object's dimensions are "overshot" by the instrument, arriving at the object's measurement from above, the object's measurement from above may differ from the measurement arrived at from below. This difference is called hysteresis, and is a form of nonlinearity that may be built into the measuring instrument. It is an undesirable measuring instrument quality, and needs to be understood and corrected for as a source of error.[9]

Bias

Bias is the amount by which a precision-calibrated instrument's measurements differ from a true value. Accuracy is often called the *bias* in the measurement.[10]

Chapter 20

E. Statistical Process Control (SPC)

1. TECHNIQUES AND APPLICATIONS

> Select appropriate control charts for
> monitoring or analyzing various processes
> and explain their construction and use.
> (Apply)
>
> **Body of Knowledge III.E.1**

One of the disadvantages of many of the tools discussed up to this point is that they have no time reference. The histogram may clearly demonstrate that a process appears to be bimodally distributed, but gives no clue as to whether the two modes are time-related, that is, a time-related shift in the process occurred. This section will illustrate a number of time-series analytical techniques culminating in control charts.

Run Chart

If a specific measurement is collected on a regular basis, say every 15 minutes, and plotted on a time scale, the resultant diagram is called a *run chart*. An example of a run chart is shown in Figure 20.1.

One of the problems of the run chart is that the natural variation in the process and in the measurement system tend to cause the graph to go up and down when no real change is occurring. One way to smooth out some of this "noise" in the process is to take readings from several consecutive parts and plot the average of the readings. The result is called an *averages* or $\bar{X}$ *chart*. An example is shown in Figure 20.2.

One of the dangers of the averages chart is that it can make the process look better than it really is. For example, note that the average of the five 3:00 readings is 0.790, which is well within the tolerance of 0.780–0.795, even though every one of the five readings is outside the tolerance. Therefore, tolerance limits should never be drawn on an averages chart. To help alert the chart user that the readings are

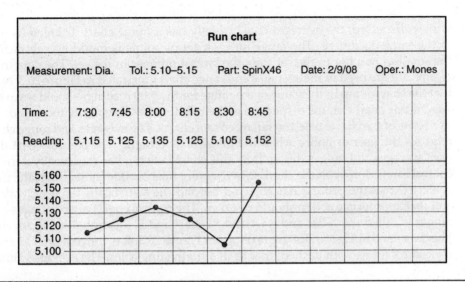

Run chart

Measurement: Dia.   Tol.: 5.10–5.15   Part: SpinX46   Date: 2/9/08   Oper.: Mones

Time:	7:30	7:45	8:00	8:15	8:30	8:45			
Reading:	5.115	5.125	5.135	5.125	5.105	5.152			

Figure 20.1   Example of run chart with first six points plotted.

Source: D. W. Benbow, A. K. Elshennawy, and H. F. Walker, *The Certified Quality Technician Handbook* (Milwaukee: ASQ Quality Press, 2003): 18.

Averages chart

Measurement: Length   Tol.: .780–.795   Part: WS4A   Date: 1/8/08   Oper.: White

Time:	Noon	1:00	2:00	3:00					
Readings:	.788	.788	.782	.775					
	.782	.792	.784	.774					
	.782	.790	.781	.798					
	.779	.794	.782	.800					
	.786	.790	.783	.801					
Average:	**.783**	**.791**	**.782**	**.790**					

Figure 20.2   Example of averages chart with first four points plotted.

Source: D. W. Benbow, A. K. Elshennawy, and H. F. Walker, *The Certified Quality Technician Handbook* (Milwaukee: ASQ Quality Press, 2003): 18.

Part III.E.1

widely dispersed, the averages chart usually has a range chart included as part of the same document. The range for each set of points is found by subtracting the smallest number in the set from the largest number in the set. The data from Figure 20.2 are used in the averages and range chart illustrated in Figure 20.3. Note that the sharp jump in the value of the range for the 3:00 readings should warn the user of this chart that the dispersion of the readings has drastically increased.

None of the charts listed so far are *control* charts. The averages and range chart requires the user to notice when the range is "too high." The control chart that uses averages and ranges differs from this chart in that it has statistically derived control limits drawn on the chart. When a point falls outside these limits, the user is alerted that the process has changed beyond the statistically expected values and that appropriate action should be taken. The control chart using averages and ranges is called the *X*-bar and *R* ($\overline{X}$ and *R*) chart and is illustrated in Figure 20.4. The data used in Figure 20.4 are taken from Figure 20.3. It is conventional to draw the control limits with dashed lines or in a contrasting color. The average value is

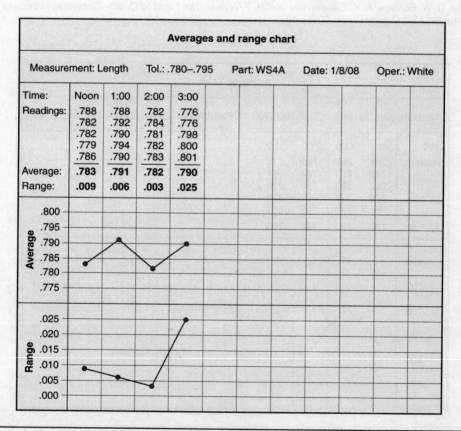

Figure 20.3   Example of an averages and range chart.

Source: D. W. Benbow, A. K. Elshennawy, and H. F. Walker, *The Certified Quality Technician Handbook* (Milwaukee: ASQ Quality Press, 2003): 19.

usually drawn with a solid line. Control limits help the user of the chart make statistically sound decisions about the process. This is because the limits are drawn so that a very high percentage of the points should fall between them. In the case of the averages chart, about 99.7 percent of the points from a *stable* process should fall between the upper and lower control limits. This means that when a point falls outside the limits, there is approximately 0.3 percent probability that this could have happened if the process has not changed. Therefore, points outside the control limits are very strong indicators that the process *has* changed. Control charts, then, can be used by process operators as real-time monitoring tools.

There are a number of events that are very unlikely to occur unless the process has changed and thus serve as statistical indicators of process change. The lists vary somewhat from textbook to textbook, but usually include something like those shown in Table 20.1.

When one of these events occurs on a control chart, the process operator needs to take appropriate action. Sometimes this may entail a process adjustment.

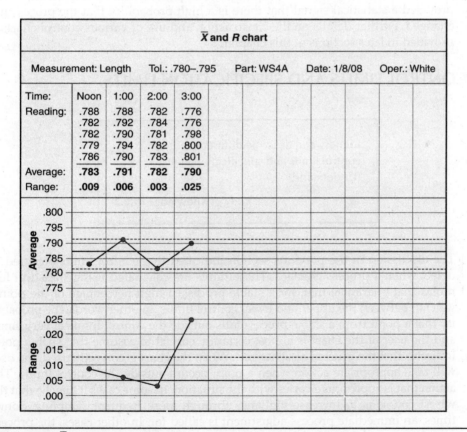

Figure 20.4 $\bar{X}$ and R chart.

Source: D. W. Benbow, A. K. Elshennawy, and H. F. Walker, *The Certified Quality Technician Handbook* (Milwaukee: ASQ Quality Press, 2003): 20.

Table 20.1 Control chart indicators of process change.

1. A point above the upper control limit or below the lower control limit

2. Seven successive points above (or below) the average line

3. Seven successive points trending up (or down)

4. Middle one-third of the chart includes more than 90% or fewer than 40% of the points after at least 25 points have been plotted

5. Nonrandom patterns

Source: D. W. Benbow, A. K. Elshennawy, and H. F. Walker, *The Certified Quality Technician Handbook* (Milwaukee: ASQ Quality Press, 2003): 20.

Sometimes the appropriate action is to stop the process. In some situations, the operator should increase watchfulness, perhaps taking readings every few minutes instead of every hour, for instance. The important issue is that the chart has provided a statistical signal that there is a high probability that the process has changed. Further details on the construction and use of various control charts is provided in later sections of this chapter.

2. CONTROL LIMITS AND SPECIFICATION LIMITS

> Identify and describe different uses of control limits and specification limits. (Understand)
>
> **Body of Knowledge III.E.2**

The discussion in the previous section indicated that control limits are used to detect when a process changes. These limits are calculated using statistical formulas that guarantee that, for a stable process, a high percentage of the points will fall between the upper and lower control limits. In other words, the probability that a point from a stable process falls outside the control limits is very small and the user of the chart is almost certainly correct to assume that such a point indicates that the process has changed. When the chart is correctly used, the user will take appropriate action when a point occurs outside the control limits. The action that is appropriate varies with the situation. In some cases it may be that the wisest course is to increase vigilance through more frequent sampling. Sometimes, an immediate process adjustment is called for. In other cases, the process must be stopped immediately.

One of the most common mistakes in using control charts is to use the specification limits as control limits. Since the specification limits have no statistical basis, the user of the chart has no statistical basis for assuming that the process has changed when a point occurs outside the specification limits. The statement, "I just plotted a point outside the control limits, but it is well within specification limits, so I don't have to worry about it," represents a misunderstanding of the chart. The main purpose of the control chart is to signal the user that the process has changed. If the signal is ignored, the control chart loses much of its value as an early warning tool. This misunderstanding of the significance of the control limits is especially dangerous in the case of the averages portion of the $\bar{X}$ and R or $\bar{X}$ and s chart. Recall that the points on this portion of the chart are calculated by averaging several measurements. It is possible for the average (or mean) to fall within the specification limits even though none of the actual measurements are within these limits. For example, suppose the specification limits for a dimension are 7.350 to 7.360 and a sample of five parts yields the following measurements: 7.346, 7.344, 7.362, 7.365, 7.366. The mean of these values is 7.357, well within the specification limits. In this case, the range chart would likely have shown the point above its upper control limit. Should the user take comfort in the fact that the plotted point, located at the mean, is well inside the specification limits?

Control limits are calculated based on data from the process. Formulas for control limits and examples of each are given in the following sections. The formulas are repeated in Appendix C. Several constants are needed in the formulas. These appear as subscripted capital letters such as A_2. The values of these constants are given in Appendix D. When calculating control limits, it is prudent to collect as much data as is practical. Many authorities specify at least 25 samples.

As explained in the following section, the control limits for each control chart are calculated based on data from the process. The control chart compares each new point with the distribution that was used as the basis for the control limits. The control limits enclose the vast majority of the points from the distribution, 99.72 percent if it is a normal distribution. Therefore, for the normal distribution, if the point is outside the control limits, it is common to state that there is a 99.72 percent probability that the point did not come from the distribution that was used to calculate the control limits. The percentages vary somewhat from chart to chart, but the control limits are constructed so that this probability will be quite high. It should be noted that these probabilities are somewhat theoretical because no process runs as if its output were randomly selected numbers from some historical distribution. It is enough to say that when a point falls outside the control limits, the probability is quite high that the process has changed.

When the probability is very high that a point did not come from the distribution used to calculate the control limits, the process is said to be "out of statistical control." The "out of statistical control" condition is often very subtle and would perhaps not be detected without the aid of a control chart. This, in fact, is one of the main values of the control chart: it detects changes in a process that would not otherwise be noticed. This may permit adjustment or other action on the process before serious damage is done, such as making large quantities of nonconforming product or making large quantities of product that differ significantly in some measurable quality characteristic, one part from another.

However, one of the hazards of using a control chart without proper training is the tendency to react to a point that is not right on target by adjusting the process, even though the chart does not indicate that the process has statistically changed. If an adjustment is made whenever the point is not exactly on target, it may tend to destabilize a stable process. Moreover, the operator adjusting the process becomes part of the variation. Additionally, such adjustments may instead serve to 'oversteer' the process. For instance, suppose a new student nurse wakes a patient for morning vitals at 4:00 AM and finds the body temperature to be 98.64°. The student nurse, having just covered normal body temperature in class the previous day, sees that the patient has a 0.04° fever and administers a fever reducing medicine, like aspirin. At the 5:00 AM vitals, the temperature is down to 98.4°, the aspirin having done its job. The nurse realizes that the temperature is now below target and administers a medicine to warm the patient. By 6:00 AM, the patient is warm enough to justify three aspirin tablets and by 7:00 AM, the patient needs more medicine to raise their body temperature, and so on.

In the ideal situation, a process should not need adjustment except when the chart indicates that it is out of statistical control. Dr. W. E. Deming, one of the key figures in the quality field, stated, "The function of a control chart is to minimize the net economic loss from . . . overadjustment and underadjustment."[1]

3. VARIABLES CHARTS

> Identify, select, construct, and interpret $\bar{X} - R$ and $\bar{X} - s$ charts. (Analyze)
>
> **Body of Knowledge III.E.3**

Earlier in this chapter, the $\bar{X}$ and R chart was introduced. It is called a *variables chart* because the data to be plotted result from measurement on a variable or continous scale. The $\bar{X}$ and s chart is another variables control chart. With this chart, the sample standard deviation s is used instead of the range to indicate dispersion. The standard deviation is a better measure of spread than the range, so this chart provides a somewhat more precise statistical signal than the $\bar{X}$ and R chart. Users who are comfortable with the standard deviation function on a handheld calculator will be able to calculate s almost as easily as R. The $\bar{X}$ and s chart is often the preferred chart when the chart is constructed using SPC software.

Control Limits for the $\bar{X}$ and R Chart

Upper control limit for the averages chart: $\text{UCL}_{\bar{x}} = \bar{\bar{X}} + A_2\bar{R}$

Lower control limit for the averages chart: $\text{LCL}_{\bar{x}} = \bar{\bar{X}} - A_2\bar{R}$

Upper control limit for the range chart: $\text{UCL}_R = D_4\bar{R}$

Lower control limit for the range chart: $\text{LCL}_R = D_3\bar{R}$

Example: Data are collected in a face-and-plunge operation done on a lathe. The dimension being measured is the groove inside diameter (ID), which has a tolerance of 7.125 ± 0.010. Four parts are measured every hour. The data are shown in Table 20.2. Since the formulas use the values of $\bar{X}$ and R, the next step is to calculate the average ($\bar{X}$) and range (R) for each time and then calculate the average of the averages ($\bar{\bar{X}}$) and the average range ($\bar{R}$). These calculations have been completed in Table 20.2. The values of $\bar{\bar{X}}$ and $\bar{R}$ are listed in the last row of Table 20.2. Notice that the value of $\bar{\bar{X}}$ is 7.125, which happens to be at the center of the tolerance in this example. This means that the process is centered at the same point as the center of the specifications.

$$\text{UCL}_{\bar{x}} = \bar{\bar{X}} + A_2\bar{R} = 7.125 + 0.729 \times 0.004 = 7.128$$

$$\text{LCL}_{\bar{x}} = \bar{\bar{X}} - A_2\bar{R} = 7.125 - 0.729 \times 0.004 = 7.122$$

$$\text{UCL}_R = D_4\bar{R} = 2.282 \times 0.004 \approx 0.009$$

$$\text{LCL}_R = D_3\bar{R} = 0 \times 0.004 = 0$$

In these calculations, the values A_2, D_3, and D_4 are found in Appendix D. The row for subgroup size $n = 4$ is used because each hourly sample has four readings.

The goal now is to construct the chart and use it to monitor future process activity. The first step is to choose scales for the $\bar{X}$ and R sections of the chart in such a way that the area between the control limits covers most of the chart area. One such choice is shown in Figure 20.5. Figure 20.5 also shows the chart with several points plotted. As each point is plotted, the user should check to determine whether any of the indicators of process change listed in Table 20.1, page 176, have occurred.

If none have occurred, continue with the process. If one of the indicators has occurred, appropriate action should be taken as spelled out in the process operating instructions.

Control Limits for the $\bar{X}$ and s Chart

Table 20.2 also shows the calculated values for s from the data. The formula calculations for the control limits for the $\bar{X}$ and s chart are:

Upper control limit for the averages chart: $\text{UCL}_{\bar{x}} = \bar{\bar{X}} + A_3\bar{s}$

Lower control limit for the averages chart: $\text{LCL}_{\bar{x}} = \bar{\bar{X}} - A_3\bar{s}$

Upper control limit for the range chart: $\text{UCL}_s = B_4\bar{s}$

Lower control limit for the range chart: $\text{LCL}_s = B_3\bar{s}$

$$\text{UCL}_{\bar{x}} = \bar{\bar{X}} + A_3\bar{s} = 7.125 + 1.628 \times 0.0020 = 7.128$$

$$\text{LCL}_{\bar{x}} = \bar{\bar{X}} - A_3\bar{s} = 7.125 - 1.628 \times 0.002 = 7.122$$

$$\text{UCL}_s = B_4\bar{s} = 2.266 \times 0.0020 \approx 0.004$$

$$\text{LCL}_s = B_3\bar{s} = 0 \times 0.0020 = 0$$

Construction and use of the $\bar{X}$ and s chart with these control limits is quite similar to that for the $\bar{X}$ and R chart and is not shown here.

Table 20.2   Data for $\bar{X}$ and R chart control limit calculations.

Time	1st Meas.	2nd Meas.	3rd Meas.	4th Meas.	Average	Range	Std. Dev.
4 PM	7.124	7.122	7.125	7.125	7.124	0.003	0.0014
5	7.123	7.125	7.125	7.128	7.125	0.005	0.0021
6	7.126	7.128	7.128	7.125	7.127	0.003	0.0015
7	7.127	7.123	7.123	7.126	7.125	0.004	0.0021
8	7.125	7.126	7.121	7.122	7.124	0.005	0.0024
9	7.123	7.129	7.129	7.124	7.126	0.006	0.0032
10	7.122	7.122	7.124	7.125	7.123	0.003	0.0015
11	7.128	7.125	7.126	7.123	7.126	0.005	0.0021
12 AM	7.125	7.125	7.121	7.122	7.123	0.004	0.0021
1	7.126	7.123	7.123	7.125	7.124	0.003	0.0015
2	7.126	7.126	7.127	7.128	7.127	0.002	0.0010
3	7.127	7.129	7.128	7.129	7.128	0.002	0.0010
4	7.128	7.123	7.122	7.124	7.124	0.006	0.0026
5	7.124	7.125	7.129	7.127	7.126	0.005	0.0022
6	7.127	7.127	7.123	7.125	7.126	0.004	0.0019
7	7.128	7.122	7.124	7.126	7.125	0.006	0.0026
8	7.123	7.124	7.125	7.122	7.124	0.003	0.0013
9	7.122	7.121	7.126	7.123	7.123	0.005	0.0022
10	7.128	7.129	7.122	7.129	7.127	0.007	0.0034
11	7.125	7.125	7.124	7.122	7.124	0.003	0.0014
12 PM	7.125	7.121	7.125	7.128	7.125	0.007	0.0029
1	7.121	7.126	7.12	7.123	7.123	0.006	0.0026
2	7.123	7.123	7.123	7.123	7.123	0	0.0000
3	7.128	7.121	7.126	7.127	7.126	0.007	0.0031
4	7.129	7.127	7.127	7.124	7.127	0.005	0.0021
Average	7.125	7.125	7.125	7.125	7.125	0.004	0.0020

Source: Adapted from D. W. Benbow, A. K. Elshennawy, and H. F. Walker, *The Certified Quality Technician Handbook* (Milwaukee: ASQ Quality Press, 2003): 53.

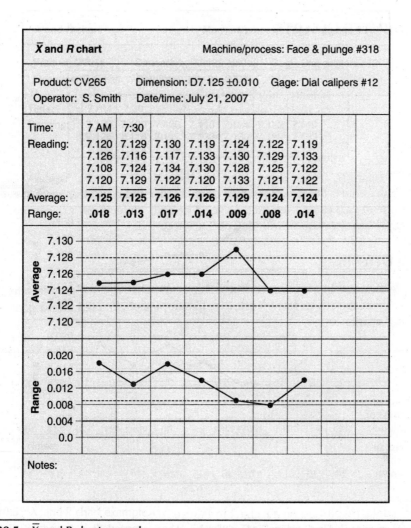

Figure 20.5 $\bar{X}$ and *R* chart example.

Source: D. W. Benbow, A. K. Elshennawy, and H. F. Walker, *The Certified Quality Technician Handbook* (Milwaukee: ASQ Quality Press, 2003): 64.

4. ATTRIBUTES CHARTS

> Identify, select, construct, and interpret *p*,
> *np*, *c*, and *u* charts. (Analyze)
>
> **Body of Knowledge III.E.4**

Attributes charts are used for count data. As discussed in Chapter 17, if every item is in one of two categories such as good or bad, "defectives" are counted. If each part may have several flaws, "defects" are counted.

Charting Defectives

If defectives are being counted, the *p* chart can be used. For example, a test for the presence of the Rh negative factor in 13 samples of donated blood gives the following results:

	Test number												
	1	2	3	4	5	6	7	8	9	10	11	12	13
No. of units of blood	125	111	113	120	118	137	108	110	124	128	144	138	132
No. of Rh– units	14	18	13	17	15	15	16	11	14	13	14	17	16

These data are plotted on a *p* chart in Figure 20.6.

Note that the *p* chart in Figure 20.6 has two points that are outside the control limits. These points indicate that the process was out of statistical control, which is sometimes referred to as *out of control*. It means that there is a very low probability that these points came from the same distribution as the one used to calculate the control limits. It is therefore very probable that the distribution has changed. These out-of-control points are a statistical signal that the process needs attention of some type. People familiar with the process need to decide how to react to various points that are outside the control limits. In the situation in this example, an unusually high number of units of blood test Rh negative. This could indicate a different population of donors or possibly a malfunction of the testing equipment or procedure.

If defectives are being counted and the sample size remains constant, the *np* chart can be used.

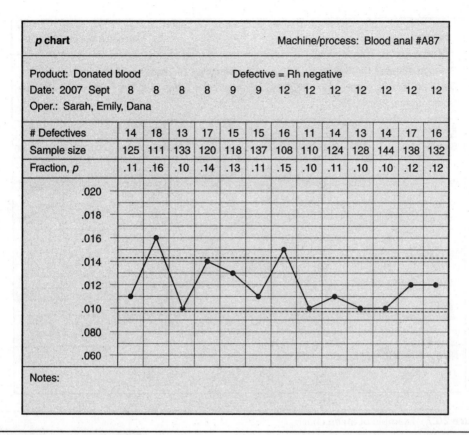

p chart													Machine/process: Blood anal #A87

Product: Donated blood   Defective = Rh negative
Date: 2007 Sept   8   8   8   8   9   9   12   12   12   12   12   12   12
Oper.: Sarah, Emily, Dana

# Defectives	14	18	13	17	15	15	16	11	14	13	14	17	16
Sample size	125	111	133	120	118	137	108	110	124	128	144	138	132
Fraction, p	.11	.16	.10	.14	.13	.11	.15	.10	.11	.10	.10	.12	.12

Notes:

Figure 20.6 *p* chart example.

Source: D. W. Benbow, A. K. Elshennawy, and H. F. Walker, *The Certified Quality Technician Handbook* (Milwaukee: ASQ Quality Press, 2003): 47.

Part III.E.4

Example: Packages containing 1000 light bulbs are randomly selected and all 1000 bulbs are light-tested. The *np* chart is shown in Figure 20.7. Note that on March 25, the point is outside the control limits. This means that there is a high probability that the process was different on that day than on the days that were used to construct the control limits. In this case, the process was different in a good way. It would be advisable to pay attention to the process to see what went right and to see if the conditions could be incorporated into the standard way of running the process. Notice the operator note at the bottom of the chart.

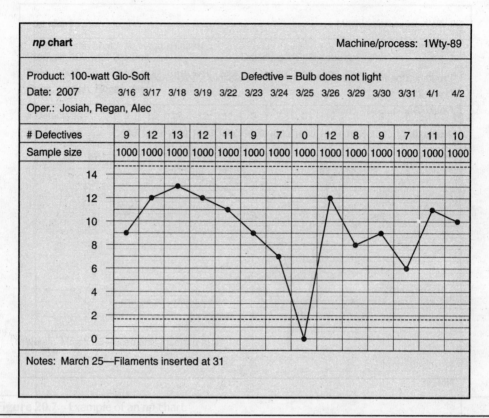

Figure 20.7 Example of an *np* chart.

Source: D. W. Benbow, A. K. Elshennawy, and H. F. Walker, *The Certified Quality Technician Handbook* (Milwaukee: ASQ Quality Press, 2003): 48.

The *u* chart and *c* chart are used when defects rather than defectives are being counted. If the sample size varies, the *u* chart is used. If the sample size is constant, the *c* chart may be used. An example of a *u* chart is shown in Figure 20.8. A *c* chart would look much like the *np* chart illustrated in Figure 20.7 and is not shown here.

To decide which attributes chart to use:

- For defectives, use *p* or *np*.

  – Use *p* for varying sample size.

  – Use *np* for constant sample size.

- For defects, use *u* or *c*.

  – Use *u* for varying sample size.

  – Use *c* for constant sample size.

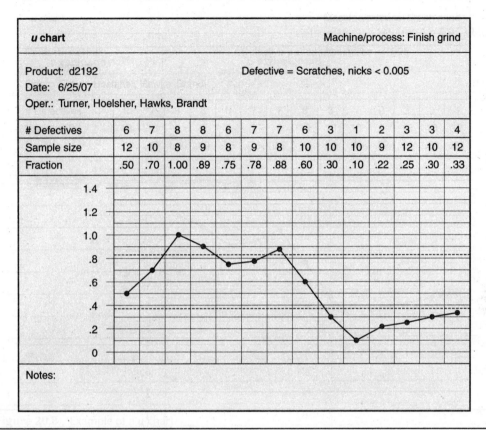

Figure 20.8   Example of a *u* chart.

Source: D. W. Benbow, A. K. Elshennawy, and H. F. Walker, *The Certified Quality Technician Handbook* (Milwaukee: ASQ Quality Press, 2003): 49.

Control Limits for *p* Charts

As indicated earlier, *p* charts are used when the data indicate the number of defective units in each sample. An example of this type of data is shown in Table 20.3, where the column labeled "No. discrepancies *d*" shows the number of defective units.

The formulas for the the control limits for the *p* chart are:

$$\mathrm{UCL}_p = \bar{p} + 3\sqrt{\frac{\bar{p}(1-\bar{p})}{\bar{n}}}$$

$$\mathrm{LCL}_p = \bar{p} - 3\sqrt{\frac{\bar{p}(1-\bar{p})}{\bar{n}}}$$

Part III.E.4

Table 20.3   Attributes data with totals.

Time	Sample size n	No. discrepancies d
8 AM	48	3
8:30	45	6
9	47	2
9:30	51	3
10	48	5
10:30	47	4
11	48	1
11:30	50	0
12 PM	46	1
12:30	45	0
1	47	2
1:30	48	5
2	50	3
2:30	50	6
3	49	2
3:30	46	4
4	50	1
4:30	52	1
5	48	6
5:30	47	5
6	49	6
6:30	49	4
7	51	3
7:30	50	4
8	48	1
8:30	47	2
9	47	5
9:30	49	0
10	49	0
Totals	1401	85

Source: D. W. Benbow, A. K. Elshennawy, and H. F. Walker, *The Certified Quality Technician Handbook* (Milwaukee: ASQ Quality Press, 2003): 57.

where

$$\bar{n} = \frac{\text{Sum of the sample sizes}}{\text{Number of samples}}$$

$$\bar{p} = \frac{\text{Sum of the discrepancies}}{\text{Sum of the sample sizes}} = \frac{\Sigma \text{discrepancies}}{\Sigma n}$$

The first step in calculating these control limits is to find $\bar{n}$ and $\bar{p}$. The totals needed for these values are calculated in Table 20.3.

$$\bar{n} = \frac{\Sigma n}{\text{Number of samples}} = \frac{1401}{29} = 48.3$$

$$\bar{p} = \frac{\Sigma d}{\Sigma n} = \frac{85}{1401} = 0.061$$

Using these values in the formula for the upper control limit:

$$\text{UCL}_p = 0.061 + 3\sqrt{\frac{0.061(1-0.061)}{48.3}} \approx 0.061 + 3\sqrt{0.1186} \approx 0.061 + 0.103 = 0.164$$

Using the formula for the lower control limit and taking advantage of some of the values just calculated: $\text{LCL}_p = 0.061 - 0.103 = -0.042$

Since this value is negative, there is no lower control limit.

It is important to note that zero is commonly used as the lower control limit. It is equally important to note that using zero as the lower control limit ignores the statistical signficance of a grouping of points near zero, wherein such a grouping does not mean the same thing statistically as a grouping of points near a lower control limit (LCL).

Control Limits for *u* Charts

If the values in the "d" column of Table 20.3 represent the number of defects rather than the number of defectives, then the *u* chart is appropriate. The formulas for control limits for the *u* chart are:

$$\text{UCL}_u = \bar{u} + 3\sqrt{\frac{\bar{u}}{n}}$$

$$\text{LCL}_u = \bar{u} - 3\sqrt{\frac{\bar{u}}{n}}$$

where

$$\bar{u} = \frac{\Sigma d}{\Sigma n}$$

$$\text{and } \bar{n} = \frac{\Sigma n}{\text{Number of samples}}$$

Using the data from Table 20.3:

$$\bar{u} = \frac{\Sigma d}{\Sigma n} = \frac{85}{1401} = 0.061$$

$$\text{UCL}_u = 0.061 + 3\sqrt{\frac{0.061}{48.3}} \approx 0.061 + 0.107 \approx 0.168$$

$$\text{LCL}_u = 0.061 - 3\sqrt{\frac{0.061}{48.3}} \approx 0.061 - 0.107 \approx -0.046$$

Again, since the lower control limit formula returns a negative value, there is no lower control limit. (Some authors use zero as the lower control limit.)

Figures 20.7 and 20.8 illustrate np and u charts. The other charts, p and c, are similar and are not shown. Analysis of attributes charts follows the same procedures as those for the $\bar{X}$ and R charts outlined earlier in the section on variables charts.

5. RATIONAL SUBGROUPS

> Define and describe the principles of rational subgroups. (Understand)
>
> **Body of Knowledge III.E.5**

The method used to select samples for a control chart must be logical or "rational." In the case of the $\bar{X}$ and R chart, in order to assign "rational" subgroup or sample size, there should be a high probability of variation between successive samples while the variation within the sample is kept small. Therefore, samples frequently consist of parts that are produced successively by the same process to minimize the within-sample variation. The next sample is chosen somewhat later so that any process shifts that have occurred will be displayed on the chart. Choosing the rational subgroup requires care to make sure that the same process is producing each item. For example, suppose a candy-making process uses 40 pistons to deposit 40 gobs of chocolate on a moving sheet of waxed paper in a 5 × 8 array as shown in Figure 20.9. How should rational subgroups of size five be selected? The choice consisting of the first five chocolates in each row, as shown in Figure 20.9(a), would have the five units of the sample produced by five different processes (the five different pistons). A better choice would be to select the upper left-hand chocolate in five consecutive arrays as shown in Figure 20.9(b) because they are all formed by the same piston.

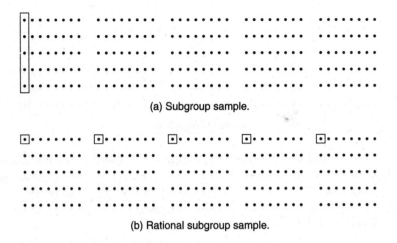

(a) Subgroup sample.

(b) Rational subgroup sample.

Figure 20.9 Conveyor belt in chocolate making process. Possible subgroup sampling.
Source: D. W. Benbow, A. K. Elshennawy, and H. F. Walker, *The Certified Quality Technician Handbook* (Milwaukee: ASQ Quality Press, 2003): 62.

If the original sampling plan had been used and one piston were functioning improperly, the data might look like this:

Piston 1	.25	.24	.26	.24	.24	.25	.26	.26	.23	.24
Piston 2	.36	.38	.33	.34	.36	.35	.38	.35	.34	.34
Piston 3	.24	.26	.25	.25	.24	.23	.24	.25	.26	.24
Piston 4	.24	.21	.25	.24	.23	.21	.22	.22	.24	.21
Piston 5	.25	.26	.29	.24	.23	.25	.26	.26	.24	.29
Average	.248	.270	.276	.262	.260	.258	.272	.268	.262	.264
Range	.12	.17	.08	.10	.13	.14	.16	.13	.11	.13

If the upper and lower control limits for the averages chart are based on these data, they would be:

$$UCL \approx 0.34 \text{ and } LCL \approx 0.19$$

and the averages would all fall ridiculously near the centerline of the chart, 0.264. The chart would be worthless as a tool to signal process change. The reason that this happened is that the errant piston #2 caused the ranges to be very high, which in turn caused the control limit values to be spread too far apart.

Another mistake sometimes made in selecting a rational subgroup is to measure what is essentially the same thing multiple times and use those values to make up a sample. One example of this might be to check the hardness of a heat-treated shaft at three different points or to measure the temperature of a well-stirred vat at three different points. Data such as these might result:

	7 AM	8 AM	9 AM	10 AM	11 AM	Noon	1 PM	2 PM	3 PM	4 PM	5 PM
	21	33	19	30	22	25	31	31	22	29	28
	22	30	19	29	20	26	33	32	22	28	29
	23	31	17	29	22	28	29	33	19	30	28
Average	22.0	31.3	18.3	29.3	21.3	26.3	31.0	32.0	21.0	29.0	28.3
Range	2	3	2	1	2	3	4	2	3	2	1

For the averages chart, the control limits are 29.6 and 25.6. In this situation, the readings in each sample are so close to each other that the range is very small, which places the UCL and LCL so close together that only two points on the chart will be within the control limits. Again, the chart is worthless as a tool for signaling process change.

6. PROCESS CAPABILITY MEASURES

> Define the prerequisites for measuring capability, and calculate and interpret C_p, C_{pk}, P_p, P_{pk}, and CR in various situations. (Analyze)
>
> **Body of Knowledge III.E.6**

The data in Table 20.2 were used to calculate the control limits for the $\bar{X}$ and R control chart. If the control chart does not exhibit any of the indicators of instability as discussed in the section on control limits and specification limits, it is assumed that the process is stable. If the process is stable, it is appropriate to conduct a process capability analysis. The purpose of a capability analysis is to estimate the percent of the population that meets specifications. If the process continues to run exactly the same way, this estimate can be used as a prediction for future production. Note that such a study is not valid if the process is not stable (that is, out of control) because the analysis produces a prediction of the ability of the process to meet specifications. If the process is unstable, then it is unpredictable.

The process capability study answers the question, "is my process good enough?" This is quite different from the question answered by a control chart, which is, "has my process changed?"

Properly, use of a control chart to establish that a process is stable and predictable precedes the use of a capability study to see if items produced by the process are good enough.

Five indices are produced by a capability study—the C_p index, P_p index, C_{pk} index, P_{pk} index, and CR index. The most optimistic of these is C_p, which disregards centering and is insensitive to "shifts and drifts" (special causes) in the data. It indicates what the capability of the process would be if no such problems existed. It is also sometimes called the process potential because it represents the ideal number of times that the process distribution fits within the specification limits, assuming that the process is perfectly centered. The most realistic estimate is P_{pk}, which indicates how the process really is. Because of their different properties, experienced users can compare these indices and know what type of remedial action a process needs: removal of a special cause, centering, or reduction of natural variation.

In the case where $\bar{R}/d_2 \approx s$, the sample standard deviation, the same formula that produces C_p also produces P_p. The difference is in how standard deviation is estimated. The same is true for P_{pk} and the C_{pk} formula.

Generally, P_{pk} numbers below 1.0 indicate a process in need of work, while numbers as high as 1.5 indicate an excellent process.

If a process is stable and predictable and has consistently provided high P_{pk} numbers for a long period, the process is a candidate for removing final inspection, which is likely to produce more defects than it finds. The process should then have the input variables controlled and undergo periodic audits.

Perhaps the most common capability study error is requesting suppliers to provide just C_{pk} along with shipped goods. C_{pk} can easily be manipulated by selectively changing the order of the data. To get spectacularly inflated values, simply sort the data. If you are to rely on a single index, then P_{pk} is the appropriate choice. Authoritative texts allow the substitution of C_{pk} for P_{pk} if the process is stable and predictable. This is allowed because in that case they are equal.

The following discussion shows how to calculate five different capability indices: C_{pk}, C_p, P_{pk}, P_p, and CR. These indices were developed to provide a single number to quantify process capability.

Part III.E.6

Calculating C_{pk}

Probably the most useful capability index is C_{pk}. The first step in calculating C_{pk} is to find the values of Z_U and Z_L using the following formulas:

$$Z_U = \frac{\text{USL} - \bar{\bar{X}}}{\sigma}$$

$$Z_L = \frac{\bar{\bar{X}} - \text{USL}}{\sigma}$$

where USL and LSL are the upper and lower specification limits and σ is the process standard deviation, which may be approximated by

$$\hat{\sigma} = \frac{\bar{R}}{d_2} \text{ (values of } d_2 \text{ are given in Appendix C)}$$

For example, suppose the tolerance on a dimension is 1.000 ± 0.005 and a control chart shows that the process is stable with $\bar{\bar{X}}$ = 1.001 and $\bar{R}$ = 0.003 with sample size n = 3.

For this example USL = 1.005, LSL = 0.995, $\bar{\bar{X}}$ = 1.001, $\bar{R}$ = 0.003 and the estimated value of σ is

$$\frac{0.003}{1.693} \approx 0.0018$$

Substituting these values into the Z formulas:

$$Z_U = \frac{1.005 - 1.001}{0.0018} \approx 2.22$$

$$Z_L = \frac{1.001 - 0.995}{0.0018} \approx 3.33$$

The formula for C_{pk}:

$$C_{pk} = \frac{\text{Min}(Z_U, Z_L)}{3}$$

where Min means select the smaller of the values in the parentheses. In this example,

$$C_{pk} = \frac{\text{Min}(2.22, 3.33)}{3} \approx 0.74$$

The higher the value of C_{pk}, the more capable the process. A few years ago, a C_{pk} value of one or more was considered satisfactory. This would imply that there is at least three σ between the process mean ($\bar{\bar{X}}$) and the nearest specification limit. More recently, many industries require four, five, or even six standard deviations between the process mean and the nearest specification limit. This would correspond to a C_{pk} of 1.33, 1.67, and 2, respectively. The push for six standard deviations was the origin of the Six Sigma methodology.

If the process data are normally distributed, the Z-values may be used in a standard normal table (Appendix B) to find the approximate values for the percent of production that lies outside the specification. In this example, the upper Z of 2.22 corresponds to 0.0132 in Appendix B. This means that approximately 1.32 percent of the production of this process violates the upper specification limit. Similarly, the lower Z of 3.33 corresponds to 0.0004 in Appendix B. This means that approximately 0.04 percent of the production of this process violates the lower specification limit. Note that the use of the standard normal table in Appendix B is only appropriate if the process data are normally distributed. Since no real-world process data are exactly normally distributed, it is best to state these percentages as estimates. The percentages can be useful in estimating return on investment for quality improvement projects.

When the two Z-values are not equal, this means that the process average is not centered within the specification limits and some improvement to the percentages can be made by centering. Suppose, in the above example, that a process adjustment can be made that would move $\bar{\bar{X}}$ to a value of 1.000. Then Z_U and Z_L would each be 2.77, and from the table in Appendix B, the two percentages would be 0.28 percent.

C_{pk} is a type of process capability index. It is sensitive to whether the process is centered, but insensitive to special causes.

In simplest terms, C_{pk} indicates how many times you can fit three standard deviations of the process between the mean of the process and the nearest specification limit. Assuming that the process is stable and predictable, if you can do this once, C_{pk} is 1.0, and your process probably needs attention. If you can do it 1.5 times, your process is excellent, and you are on the path to being able to discontinue final inspection. If you can do it two times, you have an outstanding process. If C_{pk} is negative, the process mean is outside the specification limits.

There is a more sophisticated way to calculate C_{pk}, using a sum of squares estimate of within-subgroup standard deviation. In the more common case shown here, the same formula is used to calculate both C_{pk} and P_{pk}. The difference is in how standard deviation is estimated. For C_{pk}, the moving range method is used, and for P_{pk} the sum of squares for all the data is used.

Since the moving range method of estimating standard deviation is insensitive to shifts and drifts (special causes), C_{pk} tends to be an estimate of the capability of the process assuming special causes are not present. Anyone who wants you to accept C_{pk} as an indicator of process capability automatically owes you a control chart (process behavior chart) demonstrating that the process is stable and predictable. (P_{pk} will equal C_{pk}.) Absent that, P_{pk} is the correct indicator of capability.

The confidence interval around an estimate of standard deviation tends to be larger than many users expect. This uncertainty becomes part of the calculation. Even with a few dozen samples, the estimates of any of the quality indices may not be very precise.

While normality of the data is not a concern in control charts,[2] it is of concern in interpreting the results of a process capability study. Data should be tested for nonnormality. If the data are nonnormal, estimates of defective parts per million may be improved by applying a transformation to the data.

Calculating C_p

The formula for the capability index C_p is:

$$C_p = \frac{USL - LSL}{6\sigma}$$

Using the values from the previous example:

$$C_p = \frac{1.005 - 0.995}{6(0.0018)} \approx 0.93$$

Note that the formula for this index does not make use of the process average $\overline{\overline{X}}$. Therefore this index does not consider whether the process average is centered within the specification limits or, indeed, whether it is even between the two limits. In reality, C_p tells what the process could potentially do if it were centered. For centered processes, C_p and C_{pk} have the same value. For processes that are not centered, C_p is greater than C_{pk}.

C_p is a statistical measure of process capability. This means that it is a measure of how capable a process is of producing output within specification limits. This measurement only has meaning when the process being examined is in a state of statistical control. C_p disregards centering, and is insensitive to shifts and drifts (special causes) in the data. If the process mean is not centered within the specification limits, this value may therefore be misleading, and the C_{pk} index should be used instead for analyzing process capability.[3]

Calculating CR

This index is also referred to as the *capability ratio*. It is the reciprocal of C_p.

$$CR = \frac{6\sigma}{USL - LSL} \quad \text{or} \quad \frac{1}{C_p}$$

Using the data from the previous example,

$$CR = \frac{6(0.0018)}{1.005 - 0.995} \approx 1.08 \text{ which, not surprisingly, is approximately } \frac{1}{0.093}.$$

Of course, lower values of CR imply more capable processes.

Calculating P_{pk}

P_{pk} is a type of process performance index, specifically a measure of historical variation in a process. As opposed to the related C_{pk} index, P_{pk} cannot be used to predict the future and is used when a process is not in statistical control. It is often used for preproduction runs where the population standard deviation is unknown. First compute the Z-scores or number of standard deviations the process average $\bar{\bar{X}}$ is from each specification, substituting the sample standard deviation s for an estimate of the process standard deviation σ.

$$Z_U = \frac{USL - \bar{\bar{X}}}{s}$$

$$Z_L = \frac{\bar{\bar{X}} - USL}{s}$$

with s computed as

$$s = \sqrt{\frac{\sum_i \left(X_i - \bar{\bar{X}}\right)^2}{(n-1)}}$$

As a formula,

$$P_{pk} = \frac{Min\left(Z_U, Z_L\right)}{3}$$

and USL is upper specification limit and LSL is lower specification limit.[4,5]

Calculating P_p

Like the formula for C_p, P_p does not make use of the process average $\bar{\bar{X}}$. However, like the formula for P_{pk}, it does use the sample standard deviation s instead of $\bar{R}/d_2$. The formula is:

$$P_p = \frac{USL - LSL}{6s}$$

Machine Capability

Process capability analysis was discussed earlier in this section. Machine capability analysis is calculated mathematically in the same way, that is, the same formulas are used. The difference is that machine capability analysis attempts to isolate and analyze the variation due to an individual machine rather than the entire process. If a process has several machines linked together either in series or parallel, a machine capability analysis should be conducted on each machine. It is usually best to provide a very consistent input to the machine and very precise and accurate measurement of the output. This helps assure that the analysis does not consider variation from other sources or from the measurement system. Machine capability analyses can be very useful in efforts to reduce process variation by identifying points in the process where excess variation is caused.

7. PRE-CONTROL CHART

> Define the concept and use of pre-control charts. (Understand)
>
> **Body of Knowledge III.E.7**

Pre-Control Chart

Pre-control, also called stoplight control, is sometimes used in place of control charts or until sufficient data are collected to construct a control chart. Upper and lower pre-control limits, called PC limits, are calculated based on the tolerance limits. The value of the tolerance (upper specification limit – lower specification limit) is multiplied by 0.25. The resulting value is subtracted from the upper specification limit, forming the upper PC limit, and added to the lower specification limit, forming the lower PC limit. As parts are measured, their values are compared to the PC limits and appropriate action is taken based on rules such as these:

1. If the first part is outside the specification limits, adjust the process.

2. If a part is inside the specification limits, but outside PC limits, measure the next part.

3. If two successive parts are outside PC limits, adjust the process.

4. If five successive parts are inside PC limits, switch to less frequent monitoring.

Various authors provide additional rules. The principle advantage of pre-control is that it is simpler. The main disadvantage is that it is not statistically based. When the pre-control rules indicate that the process should be adjusted, there is not necessarily a high probability that the process has changed. This may lead to overadjustment and decreased stability of the process. For this reason, there is some controversy over the use of pre-control, with Montgomery stating that, "This author believes that pre-control is a poor substitute for standard control charts and would never recommend it in practice."[6]

8. COMMON AND SPECIAL CAUSE VARIATION

Interpret various control chart patterns (runs, hugging, and trends) to determine process control, and use rules to distinguish between common cause and special cause variation. (Analyze)

Body of Knowledge III.E.8

The variation of a process that is in statistical control is called *common cause variation*. This variation is inherent in the process and can only be reduced through changes in the process itself. Therefore, it is important that process operators not respond to changes attributed to common cause variation.

When additional variation occurs, it is referred to as *special cause variation*. This variation can be assigned to some outside change that affects the process output. Hence it is also called *assignable cause variation*. It is important that the process operator respond to this variation.

The purpose of a control chart is to distinguish between these two types of variation. It does this by providing a statistical signal that a special or assignable cause is impacting the process. This permits the operator to have immediate process feedback and to take timely and appropriate action. Although the term "assignable" is used synonymously with "special" causes of variation, the root cause of the variation and its remedy may not be known but may be "assigned" to some cause, perhaps yet to be determined, other than the random noise inherent in the process.

The statistical signal consists of the occurrence of an event that is on the list of indicators of process change. See Table 20.1, page 176.

The action that is appropriate for the operator to take depends on the event and process. The action that is needed should be clearly spelled out in the operating instructions for the process. These instructions should provide for logging of the event and actions taken.

Figures 20.10 through 20.14 illustrate the use of the indicators of process change to distinguish between special cause and common cause variation. The caption for each figure explains the indicator involved.

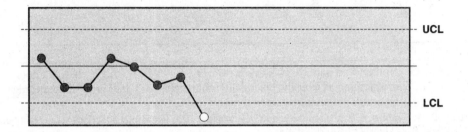

Figure 20.10   Point outside control limit. When the open dot is plotted, the operator is signaled that there is a very high probability that the process has changed.

Source: D. W. Benbow, A. K. Elshennawy, and H. F. Walker, *The Certified Quality Technician Handbook* (Milwaukee: ASQ Quality Press, 2003): 67.

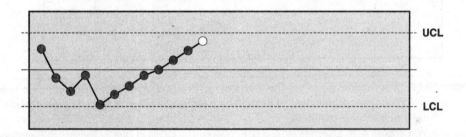

Figure 20.11   Seven successive points trending upward. When the open dot is plotted, the operator is signaled that there is a very high probability that the process has changed.

Source: D. W. Benbow, A. K. Elshennawy, and H. F. Walker, *The Certified Quality Technician Handbook* (Milwaukee: ASQ Quality Press, 2003): 67.

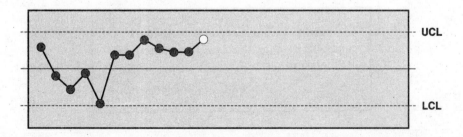

Figure 20.12   Seven successive points on one side of the average. When the open dot is plotted, the operator is signaled that there is a very high probability that the process has changed.

Source: D. W. Benbow, A. K. Elshennawy, and H. F. Walker, *The Certified Quality Technician Handbook* (Milwaukee: ASQ Quality Press, 2003): 67.

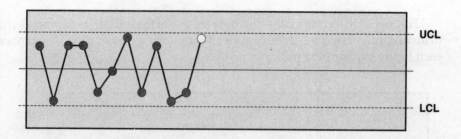

Figure 20.13   Nonrandom pattern. This rule is not very well defined. In this case, the dots are jumping all over the place rather than being clustered around the average. Operator judgment is required here.

Source: D. W. Benbow, A. K. Elshennawy, and H. F. Walker, *The Certified Quality Technician Handbook* (Milwaukee: ASQ Quality Press, 2003): 67.

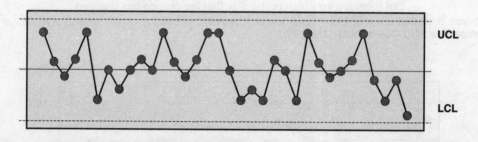

Figure 20.14   Nonrandom patterns. The V-shape seems to occur at three different times. This could be meaningless or it could be a symptom of some special cause. Operator judgment is required.

Source: D. W. Benbow, A. K. Elshennawy, and H. F. Walker, *The Certified Quality Technician Handbook* (Milwaukee: ASQ Quality Press, 2003): 68.

9. DATA PLOTTING

> Identify the advantages and limitations of
> analyzing data visually instead of numerically.
> (Understand)
>
> **Body of Knowledge III.E.9**

One of the hazards of using software for statistical analysis is the temptation to perform analysis on data without first "looking" at the data. One example would be the calculation of linear correlation coefficients. This number helps determine whether two variables have a relationship that permits the prediction of one variable using the other variable in a straight line formula. The correct procedure is to construct a scatter diagram of the data as shown in Chapter 9. If the points seem to be grouped around a straight line in the scatter diagram, it would be appropriate to calculate the coefficient. Otherwise, the calculation may be misleading. In some cases (especially with small data sets), a fairly high correlation coefficient may result from data that, when plotted, are clearly not related. Displaying data graphically cannot be overemphasized. "Always *graph* your data in some simple way—*always.*"[7]

A few graphical tools are:

- Histogram or frequency chart

- Stem-and-leaf diagram

- Box plot

- Dot plot

- Scatter diagram

- Pareto diagram

Chapter 21

F. Regression and Correlation

Describe how regression and correlation models are used for estimation and prediction. (Apply)

Body of Knowledge III.F

REGRESSION

Regression analysis is used to model relationships between variables, determine the magnitude or strength of the relationships between variables, and make predictions based on the models.

Regression analysis models the relationship between one or more response variables (also called dependent variables, explained variables, predicted variables, or regressands), (usually named Y), and the predictors (also called independent variables, explanatory variables, control variables, or regressors), (usually named $X_1, \ldots, X_p$). If there is more than one response variable, we speak of multivariate regression.

Simple and Multiple Linear Regression

Simple linear regression and multiple linear regression are related statistical methods for modeling the relationship between two or more random variables using a linear equation. Simple linear regression refers to a regression on two variables while multiple regression refers to a regression on more than two variables. Linear regression assumes that the best estimate of the response is a linear function of some parameters (though not necessarily linear on the predictors).

Nonlinear Regression Models

If the relationship between the variables being analyzed is not linear in parameters, a number of nonlinear regression techniques may be used to obtain a more accurate regression.

Linear Models

Predictor variables may be defined quantitatively, that is, as continuous variables, or qualitatively, that is, as categorical variables. Categorical predictors are sometimes called factors. Although the method of estimating the model is the same for each case, different situations are sometimes known by different names for historical reasons:

- If the predictors are all quantitative, we speak of multiple regression.

- If the predictors are all qualitative, one performs analysis of variance.

- If some predictors are quantitative and some qualitative, one performs an analysis of covariance.

The linear model usually assumes that the dependent variable is continuous. If least squares estimation is used, then if it is assumed that the error component is normally distributed, the model is fully parametric. If it is not assumed that the data are normally distributed, the model is semi-parametric. If the data are not normally distributed, there are often better approaches to fitting than least squares. In particular, if the data contain outliers, robust regression might be preferred.

If two or more independent variables are correlated, we say that the variables are multicollinear. Multicollinearity results in parameter estimates that are unbiased and consistent, but which may have relatively large variances.

If the regression error is not normally distributed but is assumed to come from an exponential family, generalized linear models should be used. For example, if the response variable can take only binary values (for example, a Boolean or yes/no variable), logistic regression is preferred. The outcome of this type of regression is a function that describes how the probability of a given event—for example, the probability of getting "yes"—varies with the predictors.[1]

CORRELATION

In probability theory and statistics, correlation, also called *correlation coefficient*, indicates the strength and direction of a linear relationship between two random variables. In general statistical usage, correlation or co-relation refers to the departure of two variables from independence, although correlation does not imply causality. In this broad sense there are several coefficients, measuring the degree of correlation, adapted to the nature of the data.

A number of different coefficients are used for different situations. The best known is the Pearson product-moment correlation coefficient, which is obtained by dividing the covariance of the two variables by the product of their standard deviations. Despite its name, it was first introduced by Francis Galton.

The correlation $\rho_{X,Y}$ (the Greek symbol ρ, pronounced "row"), also symbolized as r between two random variables X and Y with expected values μ_X and μ_Y and standard deviations σ_X and σ_Y, is defined as:

$$\rho_{X,Y} = \frac{\text{cov}(X,Y)}{\sigma_X \sigma_Y} = \frac{E\big((X - \mu_X)(Y - \mu_Y)\big)}{\sigma_X \sigma_Y}$$

where E is the expected value of the variable and cov means covariance. Since

$$\mu_X = E(X),$$

$$\sigma_X^2 = E(X^2) - E^2(X) \text{ and likewise for } Y.$$

We may also write

$$\rho_{X,Y} = \frac{E(XY) - E(X)E(Y)}{\sqrt{E(X^2) - E^2(X)}\sqrt{E(Y^2) - E^2(Y)}}$$

The correlation is 1 in the case of an increasing linear relationship, –1 in the case of a decreasing linear relationship, and some value in between in all other cases, indicating the degree of linear dependence between the variables. The closer the coefficient is to either –1 or 1, the stronger the correlation between the variables.

If the variables are independent, then the correlation is zero, but the converse is not true because the correlation coefficient detects only linear dependencies between two variables. Here is an example: Suppose the random variable X is uniformly distributed on the interval from –1 to 1, and $Y = X^2$. Then Y is completely determined by X, so that X and Y are dependent, but their correlation is zero; they are uncorrelated. However, in the special case when X and Y are jointly normal, independence is equivalent to uncorrelatedness.

A correlation between two variables is diluted in the presence of measurement error around estimates of one or both variables, in which case disattenuation provides a more accurate coefficient.

Interpretation of the Size of a Correlation

Several authors have offered guidelines for the interpretation of a correlation coefficient. Cohen, for example, has suggested the following interpretations for correlations in psychological research:[2]

Correlation	Negative	Positive
Small	−0.29 to −0.10	0.10 to 0.29
Medium	−0.49 to −0.30	0.30 to 0.49
Large	−1.00 to −0.50	0.50 to 1.00

As Cohen himself has observed, however, all such criteria are in some ways arbitrary and should not be observed too strictly. This is because the interpretation of a correlation coefficient depends on the context and purposes. A correlation of 0.9 may be very low if one is verifying a physical law using high-quality instruments but may be regarded as very high in the social sciences where there may be a greater contribution from complicating factors.[3]

A scatter diagram is helpful in detecting evident nonlinearity, bimodal patterns, mavericks, and other nonrandom patterns. The pattern of experimental data should be that of an "error ellipse": one extreme, representing no correlation,

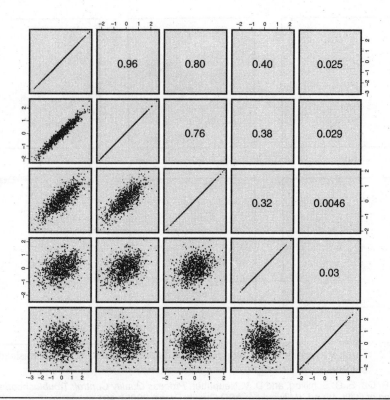

Figure 21.1   Example of linear correlation as demonstrated by a series of scatter diagrams. The data are graphed on the lower left and their correlation coefficients listed on the upper right. Each square in the upper right corresponds to its mirror-image square in the lower left, the "mirror" being the diagonal of the whole array.

would be circular; the other extreme, representing perfect correlation, would be a perfectly straight line. Figure 21.1 illustrates various correlations and their respective coefficients

Whether the data are not normally distributed is not entirely answered by a histogram alone; some further evidence on the question can be had by plotting accumulated frequencies Σf_i on normal-probability paper, and checking for mavericks or outliers.

Some Interpretations of r, *under the Assumptions.* It can be proved that the maximum attainable absolute value of the correlation coefficient r is + 1; this corresponds to a straight line with positive slope. (Its minimum attainable value of –1 corresponds to a straight line with negative slope.) Its minimum attainable absolute value is zero; this corresponds to a circular pattern of random points.

Very good predictability of Y from X in the region of the data is indicated by a value of r near +1 or –1.

Very poor predictability of Y from X is indicated by values of the correlation coefficient r close to zero. Thus $r = 0$ and $r = \pm 1$ represent the extremes of no predictability and perfect predictability. Values of r greater than 0.8 or 0.9 (r^2 values of 0.64 and 0.81, respectively) computed from production data are uncommon.

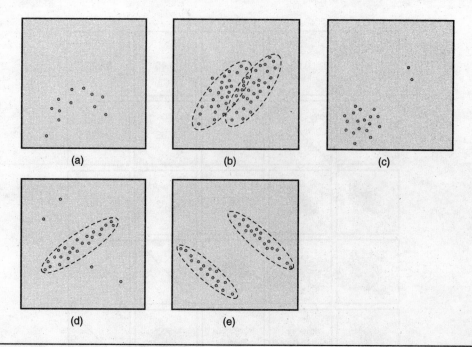

Figure 21.2 Some frequently occurring patterns of data that lead to seriously misleading values of *r* and are not recognized as a consequence.

Source: E. R. Ott, E. G. Schilling, and D. V. Neubauer, *Process Quality Control: Troubleshooting and Interpretation of Data,* 4th ed. (Milwaukee: ASQ Quality Press, 2005): 401.

Misleading Values of r. The advantage of having a scatter diagram is important in the technical sense of finding causes of trouble. There are different patterns of data that will indicate immediately the inadvisability of computing *r* until the data provide reasonable agreement with the required assumptions.

Some patterns producing seriously misleading values of *r* when a scatter diagram is not considered are as follows:

- Figure 21.2(a); a curvilinear relationship will give a low value of *r*.

- Figure 21.2(b); two populations each with a relatively high value of *r*. When a single *r* is computed, a low value of *r* results. It is important to recognize the existence of two sources in such a set of data.

- Figure 21.2(c); a few mavericks, such as shown, give a spuriously high value of *r*.

- Figure 21.2(d); a few mavericks, such as shown, give a spuriously low value of r, and a potentially opposite sign of *r*.

- Figure 21.2(e); two populations distinctly separated can give a spuriously high value of *r*, and a potentially opposite sign of *r*.

All the patterns shown in Figure 21.2 are typical of sets of production and experimental data. Hence the value of plotting the data using a scatter diagram cannot be overemphasized.

Chapter 22

G. Hypothesis Testing

> Determine and calculate confidence intervals
> using t tests and the *Z* statistic, and determine
> whether the result is significant. (Analyze)
>
> [Note: The *F* test is covered in Chapter 24.]
>
> **Body of Knowledge III.G**

POPULATION PARAMETER VERSUS SAMPLE STATISTIC

From sample data, we can calculate values called statistics such as the average $\bar{X}$, the standard deviation s, or the range R. These are called descriptive statistics. Note that "sample" and "statistics" start with "s." These statistics are denoted with English letters. We use statistics to *infer* population parameters, hence the branch of statistics called *inferential statistics*. Note that "population" and "parameter" start with the letter "p." These are related to "sample" and "statistics," respectively. Parameters are signified by lowercase Greek letters. The population mean μ, pronounced "mew," is related to the sample mean $\bar{X}$. The population standard deviation σ (sigma) is related to the sample standard deviation s. One can never really truly know a parameter unless it is possible to measure the entire population under consideration. Hence the purpose of inferential statistics is to make inferences about a population parameter based on a sample statistic.

Inferential statistics falls into two categories:

1. Estimating (predicting) the value of a population parameter from a sample statistic

2. Testing a hypothesis about the value of a population parameter, answering questions such as, "Is the population parameter equal to this specific value?"

POINT ESTIMATE

An example of a point estimate would be a sample mean $\bar{X}$ for μ or s for σ. The sample mean may also be used to form an interval estimate, known as a *confidence interval*, described next.

CONFIDENCE INTERVAL

In statistics, a confidence interval (CI) for a population parameter is an interval between two numbers with an associated probability p that is generated from a random sample of an underlying population such that if the sampling were repeated numerous times and the confidence interval recalculated from each sample according to the same method, a proportion p of the confidence intervals would contain the population parameter in question. Confidence intervals are the most prevalent form of interval estimation.

If U and V are statistics (that is, observable random variables, like sample averages) whose probability distribution depends on some unobservable parameter μ, and

$$P(U \le \mu \le V) = x \text{ (where } x \text{ is a number between 0 and 1)}$$

then the random interval (U, V) is a "$(100x)\%$ confidence interval for μ." The number x (or $100x\%$) is called the confidence level or confidence coefficient. In modern applied practice, almost all confidence intervals are stated at the 95% level.[1]

Example   A machine fills cups with margarine and is supposed to be adjusted so that the mean content of the cups is close to 250 grams of margarine. Of course it is not possible to fill every cup with exactly 250 grams of margarine. Hence the weight of the filling can be considered to be a random variable X. The distribution of X is assumed here to be a normal distribution with unknown expectation or average μ and (for the sake of simplicity) known standard deviation $\sigma = 2.5$ grams. To check if the machine is adequately adjusted, a sample of $n = 25$ cups of margarine is chosen at random and the cups weighed. The weights of margarine are $X_1, X_2, \ldots, X_{25}$, a random sample from X.

To get an impression of the population average μ, it is sufficient to give an estimate. The appropriate estimator is the sample mean $\bar{X}$:

$$\hat{\mu} = \bar{X} = \frac{1}{n}\sum_{i=1}^{n} X_i$$

The sample shows actual weights $X_1, X_2, \ldots, X_{25}$, with mean

$$\bar{X} = \frac{1}{25}\sum_{i=1}^{25} X_i = 250.2 \text{ (grams)}$$

If we take another sample of 25 cups, we could easily expect to find values like 250.4 or 251.1 grams. A sample mean value of 280 grams, however, would be extremely rare if the mean content of the cups is in fact close to 250g. There is a whole interval around the observed value 250.2 of the sample mean within which, if the whole population mean actually takes a value in this range, the observed data would not be considered particularly unusual. Such an interval is called a confidence interval for the parameter μ. How do we calculate such an interval? The endpoints of the interval have to be calculated from the sample, so they are statistics, functions of the sample $X_1, X_2, \ldots, X_{25}$ and hence random variables themselves.

In our case, we may determine the endpoints by considering that the sample mean $\bar{X}$ from a normally distributed sample is also normally distributed, with the same expectation μ, but with standard deviation $\sigma / \sqrt{n}$ = 0.5 (grams). The value $\sigma / \sqrt{n}$ is called the standard error for $\bar{X}$, denoted by $\sigma_{\bar{x}}$. By standardizing we get a random variable

$$Z = \frac{\bar{X} - \mu}{\sigma / \sqrt{n}} = \frac{\bar{X} - \mu}{0.5}$$

dependent on μ, but with a standard normal distribution independent of the parameter μ to be estimated. Hence it is possible to find numbers $-z$ and z, independent of μ, where Z lies in between with probability $1 - \alpha$, a measure of how confident we want to be. We take $1 - \alpha = 0.95$. So we have

$$P(-z \leq Z \leq z) = 1 - \alpha = 0.95$$

The number z follows from

$$\Phi(z) = P(Z \leq z) = 1 - \frac{\alpha}{2} = 0.975$$

Hence the value of z, known as the *z score*, can be found by using a z-table (Refer to Areas under Standard Normal Curve, Appendix B)

$$z = \Phi^{-1}(\Phi(z)) = \Phi^{-1}(0.975) = 1.96$$

Thus

$$0.95 = 1 - \alpha = P(-z \leq Z \leq z) = P\left(-1.96 \leq \frac{\bar{X} - \mu}{\sigma / \sqrt{n}} \leq 1.96\right)$$

$$= P\left(\bar{X} - 1.96 \frac{\sigma}{\sqrt{n}} \leq \mu \leq \bar{X} + 1.96 \frac{\sigma}{\sqrt{n}}\right)$$

$$= P\left(\bar{X} - 1.96 \times 0.5 \leq \mu \leq \bar{X} + 1.96 \times 0.5\right)$$

$$= P\left(\bar{X} - 0.98 \leq \mu \leq \bar{X} + 0.98\right)$$

This might be interpreted as: with probability 0.95 one will find the parameter μ between the endpoints

$$\bar{X} - 0.98 \text{ and } \bar{X} + 0.98$$

Every time the measurements are repeated, there will be another value for the mean $\bar{X}$ of the sample. In 95 percent of the cases, μ will be between the endpoints calculated from this mean, but in five percent of the cases it will not be. The actual confidence interval is calculated by entering the measured weights into the formula. Our 0.95 confidence interval becomes

$$(\bar{X}-0.98;\ \bar{X}+0.98)=(250.2-0.98;\ 250.2+0.98)=(249.2;\ 251.18)$$

This interval has fixed endpoints, where μ might be in between (or not). There is no probability of such an event. We cannot say "with probability $1-\alpha$ the parameter μ lies in the confidence interval." We only know that by repetition in $100(1-\alpha)\%$ of the cases μ will be in the calculated interval. In $100\alpha\%$ of the cases, however, it does not. And unfortunately we do not know in which of the cases this happens. That is why we say "with confidence level $100(1-\alpha)\%$, μ lies in the confidence interval."

Figure 22.1 shows 50 realizations of a confidence interval for μ.

As seen in the figure, there was a fair chance that we chose an interval containing μ; however, we may be unlucky and have picked the wrong one. We will never know. We are stuck with our interval.

Theoretical Example   Here is a familiar example. Suppose $X_1, \ldots, X_n$ are an independent sample from a normally distributed population with mean μ and variance σ^2. Let

$$\bar{X} = \frac{X_1 + X_2 + \ldots + X_n}{n}$$

$$S^2 = \frac{1}{n-1}\sum_{i=1}^{n}(X_i - \bar{X})^2$$

then

$$T = \frac{\bar{X} - \mu}{S/\sqrt{n}}$$

has a Student's t-distribution with $n-1$ degrees of freedom. Note that the distribution of T does not depend on the values of the unobservable parameters μ and σ^2. If c is the 95th percentile of this distribution, then

$$P(-c < T < c) = 0.9$$

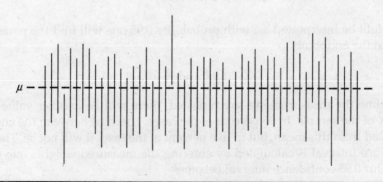

Figure 22.1   Fifty realizations of a 95 percent confidence interval for μ.

(Note: "95th" and "0.9" are correct in the preceding expressions. There is a five percent chance that T will be less than $-c$ and a five percent chance that it will be larger than $+c$. Thus, the probability that T will be between $-c$ and $+c$ is 90 percent.)

Consequently

$$P\left(\overline{X} - cS / \sqrt{n} < \mu < \overline{X} + cS / \sqrt{n}\right) = 0.9$$

and we have a theoretical 90 percent confidence interval for μ.

After actually observing the sample, we find values for $\overline{x}$ to replace the theoretical $\overline{X}$ and s to replace the theoretical S, from which we compute the confidence interval

$$\left[\overline{x} - cs / \sqrt{n} < \mu < \overline{x} + cs / \sqrt{n}\right],$$

an interval with fixed numbers as endpoints, of which we can no longer say with a certain probability that it contains the parameter μ. Either μ is in this interval or it is not.

CONFIDENCE INTERVALS IN MEASUREMENT

More concretely, the results of measurements are often accompanied by confidence intervals. For instance, suppose a scale is known to yield the actual mass of an object plus a normally distributed random error with mean zero and known standard deviation σ. If we weigh 100 objects of known mass on this scale and report the values $\pm\sigma$, then we can expect to find that around 68 of the reported ranges include the actual mass. This is an application of the *empirical rule*:

If a set of measurements can be illustrated by a roughly mound-shaped histogram, then:

- *The interval $\overline{X} \pm s$ contains approximately 68 percent of the measurements,*

- *The interval $\overline{X} \pm 2s$ contains approximately 95 percent of the measurements,*

- *The interval $\overline{X} \pm 3s$ contains approximately all of the measurements.*[2]

If we wish to report values with a smaller standard error value, then we repeat the measurement n times and average the results. Then the 68.2 percent confidence interval is $\pm\ \sigma / \sqrt{n}$. For example, repeating the measurement 100 times reduces the confidence interval to 1/10 of the original width.

Note that when we report a 68.2 percent confidence interval (usually termed *standard error*) as $v \pm \sigma$, this does not mean that the true mass has a 68.2 percent chance of being in the reported range. In fact, the true mass is either in the range or not. How can a value outside the range be said to have any chance of being in the range? Rather, our statement means that 68.2 percent of the ranges we report are likely to include the true mass.

This is not just a quibble. Under the incorrect interpretation, each of the 100 measurements described above would be specifying a different range, and the true mass supposedly has a 68 percent chance of being in each and every range. Also, it supposedly has a 32 percent chance of being outside each and every range. If two of the ranges happen to be disjoint, the statements are obviously inconsistent. Say one range is one to two, and the other is two to three. Supposedly, the true mass has a 68 percent chance of being between one and two, but only a 32 percent chance of being less than two or more than three. The incorrect interpretation reads more into the statement than is meant.

On the other hand, under the correct interpretation, each and every statement we make is really true, because the statements are not about any specific range. We could report that one mass is 10.2 ± 0.1 grams, while really it is 10.6 grams, and not be lying. But if we report fewer than 1000 values and more than two of them are that far off, we will have some explaining to do.

HOW LARGE A SAMPLE IS NEEDED TO ESTIMATE A PROCESS AVERAGE?

There are many things to consider when answering this question. In fact, the question itself requires modification before answering. Is the test destructive, nondestructive, or semidestructive? How expensive is it to obtain and test a sample of n units? How close an answer is needed? How much variation among measurements is expected? What level of confidence is adequate? All these questions must be considered.

We begin by turning to the discussion of variation expected in averages of random samples of n items around the true but unknown process average μ. The expected variation of sample averages about μ is

$$\pm 2 \frac{\sigma}{\sqrt{n}} \quad \text{(confidence about 95\%)}$$

and

$$\pm 3 \frac{\sigma}{\sqrt{n}} \quad \text{(confidence about 99.7\%)}$$

Now let the allowable deviation (error) in estimating μ be $\pm \Delta$ (read "delta"); also let an estimate or guess of σ be $\hat{\sigma}$; then

$$\Delta \cong \pm 2 \frac{\hat{\sigma}}{\sqrt{n}} \quad \text{and} \quad n \cong \left(\frac{2\hat{\sigma}}{\Delta} \right)^2 \quad \text{(about 95\% confidence)}$$

Also

$$\Delta \cong \pm 3 \frac{\hat{\sigma}}{\sqrt{n}} \quad \text{and} \quad n \cong \left(\frac{3\hat{\sigma}}{\Delta} \right)^2 \quad \text{(about 99.7\% confidence)}$$

Confidence levels other than the two shown above can be used by referring to a table for areas under the normal curve (z-table) (Appendix A). When our estimate or guess of a required sample size n is even as small as $n = 4$, then the distribution of averages of random samples drawn from almost any shaped parent population (with a finite variance) will be essentially normal.

STUDENT'S t-DISTRIBUTION

In probability and statistics, the t-distribution, or Student's t-distribution, is a probability distribution that arises in the problem of estimating the mean of a normally distributed population when the sample size is small. It is the basis of the popular Student's t-tests for the statistical significance of the difference between two sample means and for confidence intervals for the difference between two population means. Student's t-distribution is a special case of the generalized hyperbolic distribution.

The derivation of the t-distribution was first published in 1908 by William Sealy Gosset while he worked at a Guinness brewery in Dublin. He was not allowed to publish under his own name, so the paper was written under the pseudonym Student. The t-test and the associated theory became well known through the work of R. A. Fisher, who called the distribution "Student's distribution."

Student's distribution arises when (as in nearly all practical statistical work) the population standard deviation is unknown and has to be estimated from the data. Textbook problems treating the standard deviation as if it were known are of two kinds: (1) those in which the sample size is so large that one may treat a data-based estimate of the variance as if it were certain, and (2) those that illustrate mathematical reasoning, in which the problem of estimating the standard deviation is temporarily ignored because that is not the point that the author or instructor is then explaining.

The t-distribution is related to the F-distribution as follows: the square of a value of t with ν degrees of freedom is distributed as F with 1 and ν degrees of freedom.

The overall shape of the probability density function of the t-distribution resembles the bell shape of a normally distributed variable with mean zero and variance 1, except that it is a bit lower and wider. As the number of degrees of freedom grows, the t-distribution approaches the normal distribution with mean zero and variance 1.

Figure 22.2 shows the density of the t-distribution for increasing values of ν (degrees of freedom). The normal distribution is shown as a dotted line for comparison. Note that the t-distribution (darker line) becomes closer to the normal distribution as ν increases. For $\nu = 30$ the t-distribution is almost the same as the normal distribution.

Table of Selected Values

Table 22.1 lists a few selected values for distributions with ν degrees of freedom for the 90%, 95%, 97.5%, and 99.5% confidence intervals. These are "one-sided," that is, where we see "90%," "4 degrees of freedom," and "1.533,"

it means $P(T < 1.533) = 0.9$;

it does not mean $P(-1.533 < T < 1.533) = 0.9$.

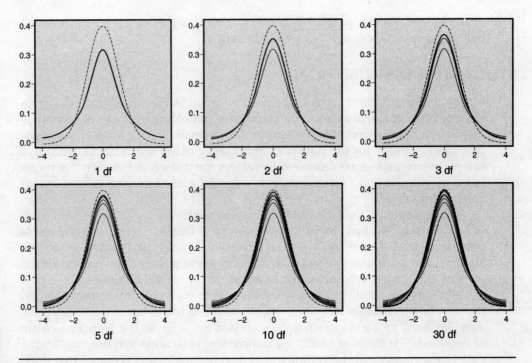

Figure 22.2   Density of the t-distribution for 1, 2, 3, 5, 10, and 30 df compared to the normal distribution.

Table 22.1   t-table.

ν	75%	80%	85%	90%	95%	97.5%	99%	99.5%	99.75%	99.9%	99.95%
1	1.000	1.376	1.963	3.078	6.314	12.71	31.82	63.66	127.3	318.3	636.6
2	0.816	1.061	1.386	1.886	2.920	4.303	6.965	9.925	14.09	22.33	31.60
3	0.765	0.978	1.250	1.638	2.353	3.182	4.541	5.841	7.453	10.21	12.92
4	0.741	0.941	1.190	1.533	2.132	2.776	3.747	4.604	5.598	7.173	8.610
5	0.727	0.920	1.156	1.476	2.015	2.571	3.365	4.032	4.773	5.893	6.869
6	0.718	0.906	1.134	1.440	1.943	2.447	3.143	3.707	4.317	5.208	5.959
7	0.711	0.896	1.119	1.415	1.895	2.365	2.998	3.499	4.029	4.785	5.408
8	0.706	0.889	1.108	1.397	1.860	2.306	2.896	3.355	3.833	4.501	5.041
9	0.703	0.883	1.100	1.383	1.833	2.262	2.821	3.250	3.690	4.297	4.781

Continued

Table 22.1   t-table. *Continued*

v	75%	80%	85%	90%	95%	97.5%	99%	99.5%	99.75%	99.9%	99.95%
10	0.700	0.879	1.093	1.372	1.812	2.228	2.764	3.169	3.581	4.144	4.587
11	0.697	0.876	1.088	1.363	1.796	2.201	2.718	3.106	3.497	4.025	4.437
12	0.695	0.873	1.083	1.356	1.782	2.179	2.681	3.055	3.428	3.930	4.318
13	0.694	0.870	1.079	1.350	1.771	2.160	2.650	3.012	3.372	3.852	4.221
14	0.692	0.868	1.076	1.345	1.761	2.145	2.624	2.977	3.326	3.787	4.140
15	0.691	0.866	1.074	1.341	1.753	2.131	2.602	2.947	3.286	3.733	4.073
16	0.690	0.865	1.071	1.337	1.746	2.120	2.583	2.921	3.252	3.686	4.015
17	0.689	0.863	1.069	1.333	1.740	2.110	2.567	2.898	3.222	3.646	3.965
18	0.688	0.862	1.067	1.330	1.734	2.101	2.552	2.878	3.197	3.610	3.922
19	0.688	0.861	1.066	1.328	1.729	2.093	2.539	2.861	3.174	3.579	3.883
20	0.687	0.860	1.064	1.325	1.725	2.086	2.528	2.845	3.153	3.552	3.850
21	0.686	0.859	1.063	1.323	1.721	2.080	2.518	2.831	3.135	3.527	3.819
22	0.686	0.858	1.061	1.321	1.717	2.074	2.508	2.819	3.119	3.505	3.792
23	0.685	0.858	1.060	1.319	1.714	2.069	2.500	2.807	3.104	3.485	3.767
24	0.685	0.857	1.059	1.318	1.711	2.064	2.492	2.797	3.091	3.467	3.745
25	0.684	0.856	1.058	1.316	1.708	2.060	2.485	2.787	3.078	3.450	3.725
26	0.684	0.856	1.058	1.315	1.706	2.056	2.479	2.779	3.067	3.435	3.707
27	0.684	0.855	1.057	1.314	1.703	2.052	2.473	2.771	3.057	3.421	3.690
28	0.683	0.855	1.056	1.313	1.701	2.048	2.467	2.763	3.047	3.408	3.674
29	0.683	0.854	1.055	1.311	1.699	2.045	2.462	2.756	3.038	3.396	3.659
30	0.683	0.854	1.055	1.310	1.697	2.042	2.457	2.750	3.030	3.385	3.646
40	0.681	0.851	1.050	1.303	1.684	2.021	2.423	2.704	2.971	3.307	3.551
50	0.679	0.849	1.047	1.299	1.676	2.009	2.403	2.678	2.937	3.261	3.496
60	0.679	0.848	1.045	1.296	1.671	2.000	2.390	2.660	2.915	3.232	3.460
80	0.678	0.846	1.043	1.292	1.664	1.990	2.374	2.639	2.887	3.195	3.416
100	0.677	0.845	1.042	1.290	1.660	1.984	2.364	2.626	2.871	3.174	3.390
120	0.677	0.845	1.041	1.289	1.658	1.980	2.358	2.617	2.860	3.160	3.373
∞	0.674	0.842	1.036	1.282	1.645	1.960	2.326	2.576	2.807	3.090	3.291

Part III.G

Consequently, by the symmetry of this distribution, we have

$$P(T < -1.533) = P(T > 1.533) = 1 - 0.9 = 0.1,$$

and consequently

$$P(-1.533 < T < 1.533) = 1 - 2(0.1) = 0.8.$$

Note that the last row of the table also gives critical points: a t-distribution with infinitely many degrees of freedom is normally distributed.

For example, given a sample with a sample variance 2 and sample mean of 10, taken from a sample set of 11 (10 degrees of freedom), using the formula[3]

$$\bar{X}_n \pm A \frac{S_n}{\sqrt{n}}$$

we can determine that at 90% confidence, we have a true mean lying below

$$10 + 1.372 \frac{2}{\sqrt{11}} = 10.585$$

And, still at 90% confidence, we have a true mean lying over

$$10 - 1.372 \frac{2}{\sqrt{11}} = 9.415$$

so that at 80% confidence, we have a true mean lying between

$$10 \pm 1.372 \frac{2}{\sqrt{11}} = \left[9.415, 10.585 \right]$$

TESTING A HYPOTHESIS

One may be faced with the problem of making a definite decision with respect to an uncertain hypothesis that is known only through its observable consequences. A statistical hypothesis test, or more briefly, hypothesis test, is an algorithm to state the alternative (for or against the hypothesis) that minimizes certain risks.

There are several preparations we make before we observe the data:

1. The hypothesis must be stated in mathematical/statistical terms that make it possible to calculate the probability of possible samples, assuming the hypothesis is correct. For example: The mean response to treatment being tested is equal to the mean response to the placebo in the control group. Both responses have the normal distribution with this unknown mean and the same known standard deviation.

2. A test statistic must be chosen that will summarize the information in the sample that is relevant to the hypothesis. In the example above, it might be the numerical difference between the two sample means, $\bar{x}_1 - \bar{x}_2$.

3. The distribution of the test statistic is used to calculate the probability sets of possible values (usually an interval or union of intervals). In this example, the difference between sample means would have a normal distribution with a standard deviation equal to the common standard deviation times the factor

$$\sqrt{\frac{1}{n_1} + \frac{1}{n_2}}$$

where n_1 and n_2 are the sample sizes.

4. Among all the sets of possible values, we must choose one that we think represents the most extreme evidence against the hypothesis. This is called the critical region of the test statistic. The probability of the test statistic falling in the critical region when the hypothesis is correct is called the alpha value (or size) of the test.

5. The probability that a sample falls in the critical region when the parameter is θ, where θ is for the alternative hypothesis, is called the power of the test at θ. The power function of a critical region is the function that maps θ to the power of θ.

After the data are available, the test statistic is calculated and we determine whether it is inside the critical region.

If the test statistic is inside the critical region, then our conclusion is one of the following:

1. The hypothesis is incorrect, therefore reject the null hypothesis. (Thus the critical region is sometimes called the rejection region, while its complement is the acceptance region.)

2. An event of probability less than or equal to alpha has occurred.

The researcher has to choose between these logical alternatives. In the example, we would say, "The observed response to treatment is statistically significant."

If the test statistic is outside the critical region, the only conclusion is that, "There is not enough evidence to reject the hypothesis."

This is not the same as evidence in favor of the hypothesis, since lack of evidence against a hypothesis is not evidence for it.[4]

NULL HYPOTHESIS

In statistics, a null hypothesis is a hypothesis set up to be nullified or refuted in order to support an alternative hypothesis. When used, the null hypothesis is presumed true until statistical evidence in the form of a hypothesis test indicates otherwise.

Although it was originally proposed to be any hypothesis, in practice it has come to be identified with the "nil hypothesis," which states that "there is no phenomenon" and that the results in question could have arisen through chance.

For example. if we want to compare the test scores of two random samples of men and women, a null hypothesis would be that the mean score of the male population was the same as the mean score of the female population, and therefore there is no significant statistical difference between them:

$$H_0 : \mu_1 = \mu_2$$

$$H_A : \mu_1 \neq \mu_2$$

where

H_0 = the null hypothesis

H_A = the alternative hypothesis

μ_1 = the mean of population 1

μ_2 = the mean of population 2

Alternatively, the null hypothesis can postulate that the two samples are drawn from the same population:

$$H_0 : \mu_1 - \mu_2 = 0$$

$$H_A : \mu_1 - \mu_2 \neq 0$$

Formulation of the null hypothesis is a vital step in statistical significance testing. Having formulated such a hypothesis, one can establish the probability of observing the obtained data or data different from the prediction of the null hypothesis if the null hypothesis is true. That probability is what is commonly called the "significance level" of the results.

When a null hypothesis is formed, it is always in contrast to an implicit alternative hypothesis, which is accepted if the observed data values are sufficiently improbable under the null hypothesis. The precise formulation of the null hypothesis has implications for the alternative. For example, if the null hypothesis is that sample A is drawn from a population with the same mean as sample B, the alternative hypothesis is that they come from populations with different means, which can be tested with a two-tailed test of significance. But if the null hypothesis is that sample A is drawn from a population whose mean is lower than the mean of the population from which sample B is drawn, the alternative hypothesis is that sample A comes from a population with a higher mean than the population from which sample B is drawn, which can be tested with a one-tailed test.

LIMITATIONS OF HYPOTHESIS TESTING

A null hypothesis is only useful if it is possible to calculate from it the probability of observing a data set with particular parameters. In general it is much harder to be precise about how probable the data would be if the alternative hypothesis is true.

If experimental observations contradict the prediction of the null hypothesis, it means that either the null hypothesis is false or we have observed an event with very low probability. This gives us high confidence in the falsehood of the null hypothesis, which can be improved by increasing the number of trials. However, accepting the alternative hypothesis only commits us to a difference in observed parameters; it does not prove that the theory or principles that predicted such a difference is true, since it is always possible that the difference could be due to additional factors not recognized by the theory.

For example, rejection of a null hypothesis (that, say, rates of symptom relief in a sample of patients who received a placebo and a sample who received a medicinal drug will be equal) allows us to make a non-null statement (that the rates differed); it does not prove that the drug relieved the symptoms, though it gives us more confidence in that hypothesis.

The formulation, testing, and rejection of null hypotheses is methodologically consistent with the falsificationist model of scientific discovery formulated by Karl Popper and widely believed to apply to most kinds of empirical research. However, concerns regarding the high power of statistical tests to detect differences in large samples have led to suggestions for redefining the null hypothesis, for example as a hypothesis that an effect falls within a range considered negligible. This is an attempt to address the confusion among non-statisticians between significant and substantial, since large enough samples are likely to be able to indicate differences, however minor.[5]

STATISTICAL SIGNIFICANCE

In statistics, a result is significant if it is unlikely to have occurred by chance. Technically, the significance level of a test is the maximum probability, assuming the null hypothesis, that the observed statistic would be observed and still be considered consistent with chance variation (consistent with the the null hypothesis). Hence, if the null hypothesis is true, the significance level is the probability that it will be rejected in error (a decision known as a Type I error). The significance of a result is also called its p-value; the smaller the p-value, the more significant the result is said to be.

Significance is represented by the Greek symbol, α (alpha). Popular levels of significance are 10%, 5%, and 1%.

For example, if you perform a test of significance, assuming the significance level is 5% and the p-value is lower than 5%, then the null hypothesis would be rejected. Informally, the test statistic is said to be "statistically significant."

If the significance level is smaller, an observed value is less likely to be more extreme than the critical value. So a result that is "significant at the 1% level" is more significant than a result that is "significant at the 5% level." However, a test at the 1% level is more likely to fail to reject a false null hypothesis (a Type II error) than a test at the 5% level, and so will have less statistical power.

In devising a hypothesis test, the tester aims to maximize power for a given significance but ultimately has to recognize that the best that can be achieved is likely to be a balance between significance and power, in other words between the risks of Type I and Type II errors. It is important to note that a Type I error is not necessarily any better or worse than a Type II error: the severity of a type of error depends on each individual case. However, one type of error may be better or worse on average for the population under study.

Pitfalls

It is important to differentiate between the vernacular use of the term "significant"—"a major effect; important; fairly large in amount or quantity"—and the meaning in statistics, which carries no such connotation of meaningfulness. With a large enough number being sampled, a tiny and unimportant difference can still be found to be "statistically significant."[6]

In statistical hypothesis testing, the p-value is the probability of obtaining a result at least as "impressive" as that obtained by chance alone, assuming the null hypothesis to be true. The fact that p-values are based on this assumption is crucial to their correct interpretation.

More technically, the p-value of an observed value $t_{observed}$ of some random variable T used as a test statistic is the probability that, given that the null hypothesis is true, T will assume a value as or more unfavorable to the null hypothesis than the observed value $t_{observed}$. "More unfavorable to the null hypothesis" can in some cases mean greater than, in some cases less than, and in some cases farther away from a specified center.

For example, say an experiment is performed to determine if a coin flip is fair (50 percent chance of landing heads or tails), or unfairly biased toward heads (> 50 percent chance of landing heads). If we choose a two-tailed test, then the null hypothesis is that the coin is fair and that any deviations from the 50 percent rate can be ascribed to chance alone.

This can be written as:

H_0: A repeated coin toss will show heads the same number of times as tails

H_A: A repeated coin toss will show heads turning up more than tails, or vice versa

Suppose that the experimental results show the coin turning up heads 14 times out of 20 total flips. The p-value of this result would be the chance of a fair coin landing on heads at least 14 times out of 20 flips (as larger values in this case are also less favorable to the null hypothesis of a fair coin) or at most six times out of 20 flips. In this case the random variable T is a binomial distribution. The probability that 20 flips of a fair coin would result in 14 or more heads is 0.0577. Since this is a two-tailed test, the probability that 20 flips of the coin would result in 14 or more heads or six or less heads is 0.115 or 2×0.0577.

Generally, the smaller the p-value, the more people there are who would be willing to say that the results came from a biased coin.

Generally, one rejects the null hypothesis if the p-value is smaller than or equal to the significance level, often represented by the Greek letter α (alpha). If the level is 0.05, then the results are only 5% likely to be as extraordinary as just seen, given that the null hypothesis is true.

In the above example, the calculated p-value exceeds 0.05, and thus the null hypothesis—that the observed result of 14 heads out of 20 flips can be ascribed to chance alone—is not rejected. Such a finding is often stated as being "not statistically significant at the 5% level."

However, had a single extra head been obtained, that is, 15 instead of 14 heads, the resulting p-value would be 0.02. This time, the null hypothesis—that the observed result of 15 heads out of 20 flips can be ascribed to chance alone—is rejected. Such a finding would be described as being "statistically significant at the 5% level."

Critics of p-values point out that the criterion used to decide "statistical significance" is based on the somewhat arbitrary choice of level (often set at 0.05).

Here is the proper way to interpret the p-value (courtesy of faculty at University of Texas Health Science Center School of Public Health): "Assuming the null [hypothesis] is true, the probability of obtaining a test statistic as extreme or more extreme than the one observed (enter the value here) is (enter p-value)."

FREQUENT MISUNDERSTANDINGS ABOUT THE USE OF THE p-VALUE

There are several common misunderstandings about p-values. All of the following statements are *false*:

- The p-value is the probability that the null hypothesis is true, justifying the "rule" of considering as significant p-values closer to zero.

- The p-value is the probability that a finding is "merely a fluke" (again, justifying the "rule" of considering small p-values as "significant").

- The p-value is the probability of falsely rejecting the null hypothesis. This error is a version of the so-called prosecutor's fallacy: guilty until proven innocent.

- The p-value is the probability that a replicating experiment would not yield the same conclusion.

- 1 − (p-value) is the probability of the alternative hypothesis being true (see the first statement in this list).

- The significance level of the test is determined by the p-value.[7]

Chapter 23

H. Design of Experiments (DOE)

Experimental design is research design in which the researcher has control over the selection of participants in the study, and these participants are randomly assigned to treatment and control groups. The first statistician to consider a methodology for the design of experiments was Sir Ronald A. Fisher. He described how to test the hypothesis that a certain lady could distinguish by flavor alone whether the milk or the tea was first placed in the cup. While this sounds like a frivolous application, it allowed him to illustrate the most important means of experimental design:

- *Randomization.* The process of making something random for example, where a random sample of observations is taken from several populations under study.

- *Replication.* Repeating the creation of a phenomenon so that the variability associated with the phenomenon can be estimated. $N = 1$ is called "one replication."

- *Blocking.* The arranging of experimental units in groups (blocks) that are similar to one another.

- *Orthogonality.* Perpendicular, at right angles, or statistically normal.

- Use of *factorial experiments* instead of the *one factor at a time* method.

Analysis of the design of experiments was built on the foundation of the analysis of variance, discussed in Chapter 24, which is a collection of models in which the observed variance is partitioned into components due to different factors that are estimated and/or tested.[1]

Success in troubleshooting and process improvement often rests on the appropriateness and efficiency of the experimental setup and its match to the environmental situation. Design suggests structure, and it is the structure of the statistically designed experiment that gives it its meaning. Consider the simple

Table 23.1 Experimental plan.

Operator

	O_1 = Dianne	O_2 = Tom
M_1 = old	(1) X_{11}	(2) X_{12}
M_2 = new	(3) X_{21}	(4) X_{22}

Machine

Source: E. R. Ott, E. G. Schilling, and D. V. Neubauer, *Process Quality Control: Troubleshooting and Interpretation of Data,* 4th ed. (Milwaukee: ASQ Quality Press, 2005): 287.

Part III.H

2^2 factorial experiment laid out in Table 23.1 with measurements X_{11}, X_{12}, X_{21}, X_{22}. The subscripts i and j on X_{ij} simply show the machine (i) and operator (j), associated with a given measurement, where i can assume the values 1 or 2 so that x_{1j} represents the measurements using the old machine, x_{2j} the new machine, x_{i1} represents Dianne's measurements, and x_{i2} as Tony's. Here there are two factors or characteristics to be tested, operator and machine.

There are two levels of each so that operator takes on the levels Dianne and Tom and the machine used is either old or new. The designation 2^p means two levels of each of p factors. If there were three factors in the experiment, say by the addition of material (from two vendors), we would have a 2^3 experiment. In a properly conducted experiment, the treatment combinations corresponding to the cells of the table must be run at random to avoid biasing the results. Tables of random numbers or slips of paper drawn from a hat can be used to set up the order of experimentation. Thus, if we numbered the cells as shown in the diagram and drew the numbers 3, 2, 1, 4 from a hat, we would run Dianne–new first, followed by Tom–old, Dianne–old, and Tom–new in that order. This is done to insure that any external effect that might creep into the experiment while it is being run would affect the treatments in random fashion. Its effect would then appear as experimental error rather than biasing the experiment.

EFFECTS

Of course, we must measure the results of the experiment. That measurement is called the *response.* Suppose the response is 'units produced in a given time,' and that the results are as shown in Table 23.2.

The *effect* of a factor is the average change in response (units produced) brought about by moving from one level of a factor to the other. To obtain the *machine* effect

Table 23.2   Experimental results.

Source: E. R. Ott, E. G. Schilling, and D. V. Neubauer, *Process Quality Control: Troubleshooting and Interpretation of Data*, 4th ed. (Milwaukee: ASQ Quality Press, 2005): 288.

we would simply subtract the average result for the old machine from that of the new. We obtain

$$\text{Machine effect} = \overline{X}_{2\bullet} - \overline{X}_{1\bullet} = \frac{(5+15)}{2} - \frac{(20+10)}{2} = -5,$$

which says the old machine is better than the new. Notice that when we made this calculation, each machine was operated equally by both Dianne and Tom for each average. Now calculate the *operator* effect. We obtain

$$\text{Operator effect} = \overline{X}_{\bullet 2} - \overline{X}_{\bullet 1} = \frac{(15+10)}{2} - \frac{(5+20)}{2} = 0$$

The dots (•) in the subscripts simply indicate which factor was averaged out in computing $\overline{X}$. It appears that operators have no effect on the operation. Notice that each average represents an equal time on each machine for each operator and so is a fair comparison.

However, suppose there is a unique operator–machine combination that produces a result beyond the effects we have already calculated. This is called an interaction. Remember that we averaged machines out of the operator effect and operators out of the machine effect. To see if there is an interaction between operators and machines, we calculate the machine effect individually for each operator. If there is a peculiar relationship between operators and machine, it will show up as the average difference between these calculations. We obtain

$$\text{Machine effect for Dianne} = X_{21} - X_{11} = 5 - 20 = -15$$

$$\text{Machine effect for Tom} = X_{22} - X_{12} = 15 - 10 = 5$$

The average difference between these calculations is

$$\text{Interaction} = \frac{(5)-(-15)}{2} = \frac{(20)}{2} = 10$$

The same result would be obtained if we averaged the operator effect for each machine. It indicates that there is, on the average, a 10-unit reversal in effect due to the specific operator–machine combination involved.

Specifically, in going from Dianne to Tom on the new machine we get, on the average, a 10-unit increase; whereas in going from Dianne to Tom on the old machine we get, on the average, a 10-unit decrease. For computational purposes in a 2^2 design, the interaction effect is measured as the difference of the averages down the diagonals of the table. Algebraically, this gives the same result as the above calculation since

$$\text{Interaction} = \frac{(X_{22}-X_{12})-(X_{21}-X_{11})}{2}$$

$$= \frac{(X_{22}+X_{11})-(X_{21}+X_{12})}{2}$$

$$= \frac{(\text{Southeast diagonal})-(\text{Southwest diagonal})}{2}$$

$$= \frac{15+20-5-10}{2}$$

$$= 10$$

There is a straightforward method for calculating the effects in more complicated experiments. It requires that the *treatment combinations* be properly identified. If we designate operator as factor A and machine as factor B, each with two levels (– and +), the treatment combinations (cells) can be identified by simply showing the letter of a factor if it is at the + level and not showing the letter if the factor is at the – level. We show (1) if all factors are at the – level. This is illustrated in Table 23.3.

The signs themselves indicate how to calculate an effect. Thus, to obtain the A (operator) effect we subtract all those observations under the – level for A from those under the + level and divide by the number of observations that go into either the + or – total to obtain an average. The signs that identify what to add and subtract in calculating an interaction can be found by multiplying the signs of its component factors together as in Table 23.4.

And we have

$$\text{A (operators) effect} = \frac{(a+ab)-(b+(1))}{2} = \frac{-(1)+a-b+ab}{2} = 0$$

$$\text{B (machines) effect} = \frac{(b+ab)-(a+(1))}{2} = \frac{-(1)-a+b+ab}{2} = -5$$

$$\text{AB interaction} = \frac{((1)+ab)-(a+b)}{2} = \frac{+(1)-a-b+ab}{2} = 10$$

Table 23.3   The 2^2 configuration.

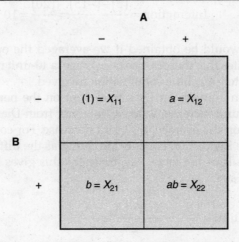

Source: E. R. Ott, E. G. Schilling, and D. V. Neubauer, *Process Quality Control: Troubleshooting and Interpretation of Data,* 4th ed. (Milwaukee: ASQ Quality Press, 2005): 290.

Table 23.4   Signs of interaction.

	A −	A +
B −	(−)(−) = +	(+)(−) = −
B +	(−)(+) = −	(+)(+) = +

Source: E. R. Ott, E. G. Schilling, and D. V. Neubauer, *Process Quality Control: Troubleshooting and Interpretation of Data,* 4th ed. (Milwaukee: ASQ Quality Press, 2005): 291.

Note that the sequence of plus (+) and minus (−) signs in the numerators on the right match those in Table 23.5 for the corresponding effects.

Tables of this form are frequently used in the collection and analysis of data from designed experiments.

Table 23.5   Treatment combination.

Treatment combination	A	B	AB	Response
(1)	–	–	+	$X_{11} = 20$
a	+	–	–	$X_{12} = 10$
b	–	+	–	$X_{21} = 5$
ab	+	+	+	$X_{22} = 15$

Source: E. R. Ott, E. G. Schilling, and D. V. Neubauer, *Process Quality Control: Troubleshooting and Interpretation of Data*, 4th ed. (Milwaukee: ASQ Quality Press, 2005): 291.

SUMS OF SQUARES

Process control and troubleshooting attempt to reduce variability. It is possible to calculate how much each factor contributes to the total variation in the data by determining the sums of squares (SS) for that factor. Sums of squares is simply the numerator in the calculation of the variance, the denominator being degrees of freedom, df. Thus, $s^2 = SS/\mathrm{df}$ and s^2 is called a mean square (*MS*) in analysis of variance. For an effect *Eff* associated with the factor, the calculation is

$$SS(Eff) = r2^{P-2}Eff^2$$

where *r* is the number of observations per cell and *p* is the number of factors. Here, $r = 1$ and $p = 2$, so

$$SS(A) = (0)^2 = 0$$

$$SS(B) = (-5)^2 = 25$$

$$SS(A \times B) = (10)^2 = 100$$

To measure the variance $\hat{\sigma}^2$ associated with an effect, we must divide the sums of squares by the appropriate degrees of freedom to obtain mean squares (*MS*). Each effect (*Eff*), or contrast, will have one degree of freedom so that

$$\hat{\sigma}^2_{Eff} = \text{Mean square } (Eff) = SS(Eff)/1 = SS(Eff)$$

The total variation in the data is measured by the sample variance from all the data taken together, regardless of where it came from. This is

$$\hat{\sigma}^2_r = s^2_r = \frac{SS(T)}{\mathrm{df}(T)} = \frac{\sum_{i=1}^{n}(X_i - \bar{X})^2}{r2^p - 1}$$

$$= \left[(15-12.5)^2 + (5-12.5)^2 + (10-12.5)^2 + (20-12.5)^2\right] / \left[1(4)-1\right]$$

$$= \frac{125}{3} = 41.67$$

Table 23.6 Analysis of variance.

Effect	SS	df	MS
Operator (A)	0	1	0
Machine (B)	25	1	25
Interaction (A × B)	100	1	100
Error	No estimate		
Total	125	3	

We can then make an analysis of variance table (Table 23.6) showing how the variation in the data is split up. We have no estimate of error since our estimation of the sums of squares for the three effects uses up all the information (degrees of freedom) in the experiment. If two observations per cell were taken, we would have been able to estimate the error variance in the experiment as well.

The F test can be used to assess statistical significance of the mean squares when a measure of error is available. Since the sums of squares and degrees of freedom add to the total, the error sum of squares and degrees of freedom for error may be determined by difference. Alternatively, sometimes an error estimate is available from previous experimentation. Suppose, for example, an outside measure of error for this experiment was obtained and turned out to be $\hat{\sigma}^2 = 10$ with 20 degrees of freedom. Then, for machines $F = 25/10 = 2.5$, the F table for $\alpha = 0.05$ and 1 and 20 degrees of freedom shows that F^* (F-critical) = 4.35 would be exceeded five percent of the time. Therefore, we are unable to declare that the machines show a significant difference from chance variation. On the other hand, interaction produces $F = 100/10 = 10$, which clearly exceeds the critical value of 4.35, so we declare interaction significant at the $\alpha = 0.05$ level of risk. Note that this is a one-tailed test.

BLOCKING

Occasionally it is impossible to run all of the experiment under the same conditions. For example, the experiment must be run with two batches of raw material or on four different days or by two different operators. Under such circumstances it is possible to "block" out such changes in conditions by confounding, or irrevocably combining, them with selected effects. It is possible to run the experiment in such a way that an unwanted change in conditions, such as operators, will be confounded with a preselected higher-order interaction in which there is no interest or which is not believed to exist. That is why experiment structure and randomization are so important.

It has been pointed out that, particularly in the screening stage of experimentation, the Pareto principle often applies to factors incorporated in a designed experiment. A few of the effects will tend to account for much of the variation observed and some of the factors may not show significance. This phenomenon has been called *factor sparsity* or *sparsity of effects*,[2] and provides some of the rationale for deliberately confounding the block effect with a preselected interaction that is deemed likely to not exist.

MAIN EFFECT AND INTERACTION PLOTS

The main effect plot simply plots the averages for the low (–) and high (+) levels of the factor of interest. The slope of the line reflects whether the effect is positive or negative on the response when the level changes from low to high. The value of the main effect is simply seen as the difference in the value of the response for the endpoints of the line. Figure 23.1 demonstrates what the main effect plots would look like for the data in Table 23.2.

The interaction plot in Figure 23.1 shows the four cell means in the case of a 2^2 design. One of the factors is shown on the horizontal axis. The other factor is represented by a line for each level, or two lines in the case of the 2^2 design. Intersecting (or otherwise nonparallel) lines indicate the presence of a possibly statistically

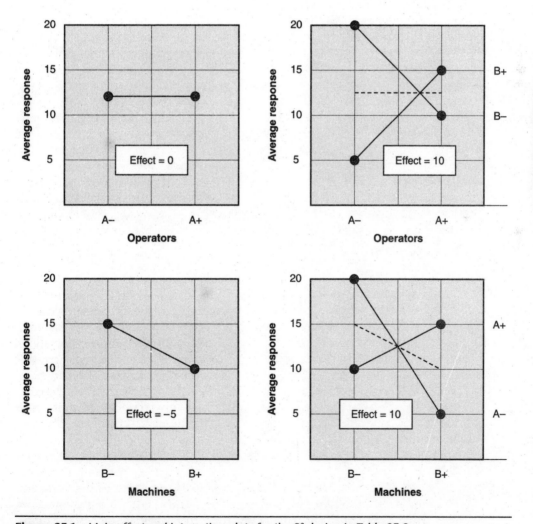

Figure 23.1   Main effect and interaction plots for the 2^2 design in Table 23.2.

Source: E. R. Ott, E. G. Schilling, and D. V. Neubauer, *Process Quality Control: Troubleshooting and Interpretation of Data*, 4th ed. (Milwaukee: ASQ Quality Press, 2005): 303.

significant interaction. The value of the interaction can be seen as the average of the difference between the lines for each level of the factor on the horizontal axis. Note that in Figure 23.1, the main effect for the factor on the horizontal axis can be seen in the interaction plot as the dotted bisecting line.

CONCLUSION

This is only a cursory discussion of design of experiments. It touches only on the most rudimentary aspects. It will serve, however, as an introduction to the concepts and content.

Experimental design is a very powerful tool in the understanding of any process—in a manufacturing facility, pilot line facility, or laboratory.

Chapter 24

I. Analysis of Variance (ANOVA)

Define and determine the applicability of ANOVAs. (Understand)

Body of Knowledge III.I

In statistics, analysis of variance (ANOVA) is a collection of statistical models and their associated procedures that compare means by splitting the overall observed variance into different parts. The initial techniques of analysis of variance were pioneered by the statistician and geneticist Sir Ronald A. Fisher in the 1920s and 1930s, and the tool is sometimes known as Fisher's ANOVA or Fisher's analysis of variance, due to the use of Fisher's F-distribution as part of the test of statistical significance.

There are three conceptual classes of such models:

- The fixed effects model assumes that the data come from normal populations that may differ only in their means. (Model 1)

- Random effects models assume that the data describe a hierarchy of different populations whose differences are constrained by the hierarchy. (Model 2)

- Mixed effects models describe situations where both fixed and random effects are present. (Model 3)

In practice, there are several types of ANOVA depending on the number of treatments and the way they are applied to the subjects in the experiment:

- One-way ANOVA is used to test for differences between three or more independent groups.

- One-way ANOVA for repeated measures is used when the subjects are repeated measures; this means that the same subjects are used for each treatment. Note that this method can be subject to carryover effects.

Part III.I

- Factorial ANOVA is used when the experimenter wants to study the effects of two or more treatment variables. The most commonly used type of factorial ANOVA is the 2 × 2 (read two by two) design, where there are two independent variables and each variable has two levels or distinct values. Factorial ANOVA can also be multilevel such as 3 × 3, or higher order such as 2 × 2 × 2, but analyses with higher numbers of factors are rarely done because the calculations are lengthy and the results are hard to interpret. Such experiments may also be costly and take a long time to complete.

- Multivariate analysis of variance (MANOVA) is used when there is more than one dependent variable.

FIXED EFFECTS MODEL

The fixed effects model of analysis of variance applies to situations in which the experimenter has subjected his experimental material to several treatments, each of which affects only the mean of the underlying normal distribution of the response variable.

An example might be comparing the gage error of three particular gages used to measure a given part characteristic.

RANDOM EFFECTS MODEL

Random effects models are used to describe situations in which incomparable differences in experimental material occur. The simplest example is that of estimating the unknown mean of a population whose individuals differ from each other. In this case, the variation between individuals is confounded with that of the observing instrument.

Contrasted with the fixed effects model, a subtle difference in the example could be first randomly selecting three gages capable of measuring a given part characteristic from all such gages available at a manufacturing facility and then comparing gage error. Inferences now not only relate to the gage error from the sampled gages, but also to what might occur by using other gages from the population. To further illustrate the difference between the fixed effects and the random effects models, consider repetition of the experiment. For the fixed effects model, we would select the same three gages, noting differences in gage errors with each gage. For the random effects model, we would again randomly sample three gages from the population of gages in order to examine the variability of the population of gages.

ASSUMPTIONS

- *Independence of cases.* This is a requirement of the design.

- *Scale of measurement.* The dependent variable is interval or ratio.

- *Normality.* The distributions in each of the groups are normal (use the Kolmogorov-Smirnov and Shapiro-Wilk tests to test normality). Note that the *F*-test is extremely non-robust to deviations from normality.[1]

- *Homogeneity of variances.* The variance of data in groups should be the same (use Levene's test for homogeneity of variances).

The fundamental technique behind an analysis of variance is a partitioning of the total sum of squares into components related to the effects in the model used. For example, we show the model for a simplified ANOVA with one type of treatment at different levels. (If the treatment levels are quantitative and the effects are linear, a linear regression analysis may be appropriate.):

$$SS_{Total} = SS_{Error} + SS_{Treatments}$$

The number of degrees of freedom (abbreviated df) can be partitioned in a similar way and specifies the chi-square distribution that describes the associated sums of squares (*SS*):

$$df_{Total} = df_{Error} + df_{Treatments}$$

DEGREES OF FREEDOM

Degrees of freedom indicates the effective number of observations that contribute to the sum of squares in an ANOVA: the total number of observations minus the number of linear constraints in the data. Generally, the degrees of freedom are the number of participants (for each group) minus one.

Example 1 Group A is given vodka, group B is given gin, and group C is given a placebo. All groups are then tested with a memory task. A one-way ANOVA can be used to assess the effect of the various treatments (that is, the vodka, gin, and placebo).

Example 2 Group A is given vodka and tested on a memory task. The same group is allowed a rest period of five days and then the experiment is repeated with gin. The procedure is repeated using a placebo. A one-way ANOVA with repeated measures can be used to assess the effect of the vodka versus the impact of the placebo.

Example 3 In an experiment testing the effects of expectations, subjects are randomly assigned to four groups:

Expect vodka–receive vodka

Expect vodka–receive placebo

Expect placebo–receive vodka

Expect placebo–receive placebo (the last group is used as the control group)

Each group is then tested on a memory task. The advantage of this design is that multiple variables can be tested at the same time instead of running two different experiments. Also, the experiment can determine whether one variable affects the other variable (known as interaction effects). A factorial ANOVA (2 × 2) can be used to assess the effect of expecting vodka or the placebo and the actual reception of either.

ANOVA measures within-group variation by a residual sum of squares $\hat{\sigma}_e^2$ whose square root $\hat{\sigma}_e$ will be found to approximate

$$\hat{\sigma} = \frac{\bar{R}}{d_2},$$

the measure of within-group variation or an estimate of inherent variability, even when all factors are thought to be held constant. This within-group variation is used as a yardstick to compare between-group variation. ANOVA compares between-group variation by a series of variance ratio tests (*F* tests).

It is important to stress that a replicate is the result of a repeat of the experiment as a whole and is not just another unit run at a single setting, which is referred to as a determination.[2]

This is just a very brief introduction to analysis of variance. In combination with the previous chapter on design of experiments, extremely powerful analytical methods may be applied to the further understanding of processes. With these tools employed, sound decisions may be made based on statistically derived, objective results.

Part III.I

Part IV

Customer–Supplier Relations

Part IV

Chapter 25

A. Internal and External Customers and Suppliers

> Define and distinguish between internal and external customers and suppliers and their impact on products and services, and identify strategies for working with them to improve products, services, and processes. (Apply)
>
> **Body of Knowledge IV.A**

Customer: *Anyone who is affected by the product or by the process used to produce the product. Customers may be external or internal.*[1]

INTERNAL CUSTOMERS

An internal customer is the recipient (person or department) of another person's or department's output (product, service, or information) within an organization.[2]

The concept of internal customers cuts across all levels of an organization—executive, managerial, and workforce. Case in point: consider how factory workers supply manufacturing data to a manager who, in turn, produces a production report for a senior executive.

Internal customers have also been defined as the next person in the work process. Ishikawa introduced the concept of internal customer as *next operation as customer*. All work-related activities of an organization can be viewed as a series of transactions between its employees or between its internal customers and internal suppliers. For example, the marketing department in any given organization might supply creative content to the Internet architecture and technology group responsible for the firm's Web site design.

Each receiver has needs and requirements. Whether the delivered service or product meets these needs and requirements impacts the customers' effectiveness and the quality of services and/or products delivered to their customers, and so on. Following are some examples of internal customer situations:

• If A delivers part X to B one hour late, B may have to apply extra effort and cost to make up the time or else perpetuate the delay by delivering late to the next customer.

- Engineering designs a product based on a salesperson's understanding of the external customer's need. Production produces the product, expending resources. The external customer rejects the product because it fails to meet the customer's needs. The provider reengineers the product and production makes a new one, which the customer accepts beyond the original required delivery date. The result is waste and possibly the last order from this customer.

- Information Technology (IT) delivers copies of a production cost report (which averages 50 pages of fine print per week) to six internal customers. IT has established elaborate quality control of the accuracy, timeliness, and physical quality of the report. However, of the six report receivers, only two still need information of this type. But neither of these finds the report directly usable for their current needs. Each has assigned clerical people to manually extract pertinent data for their specific use. All six admit that they diligently store the reports for the prescribed retention period.[3]

In every organization, each employee is a customer and a supplier. Because no work is considered an isolated activity, all internal customers are ultimately parts of a *customer–supplier value chain* (See Figure 25.1). In this chain, everyone's work is part of a process of inputs, added value, and outputs in products and services reaching the external customer. The following are some examples of how the customer–supplier value chain can be observed in every work process and management activity:

- Employees helping other employees get work done

- Supervisors leading employees to accomplish tasks

- Managers serving the needs of organizational groups in their area of responsibility

- Senior executives creating value for employees they lead

- The chief executive creating value by leading efforts to shape organizational vision, direction, philosophy, values, and priorities

A solid foundation for a successful customer-focused business is built when everyone in an organization thinks in terms of serving his or her internal customer and contributing to the ultimate value for the external end user.[4]

IMPROVING INTERNAL PROCESSES AND SERVICES

In an excerpt from Armand V. Feigenbaum and Donald S. Feigenbaum's, "The Power of Management Capital," the authors talk of the focus on enhancing quality value:

Companies can no longer focus their quality programs primarily on the reduction of defects or things gone wrong for their customers. Defect reduction has become an entry-level requirement for effective quality improvement initiatives. Companies must build their quality programs throughout the customer value chain by integrating and connecting all key quality work processes. . . .[5]

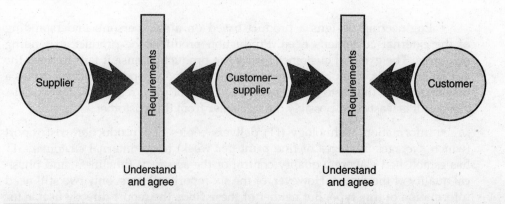

Figure 25.1 The customer–supplier value chain.

Source: ASQ's Foundations in Quality Self-Directed Learning Series, Certified Quality Manager, Module 4 (Milwaukee: ASQ Quality Press, 2001): 4-4.

The goal of the customer–supplier value chain is to integrate quality into every aspect of a work process and, ultimately, the product or service. The concept of "value added" refers to tasks or activities that convert resources into products or services consistent with external customer requirements. Strategies for working with internal customers and suppliers to improve products, processes, and services are as follows:

1. Identify internal customer interfaces (providers of services/products and receivers of their services/products)

2. Establish internal customers' service/product needs and requirements

3. Ensure that the internal customer requirements are consistent with and supportive of external customer requirements

4. Document quality-level agreements between internal customers and suppliers

5. Establish improvement goals and measurements

6. Identify and remove non-value-added work to keep work processes efficient and focused on quality

7. Implement systems for tracking and reporting performance and for supporting the continuous improvement of the process[6]

One method to help improve internal customer–supplier relations is an adaptation of Dr. Ishikawa's idea of everyone being able to talk with each other freely and frankly in order to create better internal relations and improved products and services. *Quality level agreements* (QLAs) are an adaptation of this. Internal QLAs can be used between internal suppliers and customers to document their agreed-to levels of service/product outputs.

In "Quality Level Agreements for Clarity of Expectations," Russell Westcott describes how QLAs can be used to manage internal expectations for improved results:[7]

- Quality specifications for service/product outputs are established (such as quantity, accuracy, usability, service availability and response time, attitude and cooperation of suppliers, and so on)

- Customers express performance targets with appropriate units of measure for each identified service/product output

- A periodic audit is conducted or a system for ongoing tracking, measuring, evaluating, and reporting performance compared to the QLA terms is put in place

- Identification of deficient performance levels triggers continuous quality improvement

Internal QLAs offer a sound basis for performance measurement and continual improvement. They also:

- Increase employee understanding of internal service/product expectations

- Increase employee insight about the potential impact of their contributions on a given business process

- Provide a clear link between internal product/service quality and the organization's quality management system and ultimately with external customer requirements

- Improve collaboration between internal customers and suppliers

For a successful customer-focused business, everyone in an organization must think of serving their internal customer while keeping aligned with the external customer's requirements in order to ultimately serve the external customer as the end user.

EFFECT OF TREATMENT OF INTERNAL CUSTOMERS ON THAT OF EXTERNAL CUSTOMERS

For management (and management's systems) to be inconsiderate toward internal customers (poor tools and equipment, defective or late material from a previous operation, incorrect/incomplete instructions, illegible work orders or prints, circumvention of worker safety procedures and practices, unhealthy work environment, lack of interest in internal complaints, disregard for external customer feedback, and so forth) can create careless or indifferent treatment of external customers. Continued, this indifference may generate a downward spiral that could adversely affect an organization's business. If managers want to instill a desire in their employees to care for the needs of external customers, they can't ignore the needs of internal customers.

So many organizations fail to learn, or rather, ignore, the internal customers' needs and then wonder why their management fails to stimulate internal customers to care about how and what they do for external customers. The rude and uncooperative sales representative, waitperson, housekeeping employee, healthcare provider, delivery person, or customer service representative are often a reflection of a lack of caring for internal customers.

Organizations must work constantly to address the internal customers' lament: "How do you expect me to care about the next operator, or external customer, when no one cares whether I get what I need to do my job right?"[8]

METHODS TO ENERGIZE INTERNAL CUSTOMERS

The organization should consider all or several of the following methods to energize internal customers to improve products, processes, and services:[9]

- Ensure that all employees understand who the external customers are and, generally, what the customers need and want.

- Ensure that employees know what the organization expects of them and what boundaries pertain to what they can do to satisfy the external customer.

- Involve employees in designing and implementing strategies, procedures, and tools that will facilitate customer satisfaction.

- Ensure that employees have the information, training, time, and support they need to enable them to focus on external customer needs, wants, and satisfaction.

- Ensure that employees have the tangible things they need to serve customers well, for example, supplies, tools, working equipment, clean work space, safe workplace, flexible policies and procedures.

- Ensure that the recognition and reward systems are supportive of the customer satisfaction focus.

- Demonstrate through words and actions the commitment the organization has to serving its customer to the best extent possible.

- Provide positive reinforcement to employees who are, or are trying, to do their best to satisfy customers.

- Provide employees with substantive performance feedback and assist in continual improvement.

EXTERNAL CUSTOMERS

Many times it is assumed that employees know who the external customer is. As Robin Lawton writes in his article, "8 Dimensions of Excellence," "Ask any diverse group of employees who the customer is and you are likely to hear multiple and competing answers."[10]

Terms referring to specific types of external customers include the following.

End User or Final Customer

End users or final customers purchase products/services for their own use or receive products/services as gifts. This definition can also apply to organizations, such as regulatory bodies.

Types of end users/final customers include:

• The *retail buyer* of a product. The retail buyer influences the design and usability of product features and accessories based on the volume purchased. Consumer product 'watch' organizations or regulatory bodies warn purchasers of potential problems. These can be small, independent regulatory organizations or large, established government organizations such as the FDA (Food and Drug Administration), which is one of the oldest consumer protection agencies in America.

The factors important to this type of buyer, depending on the type of product, are reasonable price, ease of use, performance, safety, aesthetics, and durability. Other influences on product offerings include easy purchase process, installation, instructions for use, post-purchase service, warranty period, packaging, friendliness of seller's personnel, and brand name.

• *Discount buyer.* The discount buyer shops primarily for price, is more willing to accept less well known brands, and is willing to buy quantities in excess of immediate needs. These buyers have relatively little influence on products, except for, perhaps, creating a market for off-brands, production surpluses, and discontinued items.

• *Employee buyer.* The employee buyer purchases the employer's products, usually at a deep discount. Often being familiar with or even a contributor to the products bought, this buyer can provide valuable feedback to the employer (both directly, through surveys, and indirectly, through volume and types of items purchased).

• *Service buyer.* The buyer of services (such as TV repair, dental work, or tax preparation) often buys by word of mouth. Word of good or poor service spreads rapidly and influences the continuance of the service provider's business.

• *Service user.* The captive service user (such as the user of electricity, gas, water, municipal services, and schools) generally has little choice as to from which supplier they receive services. Until competition is introduced, there is little incentive to providers to vary their services. Recent deregulation has resulted in a more competitive marketplace for some utilities.

• *Organization buyer.* Buyers for organizations that use a product or service in the course of their business or activity can have a significant influence on the types of products offered them as well as on the organization from which they buy. Raw materials or devices that become part of a manufactured product are especially critical in sustaining quality and competitiveness for the buyer's organization (including performance, serviceability, price, ease of use, durability, simplicity of design, safety, and ease of disposal). Other factors include flexibility in delivery, discounts, allowances for returned material, extraordinary guarantees, and so forth.

Factors that particularly pertain to purchased services are the reputation and credibility of the provider, range of services offered, degree of customization offered, timeliness, and fee structure.

Intermediate Customers

Intermediate customers are divided into two groups. One group of intermediate customers buys products and/or services and makes them available to the final customer by repackaging or reselling them or creating finished goods from components or subassemblies. The second group includes organizations that repair or modify a product or service for the end user.

Types of intermediate customers include:

• *Wholesale buyer.* Wholesalers buy what they expect they can sell. They typically buy in large quantities. They may have little direct influence on product design and manufacture, but they do influence the providers' production schedules, pricing policies, warehousing and delivery arrangements, return policies for unsold merchandise, and so forth.

• *Distributor.* Distributors are similar to wholesalers in some ways but differ in the fact that they may stock a wider variety of products from a wide range of producers. What they stock is directly influenced by their customers' demands and needs. Their customers' orders are often small and may consist of a mix of products. The distributors' forte is stocking thousands of catalog items that can be "picked" and shipped on short notice at an attractive price. Customers seeking an industry-level of quality at a good price and immediately available mainly influence distributors stocking commodity-type items, such as sheet metal, construction materials, and mineral products. *Blanket orders* for a yearly quantity delivered at specified intervals are prevalent for some materials.

• *Retail chain buyer.* Buyers for large retail chains, because of the size of their orders, place major demands on their providers, such as pricing concessions, very flexible deliveries, requirements that the providers assume warehousing costs for already purchased products, special packaging requirements, no-cost return policies, and requirements that the providers be able to accept electronically sent orders.

• *Other volume buyers.* Government entities, educational institutions, healthcare organizations, transportation companies, public utilities, cruise lines, hotel chains, and restaurant chains all represent large-volume buyers that provide services to customers. Such organizations have regulations governing their services. Each requires a wide range of products, materials, and external services in delivering its services, much of which is transparent to the consumer. Each requires high quality and each has tight limitations on what it can pay (for example, based on appropriations, cost-control mandates, tariffs, or heavy competition). Each such buyer demands much for its money but may offer long-term contracts for fixed quantities. The buying organizations' internal customers frequently generate the influences on the products required.

• *Service providers.* These buyers include plumbers, public accountants, dentists, doctors, building contractors, cleaning services, computer programmers, Web site designers, consultants, manufacturer's reps, actors, and taxi drivers, among many others. This type of buyer, often self-employed, buys very small quantities, shops for value, buys only when the product or service is needed (when the buyer has a job, patient, or client) and relies on high quality of purchases to maintain customers' satisfaction. Influences on products or services for this type of buyer may include having the provider be able to furnish service and/or replacement parts for old or obsolete equipment, be able to supply extremely small quantities of an extremely large number of products (such as those supplied by a hardware store, construction materials depot, or medical products supply house), or have product knowledge that extends to knowing how the product is to be used.

Intermediate customers must be able to understand the product and/or service requirements of the end users and know how to address product specifications or service elements with the supplier.

Customers are diverse; they have different needs and wants and certainly do not want to be treated alike. To provide a strategic focus and respond more effectively to groups of either current or prospective buyers, most businesses segment their external customers in order to better serve the needs of different types of customers.

Customers sharing particular wants or needs may be segmented by:

- Purchase volume

- Profitability (to the selling organization)

- Industry classification

- Geographic factors (such as municipalities, regions, states, countries, and continents)

- Demographic factors (such as age, income, marital status, education, and gender)

- Psychographic factors (such as values, beliefs, and attitudes)

An organization must decide whether it is interested in simply pursuing more customers (or contributors, in the case of a not-for-profit fund-raiser) or in targeting the right customers. It is not unusual for an organization, after segmenting its customer base, to find that it is not economically feasible to continue to serve a particular segment. Conversely, an organization may find that it is uniquely capable of further penetrating a particular market segment or may even discover a niche not presently served by other organizations. [11]

EXTERNAL CUSTOMERS' INFLUENCE ON PRODUCTS AND SERVICES

For an organization to assume they know a customer's needs would be very risky. Robin Lawton also writes in his article, "8 Dimensions of Excellence," "It is impossible to determine who the customer is without first identifying the product for which they are customers." With reference to his third dimension, "Product Characteristics Customers Want," he goes on to explain that when there is confusion about the product, there is guaranteed confusion about who the customer is. [12] After segmenting external customers, an organization must determine what it is about its products and services that are the most important to each group. This involves consideration of customers as people—their likes, dislikes, idiosyncrasies, biases, political orientation, and religious beliefs as well as moral codes, peer pressures, prestige factors, fads, and so on, that an organization should be aware of in order to manage the customer relationship effectively.

Chapter 26

B. Customer Satisfaction Analysis

Describe the different types of tools used
to gather and analyze customer feedback:
surveys, complaint forms, warranty analysis,
quality function deployment (QFD), etc.
(Understand)

Body of Knowledge IV.B

It is the *voice of the customer* (VOC) that an organization must be listening to in order to satisfy customer needs, improve current products and services, and generate ideas for new products and services. Customer satisfaction is the difference between perceived performance and the customer's expectations. With customer expectations continuously growing, there are a variety of tools that can be used to capture accurate customer information.

From the book, *How to Win Customers and Keep Them for Life:* "A typical business hears from only four percent of its dissatisfied customers. The other 96 percent just quietly go away and 91 percent will never come back. That represents a serious financial loss for companies whose people don't know how to treat customers, and a tremendous gain for those that do."[1]

If an organization wants to be customer-driven, it must integrate both qualitative and quantitative customer feedback data to get a more complete picture of the voice of the customer. Qualitative data complements quantitative data in that it helps companies gain a more comprehensive understanding of their customers' perceptions of quality. There are many tools used to gather and analyze customer feedback:

- To seek out customer feedback, organizations can use surveys, focus groups, order forms with comment sections, and interviews.

- To facilitate feedback, making it easy for the customer to respond, an organization can set up customer hot lines, such as '24/7' 800 numbers, pagers for contact personnel, electronic bulletin boards, and access for feedback or complaints via e-mail, fax, phone, and mail.

- Information is also available from the data provided by standard company records, such as claims, refunds, returns, repeat services, litigations, replacements, repairs, guarantee usage, warranty work, and general complaints.

Once organizations have a process, are collecting customer data, and problem-solving to improve on their product/service, they must also take action in which the use of the quality function deployment (QFD) method helps to translate customer requirements into appropriate technical requirements for each stage of the product development cycle.

SURVEYS

Companies who are customer-driven are proactive and will go out and ask the customer for feedback. Surveys are a good way to quantify customers' attitudes and perceptions of a company's products or services. By using surveys, a company goes directly to their customers to get the answers to what they want to know. What knowledge is the organization hoping to gain? Facts that they can gather and benefit from may include the answers to knowing who their customers are, why they buy their products or services, what makes them loyal and want to buy or use their services again, or what made them stop and go to a competitor. Customer satisfaction surveys can be used to drive continuous improvement activities.

Survey Types

There are a number of different types of surveys that can be used, depending on the organization's needs. Surveys can elicit a customer's evaluation of a product, a customer's purchasing experience, a company employee's feedback, and so on. See Table 26.1 for a list of various survey types, their uses, advantages, and disadvantages. Note that the types of surveys are listed in alphabetical order; the order does not imply any special significance.

Customer satisfaction and perception surveys can be conducted using one of three different formats: in-person, over the telephone, or written (traditional paper and pencil, comment cards/suggestion box, or online via e-mail or a Web site). See Table 26.2 for the advantages and disadvantages of each format.

Table 26.1 Survey types.

Type of survey	Use	Advantages	Disadvantages
Attrition analysis and reports	• Used to calculate attrition (lost-customer) rates • Used to analyze major events in the organization's environment that may have led to higher rates of attrition (for example, changes in management, changes in internal quality measures, changes in policies such as merchandise-return policies or frequent flyer policies, or the introduction of a new product or service by a competitor	• The company has a better understanding of why it is losing customers and the conditions leading up to the losses • The company can predict the kind of events likely to lead to customer attrition in the future	• May not get complete and honest answers
General customer satisfaction surveys	• Used to measure customer satisfaction and dissatisfaction with products and services • Used to track trends • Used to identify major improvement and marketing efforts	• Useful for measuring "desired" quality • General surveys, such as ratings and report cards, are easy to obtain from customers	• Because people typically do not want to write out details, the information received may be too general and therefore not very useful • Companies may work to achieve high ratings on the report cards, leading to other important things being ignored that are not on the report card. A focus on report cards can derail customer satisfaction.
Lost-customer surveys	• Used to survey (typically over the telephone) customers who have significantly reduced or have stopped purchasing company's products or services • Used to survey customers as close as possible to time of defection • Used when the organization can define who represents a lost customer (which is based on the typical buying interval) • Used when an organization can define when a customer is lost (expressed in terms of days, months, or perhaps years)	• Help identify serious customer issues • Are one of the more powerful tools for recovering customers • Useful for identifying and measuring "expected" quality • May reveal information about competitors' advancements in products and services	Require researchers who: • Deal with serious customer issues • Are very familiar with the company's products and services • Understand how things get done in the organization • Demonstrate good listening, analytical, and interpersonal skills • Retrieve customer opinions without using a sales approach

Continued

Table 26.1 *Continued.*

Type of survey	Use	Advantages	Disadvantages
Moment-of-truth inventories	• Used to collect observations from frontline employees regarding customer needs or complaints	• Employees can make observations and assessments within context of whole organization (for example, policies, systems, processes, culture, management styles, employee pay, benefits, and promotions) • Daily aggregates of observations and interactions with customers are more revealing than a once-a-year survey	• If trust is low and fear is high, employees might not feel comfortable sharing "the truth" • Not all employees view complaints as legitimate opportunities for improvement • Managers may subtly punish employees for telling the truth about customer dissatisfaction
Perceptual surveys—typically conducted as a blind survey, meaning the person surveyed does not know who is sponsoring the survey	• Used to get a customer's comparative perceptions of a company's products and services relative to the competition's • Used to get perceptions of a company's image	• The pool of sampled customers typically includes the company's customers, the competitor's customers, and, sometimes, potential customers • Useful for identifying and measuring both "expected" quality and "desired" quality	• Interviewees may not have experience using competitor's products, thus no genuine basis for comparison • The anonymity of the survey (that is, impersonal nature) often causes a reduced rate of responsiveness
Reply cards	• Used to obtain demographic information about buying customers • Used to obtain reasons why customers bought • Used to get customer reactions to the buying experience • Used to solicit information regarding other product/service needs	• Help update company's customer list • Help company research buying trends	• May be difficult to get a significant return • Not all customers will respond, thus the information received is potentially incomplete and biased
Transaction surveys—"real-time" surveys. A brief survey typically asking two key questions: "What did we do right?" and "How can we improve?"	• Used to survey customers immediately after a transaction (for example, a product delivery, an installation, or a billing) • Used to build customer relationships • Used to get new information related to products and services	• Provide quick feedback • Give employees opportunities to fix a problem on the spot • Useful for measuring "desired" quality	• If frontline employees are not empowered to resolve problems on the spot, customers might perceive the company as not being concerned about service quality issues
Win/loss analysis and reports	• Used to discover the company's relative ranking with competitors based on the decision maker's selection process • Used to assess why the company won or lost a competitive bid	• Company has new information to improve future bids • Company gains information about their competitors	• May be difficult to get an honest assessment

Source: ASQ's Foundations in Quality Self-Directed Learning Series, Certified Quality Manager, Module 4 (Milwaukee: ASQ Quality Press, 2001): 4-31, 4-32.

Table 26.2   Survey mediums.

Type of survey	Advantages	Disadvantages
In-person survey	• Enhances relationship management with key customers • Allows for in-depth probing • Offers a more flexible format • Creates opportunities to establish and maintain rapport with customers	• Is labor-intensive • Can be expensive • Can allow interviewer–interviewee bias to be introduced • Requires highly trained interviewers • Requires content analysis of open-ended questions, which is more difficult to do • Requires well-written open-ended questions • Requires excellent listening skills
Telephone survey	• Allows for greater depth of response than written surveys • Costs less than in-person surveys • Respondents are more at ease in their own environments • Creates a personal connection with the customer that may allow for follow-up	• Calls may be viewed as intrusive • The interviewee may be uncomfortable if others listen in on the conversation • Designing is more difficult than for written surveys because telephone surveys are a form of interpersonal social interaction without a visual contact. People rely on visual as well as spoken cues in a conversation. • Requires well-written open-ended questions that are easy for the interviewee to comprehend • Requires excellent listening skills
Written survey	• Typically is less expensive than in-person survey/interview • Can be self-administered • Can be made anonymous • Offers a variety of design options to choose from • Is one of the most expedient ways to have customers quantify their reactions to a limited number of features in a company's service or product (if a low response rate is acceptable)	• Requires follow-up to derive a decent response rate • Customers typically don't like answering open-ended questions (therefore, limited to asking only a few open-ended questions) • Targeted addressee may not be the one that responds • Feedback is of the "report card" nature, that is, little or no depth of information • There is less assurance that the questions are understood. (This can be minimized with field-testing or validating the survey instrument with a small group of preselected customers.)

Source: ASQ's Foundations in Quality Self-Directed Learning Series, Certified Quality Manager, Module 4 (Milwaukee: ASQ Quality Press, 2001): 4-33.

Survey Design

To maximize the quantity of responses and to get good constructive answers, the design of a survey is critical. When designing a survey the following must be addressed:

• What specific research objectives will the survey accomplish?

• How will results be used?

• How will results be communicated to customers and stakeholders?

• Which customers should be surveyed?

- What information should be sought and what kinds of questions should be posed?

- What type of survey will best achieve the research objectives?

- Will the survey selected measure what it is intended to measure?

- Will the survey be capable of producing the same results over repeated administrations?

- How will the results be analyzed (scoring format and editing policy)?

- How easy will it be to administer the selected survey?

- How easy will it be for the respondents to complete?

- Does the budget cover the total costs associated with the design, implementation, analysis, and feedback?

Survey Question Design

When designing a survey, it's like the old adage, "garbage in, garbage out"; The answers received will only be as good as the questions that were asked. The survey question design choice depends on the type of data an organization wants to receive. The responses can be qualitative or quantitative:

- *Qualitative responses.* If looking for qualitative data, questions will be designed for a nonnumeric response. The response will be text or open-ended. Open-ended questions will result in a richer understanding of the respondent's attitudes and thought processes, but there are limitations, namely analysis and control. Using tools such as cluster analysis and Pareto analysis can help quantify results. A few well-written, thought-provoking questions will get an organization more informative responses than a multitude of questions (no matter how well written!) as respondents are bound to grow tired and either skip questions or give up and not complete the survey at all.

- *Quantitative responses.* If looking for quantitative data, questions will be designed for numeric data or closed-ended. Closed-ended questions present discrete choices to the respondent to elicit a simple yes/no response. The response format determines how the closed-ended survey data can be used. The four popular scales used for these surveys are:

  - The *Likert scale* format gives customers the opportunity to respond in varying degrees. The high end represents a positive response while the low end represents a negative response. For example, if measuring level of satisfaction, a five-point scale is used for responses ranging from 'Very satisfied' to 'Very dissatisfied.' Of all scales, Likert is the most frequently used.

  - *Rank order* is used when respondents are asked to rank a series of product/service attributes according to their perceived importance.

Five or six items are optimal when rank-ordering although it is possible with up to 20 items but more difficult for most respondents. Rank-ordering is the least popular for measuring the importance of attributes.

- *Forced allocation* requires respondents to assign points across categories. Used to identify the importance of different product or service features.

- *Yes/no response* commands a simple response of a yes or no answer from the customer.

Generally, the survey results should help the organization understand customer priorities and levels of customer satisfaction. The survey data results should be communicated to all key stakeholders to ensure that the feedback is understood and used to increase customer focus and satisfaction.

COMPLAINT FORMS

From the book *How to Win Customers and Keep Them for Life:* "Seven out of 10 complaining customers will do business with you again if you resolve the complaint in their favor. If you resolve it on the spot, 95 percent will do business with you again. On average, a satisfied customer will tell five people about the problem and how it was satisfactorily resolved."[2]

Most customers who are dissatisfied will not complain, instead, they will go to the competition. When a customer does communicate a complaint, it should be regarded as an important opportunity not only to learn from that customer, but if handled quickly with positive results for the customer, it enhances that customer's loyalty. The organization should see this as a missed opportunity if they have only been proactive with other tools such as customer feedback surveys, QFD, and so on.

When a customer complains, he/she is providing information about a problem that other customers are most likely experiencing but the company is not yet aware of. The problem may be with the service received or the product itself. The customer's complaint may also be about not getting what they expected to get out of that product or service. This also needs to be captured and fed back to the service provider or manufacturer as this is valuable information providing the opportunity for improvements to a service or product. A complaint can be communicated many ways: by phone calls, comments made to an employee, a section for comments on an order form, face-to-face complaints at a customer service counter, writing a letter, customer feedback cards, or filling out and submitting a complaint form. All these formats can be captured and used for continuous improvement. The least productive and unfortunately very harmful to a company are the complaints that are lodged by customers complaining to friends and acquaintances. If organizations want customer loyalty, they must make it easy for a customer to complain and provide a quick response that satisfies the customer!

There are various formats for complaint forms whether they are interactive online forms or filled out on paper. In general, for complaint forms to be of any use for problem solving, they need to have the following information:

Part IV.B

- To be provided by the customer:
  - Customer contact information
  - A detailed description of the problem
  - When the problem occurred
  - Who was involved
  - Where the problem took place
- To be completed by organization/customer service department
  - Person investigating complaint
  - Due date for closure
  - Root cause
  - Action taken
  - Communication back to the customer

Figure 26.1 shows an example of a customer complaint form to be filled out by hand.

The complaint forms should be just a small piece of a larger picture. A *complaint management system* (CMS) needs to be in place where there is one person responsible for tracking all complaint forms. A CMS will allow for recording all complaints and is an added source for gaining information from existing customers and satisfying customers. As part of a continuous improvement process, such as PDCA (as discussed in Chapter 10), after gathering complaint data, problem solving tools (refer to Chapter 9) can be used to analyze the data, identifying opportunities for improvement.

- First, and foremost, a complaint in any form should ideally be resolved on the spot.
  - A complaining customer is likely to be emotional or angry. Any employee dealing directly with customers must have training in conflict resolution and problem solving tools. An employee must know through proper training not to take it personally. Rather than feel resentment toward an irate customer, an employee should strive to understand the problem and solve the problem, providing a win–win situation. The customer walks away knowing the company cares and the employee feels the reward of accomplishment.
  - An employee must always remember to thank the customer for bringing the problem to his/her attention.
- All complaints should be documented along with the resolution whether they were resolved on the spot or not.
- Communicate the action taken to the customers.

Customer Complaint Form

Customer name: _____

Telephone number: _____

Address: _____

Complaint: _____

Date and time of occurrence: _____

Location: _____

Name of product (if applicable): _____

Name of person(s) involved (if applicable): _____

Please give description of problem: _____

_____

_____

_____

_____

_____

_____

Findings: _____

_____

_____

_____

_____

_____

Action taken: _____

_____

_____

_____

Date customer was contacted with the results of the investigation and action taken: _____

Initials of person investigating the complaint: _____

Initials of person taking complaint: _____

Figure 26.1 Customer complaint form.

- Using problem-solving tools as discussed in Chapter 9, look for trends, Paretoize complaints, and problem-solve.

  - Recording and tracking complaints will tell a company which particular issue is not just a nuisance but indeed may be a major problem.

  - Remember the 80/20 rule (Pareto principle)—80 percent of the complaints come from 20 percent of the customers. A high number of complaints for a particular problem certainly does signify there is a problem needing resolution. However, there may only be a small number of complaints for a specific problem that has a larger impact on profits. Having already divided customers into segments, an organization should know which of their customers contribute most to their profits.

- Don't stop at problem solving. After prioritizing improvement opportunities, provide resources to a cross-functional team for implementation as part of the PDCA cycle.

Customers should be encouraged to express dissatisfaction. Training employees to handle customer complaints as opportunities can be key to a company gaining competitive advantage in the marketplace. Natalia Scriabina and Sergiy Fomichov discuss improving business performance through effective complaint handling in, "Six Ways to Benefit from Customer Complaints." They provide Table 26.3 to help identify whether an organization sees complaints as problems or opportunities:[3]

Table 26.3 Test to identify organization's attitude toward complaints.

	A	**B**
A complaint is	a problem	an opportunity
Receiving complaints is	a painful and awkward situation	a chance to retain dissatisfied customers
Above all, a complainant	wants compensation	gives important information
Employees are	defensive about complaints	open to complaints
Employees tend to	shift blame elsewhere	recognize the needs of dissatisfied customers
Complaints are resolved	with problem-solving techniques	with a systematic process linked to a continuous improvement process
When a complaint is closed,	someone will likely be punished	something should be improved
Complaints	must be reduced	are encouraged and welcomed

Source: N. Scriabina and S. Fomichov, "Six Ways to Benefit from Customer Complaints," *Quality Progress* (2005).

WARRANTY AND GUARANTEE ANALYSIS

Using data from standard company records, such as claims, refunds, returns, repeat services, litigations, replacements, repairs, guarantee usage, and warranty work, also tells an organization where to focus their problem-solving efforts. A guarantee is a promise or assurance regarding the quality of a company's product or service. A warranty is an assurance regarding the reliability of a company's product for a certain period of time. Guarantees and warrantees are legally binding contracts. They are typically offered to build customer confidence and reduce any risks associated with a purchase decision.

Having a warranty ensures that the customer seeks service using the warranty rather than replacing the product with one from a competing manufacturer. Warranty cards are often distributed with products for new owners to complete and return to the producer. The purpose of the warranty card is twofold: (1) to register the owner's product and serial number with the producer and (2) to obtain demographic information, marketing information, and/or customer feedback regarding the owner's decision to buy, the buying experience, and so on. Ultimately, warranty cards help build/update the company's customer list and provide opportunities to stay in touch with customers. A pragmatic problem with warranty cards is that many customers do not fill them out and return them.[4]

The data from tracking guarantee and warranty issues provide organizations information such as failure rates in products and services, what is unacceptable to customers, and a rough idea of what customers expect. With each guarantee and warranty problem, the organization's profits are liable through payouts and returns. Using the data to problem-solve for the root cause and taking corrective action to prevent the problem from recurring lessens future impact to a company's profits.

QUALITY FUNCTION DEPLOYMENT

The voice of the customer is the basis for setting requirements. With all the data that is being gathered from surveys, complaints, guarantee usage, and warranty issues for information about customer requirements, a company can use a planning process such as quality function deployment (QFD) to take action. QFD is a structured method in which customer requirements are translated into appropriate technical requirements for each stage of the product development cycle.

A Japanese shipbuilding company developed QFD in the early 1970s. In the 1980s, the U.S. auto industry imported QFD and it has since proven successful in virtually every industry (manufactured products and services) worldwide.

QFD, also known as the *house of quality* (see Figure 26.2), graphically displays the results of the planning process. It links organizations with consumers by translating the customer's voice into technical design requirements and customer needs into product design parameters. The QFD matrix shows customer requirements (what the customer wants and needs) as rows, and the columns are characteristics of a product or service (how to meet the customer's needs). In a typical application, a cross-functional team creates and analyzes a matrix that the company can then measure and control.

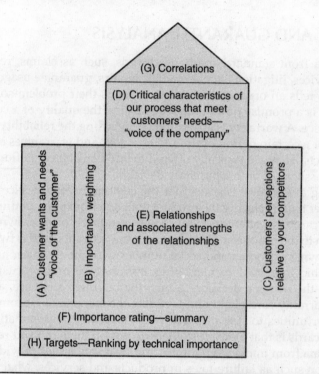

Figure 26.2 Quality function deployment matrix house of quality.
Source: R. T. Westcott, *The Certified Manager of Quality/Organizational Excellence Handbook,* 3rd ed. (Milwaukee: ASQ Quality Press, 2006): 475.

QFD is deployed through a progression of matrixes (Figure 26.3). It's an iterative process of four phases where the decisions made in one matrix drive the next. The four linked matrices convey the voice of the customer through:

- *Product planning.* Translating customer input and competitor analysis into design requirements.

- *Part deployment.* Translating design requirements into part specifications

- *Process planning.* Translating part specifications into process requirements

- *Production planning.* Translating process requirements into production requirements

Early in the planning phase, the house of quality clarifies the relationship between customer needs and product features. It correlates the customer requirements and competitor analysis with higher-level technical and product characteristics to establish the strong and weak relationships between the two.

Referring to Figure 26.2:

- Data for section (A), the customers' wants and needs, is obtained from several sources, for example, customer listening systems, surveys, and customer focus groups.

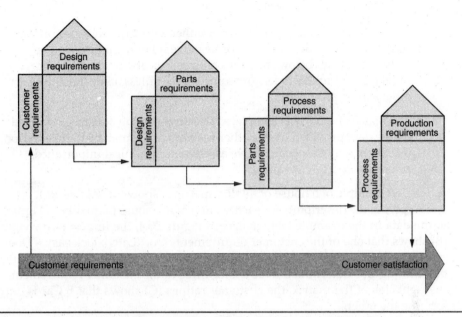

Figure 26.3   Voice of the customer deployed through an iterative QFD process.
Source: R. T. Westcott, *The Certified Manager of Quality/Organizational Excellence Handbook*, 3rd ed. (Milwaukee: ASQ Quality Press, 2006): 478.

- Each listed customer want may be given a weight as to its importance (B).

- Through research, interviews, focus groups, and so on, data are obtained to be able to rate customers' perceptions (C) of the wants listed in (A) relative to your competitors' products/services (the horizontal connection between A and C).

- The "how things are done" or critical characteristics of your process that meet the customers' needs (the voice of the company) are shown in part (D) of the matrix.

- In the relationships section of the matrix (E), where the rows and columns of the matrix intersect, the QFD team identifies the inter-relationships between each customer requirement and the technical characteristics. Relationship codes are inserted. Usually, these codes are classified as 9 = strongly related, 3 = moderately related, or 1 = weakly related.

- Section (F) is a summary of the vertical columns from the matrix (E) representing the sums of the weighted totals represented by the relationships (where coded).

- By adding a roof (G) to the house of quality, the question of whether the improvement to one requirement will impact another design requirement will be addressed. This matrix correlates the interrela-tionship between company descriptors. Usually a key is provided as to the correlation strengths to show either a positive and supporting relationship or that a trade-off or negative relationship exists. The resulting arrows below the roof's matrix indicate the direction of improvement for each design requirement (unless a key indicating otherwise is provided).

- The house of quality is completed by summarizing conclusions based on the data provided by the other sections of the matrix. In this section (H), there may be a competitive assessment, technical importance/priorities, and target values.

A detailed example of a house of quality matrix is shown in Figure 26.4.

Applying the descriptors for each section of the house of quality in Figure 26.2 to the data in the example QFD matrix in Figure 26.4, the importance weighting (B) shows that one of the customer requirements (A), "Latch/lock easy—Does not freeze," was assigned the highest importance rating of 8, and the lowest importance rating was determined to be a 2 for three of the customer requirements. On the right side of the matrix, the customer ratings (C) shows that B Car is outperforming the others.

Section (F), identified as the "absolute technical importance" row in this particular sample QFD matrix, represents the sum of the products of each column symbol value (E) and the corresponding customer requirement weight (B). Calculating the design requirement "Door close effort" is:

7 (weight of Easy close O/S) × 9 (Strong strength relationship) +
5 (weight of Easy close I/S) × 3 (Medium strength relationship) = 78.

Each entry in (F) is then divided by the sum of all the entries in that row and multiplied by 100 to convert it into a percentage giving the relative technical importance (H) rankings. The relative ranking for "Door close effort" is calculated as:

78/913 × 100% = 8.54 or rounded off = 9.0

These values represent targets of opportunity for improvement or innovation.

The QFD matrix may be changed and/or expanded according to the objectives of the analysis. If desired, customer requirements can be prioritized by importance, competitive analysis, and market potential by adding an additional matrix to the right-hand side of the QFD matrix. The company descriptors can be prioritized by importance, target values, and competitive evaluation with a matrix added to the bottom of the QFD matrix (as displayed in the sample QFD in Figure 26.4). To summarize, the QFD approach and deployment steps are:

1. Plan—determine the objective(s), determine what data are needed.

2. Collect data.

3. Analyze and understand the data using the QFD matrix to generate information.

4. Deploy the information throughout the organization (depending on the objectives).

5. Employ the information in making decisions.

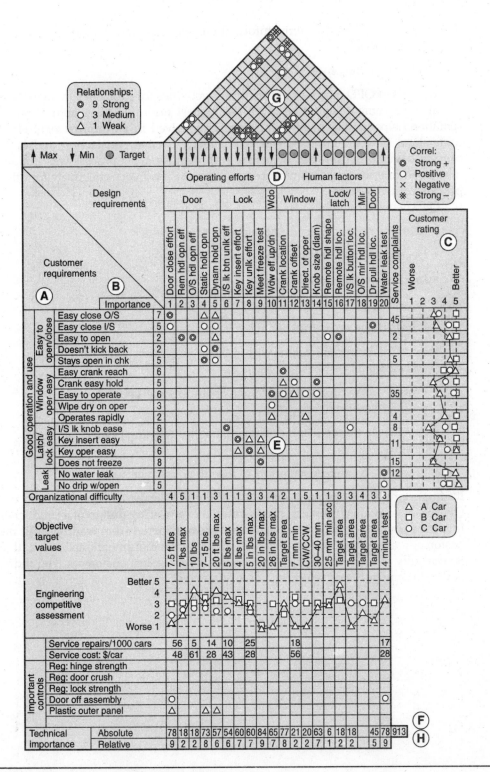

Figure 26.4   QFD house of quality matrix example.

Source: R. T. Westcott, *The Certified Manager of Quality/Organizational Excellence Handbook*, 3rd ed. (Milwaukee: ASQ Quality Press, 2006): 477.

Part IV.B

6. Evaluate the value of the information generated and evaluate the process.

7. Improve the process.

The completed QFD matrix indicates which activities for problem solving or continual improvement should be given the highest priority. This graphical representation helps management and personnel involved in an improvement effort understand how an improvement will affect customer satisfaction and helps prevent deviations from that objective. Although the QFD process was initially developed to prioritize design processes, it can be used very effectively for other process-related systems.

QFD benefits include:

- Shortens product/service development cycle time

- Decreases the number of after-launch changes

- Increases customer satisfaction

- Reduces costs and can increase profits and market share

- Breaks down cross-functional barriers

- Improves communications within the organization and with suppliers and customers

- Provides structure and a systematic planning methodology for product/service development

- Provides better data and documentation for refining product/service designs

- Leads to a more focused effort in meeting customer needs[5]

It's important that a company make use of the VOC as early as possible and use QFD to incorporate customers' needs into the engineering design of a product or into the plan for a service in order to save costs incurred from waiting until the customer receives the product and ends up using the tools mentioned above, that is, warranty costs, costs incurred from complaints or lost customers who didn't bother to complain, and so on.

Chapter 27

C. Product/Process Approval Systems

Quality function deployment (QFD) is a valuable tool for product, service, and process design. A company or organization should be using a planning process such as QFD and translating customer requirements into appropriate technical requirements. During research and development, it is important to translate the product characteristics into specifications as a basis for description and control of the product.

At the conclusion of each stage of a product development cycle (see Figure 27.1), a formal, documented, systematic, and critical review of results should be planned and conducted. A decision is then made based on the review of the results whether to move on to the next stage. This is done from the system-design level down to lower levels in the product hierarchy (breakdown). A typical breakdown would be: systems, subsystems, components, assemblies, parts, and so on. Depending on the industry, certain regulations may apply. In the medical device industry, companies must be compliant with FDA21 CFR820.30 regulation for the control of design changes to ensure that good quality assurance practices are being used for the design of medical devices.

VERIFICATION

Verification is conducted by engineering to determine if design specifications are being met. Verification ensures that the design is complete and meets design input requirements (customer needs and satisfaction, product specifications, and process specifications). Verification methods include:

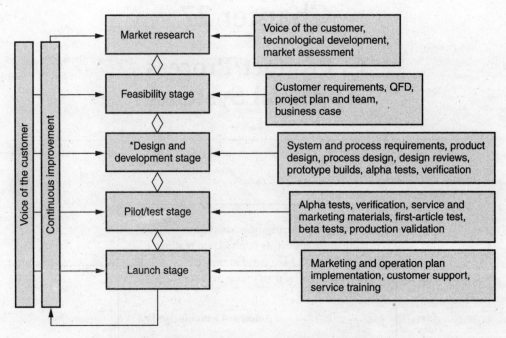

*The development stage can be lengthy and often broken up into phases, that is, definition phase, design phase, and prototype phase.

Figure 27.1 Stages of the product development cycle.

- Formal product/service design reviews*
- Qualification testing
- Burn-in/run-in tests (to detect and correct inherent deficiencies)
- Performance tests (mean time between failures, mean time to repair, and so on)
- Demonstrations
- Prototypes
- Design comparisons with similar existing designs previously proven to work

Design reviews are also used to identify critical and major characteristics so that manufacturing and quality efforts will focus on high-priority items and design for appropriate inspection and tests. In addition to the design reviews, periodic evaluations of the design involving analytical methods should be conducted at significant stages. These methods include failure mode and effects analysis (FMEA), fault tree analysis, and risk assessment.

---

* Because a lack of design reviews can cause potentially serious quality problems, quality and reliability engineers should advocate and participate in design reviews.

In Joseph M. Juran's, *Juran on Planning for Quality,* he states, "Design review originally evolved as an early warning device during product development. The method is to create a design review team, which includes specialists from those classes of customers who are heavily impacted by the design—those who are to develop the manufacturing process, produce the product, test the product, use the product, maintain the product, and so on. . . . Their responsibility is to provide early warning to the designer: 'If you design it this way, here are the consequences in my area.'"[1]

Other definitions of verification:

"The act of reviewing, inspecting, testing, checking, auditing, or otherwise establishing and documenting whether items, processes, services, or documents conform to specified requirements."[2]

"Confirmation, through the provision of objective evidence, that specified requirements have been fulfilled."[3]

VALIDATION AND QUALIFICATION METHODS

As an output of the initial planning phase, product/service quality goals are defined based on customer wants/needs. During product and process design and development, the task is to create a process that can, under operating conditions, satisfactorily produce the product or service.

Qualification testing can be used for both verification and validation. All test and evaluation results should be documented regularly throughout the qualification test cycle and any results of nonconformity and failure analysis noted. Changes indicated by the results of qualification testing get incorporated into the drawings and manufacturing planning.

Qualification process—a set of steps that demonstrate the ability of an entity to fulfill specified requirements.[4]

Other definitions from the FDA's Web site on Guidelines on General Principles of Process Validation:[5]

- *Installation qualification* of tooling and other process equipment establishes confidence that they are consistently operating within established limits and tolerances.

- *Process performance qualification* establishes confidence that the process is effective and reproducible.

- *Product performance qualification* establishes confidence through appropriate testing that the finished product produced by a specified process meets all requirements for functionality and safety.

Once manufacturing begins output of a small number of parts on production tools, and enough parts are available, a pilot assembly run is started. Various production tests are conducted to ensure that the product is being built to specifications.

Validation is the process of evaluating a system or component during or at the end of the development process, after verification, to determine whether it satisfies specified requirements. The focus is to ensure that the product functions as it was intended to. It's normally performed on the final product and conducted by the customer under defined operating conditions.

The FDA's Web site on Guidelines on General Principles of Process Validation defines process validation as:[6]

- Establishing documented evidence that provides a high degree of assurance that a specific process will consistently produce a product meeting its predetermined specifications and quality characteristics.

The ISO 9000:2000 definition for validation:[7]

- Confirmation, through the provision of objective evidence, that the requirements for a specific intended use or application have been fulfilled.

First-article testing tests a production process to demonstrate process capability and whether tooling meets quality standards. The process is tested under operating conditions but the resulting products are not salable units. Approval of first-article testing can validate the process.

A *pilot run* is a trial production run. When enough units are produced during a pilot assembly run, various production tests are conducted to ensure that the product is being built to specifications. A field test validates that the product meets performance requirements. During a pilot run, the entire manufacturing process is being analyzed to determine if there are any trouble spots. When a problem is identified and the problem solved for the root cause, a countermeasure is then implemented. The process is then validated to ensure that the processes are functioning as they should to produce a quality product that meets the customer's needs.

Alpha testing, usually done in the development stage, consists of in-house tests conducted on prototypes to ensure that the product is meeting requirements under controlled conditions. Alpha testing may go through multiple iterations. Each Pareto analysis of failure modes yields tighter results toward meeting intent.

Beta testing is the final test before the product is released to the commercial market. Beta testers are existing customers. Two criteria to be included in the beta test program: the customer must be friendly and be forthcoming with feedback. The customers in the beta program must be fully aware that they are taking part in a pre–commercial release test program. The beta program should be contained within a set time frame. Like alpha testing, beta testing may go through multiple iterations. Often, the beta product is recovered once the test is completed. This allows a further degree of testing and review.

Certain industries may have additional specific methodologies for process/product approval. In the automotive industry, the production part approval process (PPAP) is used. Documents are used to demonstrate to the customer that the new product/process is ready for full-blown production and to obtain approval to release it for production. PPAP is used to show that suppliers have met all the engineering requirements based on a sample of parts produced from a production run made with production tooling, processes, and cycle times.

Chapter 28

D. Reliability

Define basic concepts such as mean time to failure (MTTF), mean time between failures (MTBF), mean time between maintenance actions (MTBMA), and mean time to repair (MTTR), and identify failure models such as bathtub curve, prediction, growth, etc. (Remember)

Body of Knowledge IV.D

Reliability is a major area in total quality control. When a customer purchases a product, that customer expects dependability and consistency. Consumers are looking for more than flashy, innovative products; they also expect the product to work correctly every time for a long period of time.

Reliability is a considerable part of the product development cycle. If reliability is planned early in the design stages of the product lifecycle (refer back to Figure 27.1, page 260), the greater the reliability will be to satisfy the customer and reduce field warranty issues. While deploying QFD, the team may be using VOC tools such as warranty data along with SPC data on similar parts and processes but also should be using reliability tools such as process failure mode and effects analysis (PFMEA), failure rates, and maintainability, as discussed further in this chapter.

BASIC CONCEPTS

Reliability—The probability that an item can perform its intended function for a specified interval under stated conditions.[1]

The definition consists of the following concepts:

- *Probability*. Since reliability is a probability, it is a variable that can be measured. Probability theory provides a foundation to describe the properties of a physical process mathematically.

- *Performance*. To calculate probability, a product (or "unit") must exist in one of two states: performs intended function or fails to perform intended function. A unit's failure condition must be clearly defined as it could be a degree of diminishing performance.

- *Conditions.* The stated conditions specify limits within which a unit should operate. Customers have some responsibility to use a unit within these limits; however, product designers must anticipate and design for stress conditions beyond those that can occur during proper use.

- *Time.* A specified interval is a period of use, or "mission time," that must be stated or implied to calculate reliability. The interval can be measured in hours, miles, cycles, or other units of product use.

There are some devices, sometimes referred to as one-shot devices, that do not have a traditional mission time. Examples are the crash sensor for an automobile air-bag system or the igniter for a rocket engine. These devices operate only when called upon, resulting in instant success or failure.

PROBABILITY DENSITY FUNCTIONS

Reliability engineers must know how a product fails in order to calculate, predict, and improve reliability. To understand how a product fails, a reliability engineer will reference an existing probability density function that describes the failure. Probability density functions show the probability of product failure with time or some other measure of product use such as miles or cycles. There are several distributions that reliability engineers can choose from when analyzing a product's reliability. Three of the more commonly used distributions are: normal, exponential, and Weibull.

BATHTUB CURVE

The bathtub curve, named for its shape, is a model used to describe failure patterns of a population over the entire life of a product. See Figure 28.1.

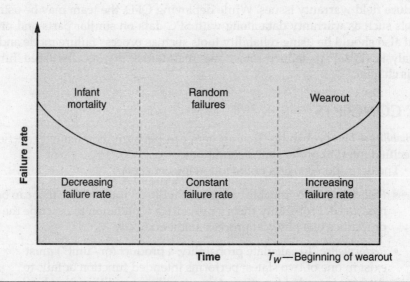

Figure 28.1 Bathtub curve.

The curve shows how fast product failures are taking place and if the failure rate is increasing, decreasing, or staying the same. The curve can be divided into three regions:

- The *infant mortality* period (also called the early life period or burn-in period) has a rapidly decreasing failure rate. These early life failures can be extracted through a process called burn-in. Burn-in can be performed at normal operating conditions or at higher than normal stress levels. Burn-in can serve two purposes: to weed out the units with nonconformities, hence improving the reliability of units delivered to the customer, and to investigate each failure to eliminate root causes and make design improvements. Burn-in is very costly. Methods to reduce early life failures include:

  - Proper vendor certification

  - Manufacturing and quality engineers participating in design reviews

  - Acceptance sampling

  - Use of process failure mode and effects analysis (PFMEA)

  - Use of continuous process capability studies and statistical process control procedures, including control charts

  - Design for manufacturability and assembly (DFMA)

- The *random failure* period (also called the useful life period) has a constant or flat failure rate. With a constant failure rate, the product's reliability is at its highest during this period, and product use and customer use are also at their highest. The exponential distribution (see Figure 28.2), having a constant failure rate, corresponds with that of the useful life period and so can be used to model a product's time to failure in this period of the lifecycle.

- The *wearout* period has an increasing failure rate signifying that there is an increasing probability of failure and that age is a factor in the probability of failure. Most items become more likely to fail with age due to accumulated wear. A good maintenance plan will require replacing a unit or appropriate parts of the unit just before it is predicted to enter the wearout period. If the shortest-lived part cannot be replaced, it becomes the determinant of the start time of the product's wearout phase. Engineers must take that into account when designing the product and determining a useful service life to the customer. Because the normal distribution has an increasing failure rate, it can be used to model a product's time to failure in the wearout period. Each region of the bathtub curve has an associated probability distribution and typical failure causes as listed in Table 28.1.

The Weibull distribution, named for Waloddi Weibull, is not associated with any single region of the bathtub curve as it could be used for any of the regions,

Part IV.D

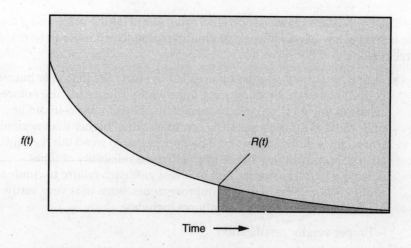

f(t)

R(t)

Time ⟶

Figure 28.2   Reliability can be modeled by the exponential distribution.
Source: R. A. Dovich, *Reliability Statistics* (Milwaukee: ASQ Quality Press, 1990): 13.

Table 28.1   Relationship between each bathtub phase, its probability distribution, and failure causes.

Bathtub curve period	Probability distribution	Typical failure causes
Infant mortality failures (Decreasing failure rate)	Weibull and decreasing exponential	Product does not meet specifications due to any number of causes: design, material, tooling, assembly, and so on
Random failures (Constant failure rate) Also known as the useful life period	Exponential and Weibull	Design factors (for example, operational stresses—situations where stress exceeds the strength of the unit)
Wearout failures (Increasing failure rate)	Normal or Weibull	Wear and tear on a product over time

depending on certain circumstances. The Weibull distribution is dependent on three parameters:

1. Shape parameter

2. Scale parameter

3. Location parameter

It is the shape parameter that provides the information to determine which phase of the lifecycle the product or system is in. It can assume the shape of distributions that have increasing, decreasing, or constant failure rates.[2]

EXPONENTIAL DISTRIBUTION

This is the most commonly used reliability distribution for predicting the probability of survival to time (t). See Figure 28.2:

The probability density function (pdf) of the exponential is:

$$f(t) = \lambda e^{-\lambda t} \text{ or } f(t) = (1/\theta)e^{-t/\theta}$$

where

$t \geq 0$

λ = failure rate

θ = mean time between failures (MTBF)

$\lambda = 1/\theta$

A reliability engineer can use the exponential distribution to calculate a product's reliability. The reliability for a given time (t) during the constant failure rate period can be calculated with the formula:

$$R_{(t)} = e^{-\lambda t}$$

where

e = base of the natural logarithms, which is 2.718281828 . . .

The *failure rate* (λ) is the rate at which failures occur for a population of units over a given time interval. It is calculated as the number of failures during the interval divided by the number of operating units at the beginning of the interval and by the length of the time interval (Dt). Several variables are involved. For example:

$$\lambda = \frac{r}{\Sigma t_i + (n-r)(T)}$$

Total number of units being tested is n

Total number of unit failures is r

Scheduled test time = T

Operating time until ith failure is t_i (i = 1 to r)

If 30 parts were being tested for 60 hours each and four parts failed at 20, 40, 45, and 50 hours,

$$\lambda = \frac{4}{20 + 40 + 45 + 50 + (26)(60) \text{ hours}}$$

$$\lambda = .002 / \text{hour}$$

Mean Time to Failure (MTTF)

This is a basic measure of system reliability, symbolized by θ, and referring to non-repairable items: the total number of life units of an item divided by the total number of failures within that population during a particular measurement interval under stated conditions.

Mean Time Between Failures (MTBF)

This is also a basic measure of system reliability, symbolized by θ, and referring to repairable items. MTBF is the mean number of life units during which all parts of the item perform within their specified limits during a particular measurement interval under stated conditions. To simplify:

$$\theta = T/r = 1/\lambda$$

where

T = total test time of all items for both failed and non-failed items

r = the total number of failures occurring during the test

λ = the failure rate

Maintainability

This is the ability of a product to be retained in, or restored to, a state in which it can perform service under the conditions for which it was designated; it is focused on reducing the duration of downtime. Maintainability is incorporated into system design during the early stages of product development.

Maintainability measures the amount of time a system is not in a functional state and is calculated based on the type of maintenance performed: corrective or preventive.

- *Corrective maintenance* consists of actions required to restore a system to its functional state after a failure occurs. It's quantified by the average time required to complete the maintenance—mean time to repair.

- *Preventive maintenance* consists of all actions required to retain a unit in its functional state and prevent a failure from occurring.

Mean Time between Maintenance Actions (MTBMA)

This is a measure of the system reliability parameter related to demand for maintenance manpower: the total number of system life units divided by the total number of maintenance actions (preventive and corrective) during a stated period of time.

Mean Time to Repair (MTTR)

This is a basic measure of maintainability: The sum of corrective maintenance times at any specific level of repair divided by the total number of failures within an item repaired at that level during a particular interval under stated conditions.

Availability

This is the probability that a unit will be ready for use at a stated time, or over a stated period of time, based on the combined aspects of reliability and maintainability. There are three types of availability:

1. *Inherent availability* is a function of reliability (MTBF) and maintainability (MTTR). To calculate inherent availability:

$$(A_I) = \text{MTBF}/(\text{MTBF} + \text{MTTR})$$

2. *Achieved availability* includes the measure of preventive and corrective maintenance. To calculate achieved availability:

$$(A_A) = \text{MTBMA}/(\text{MTBMA} + \text{MMT})$$

Where MMT = mean maintenance time = the sum of preventive and corrective maintenance times divided by the sum of scheduled and unscheduled maintenance events during a stated period of time.

3. *Operational availability* includes both the inherent and achieved availability and includes logistics and administrative downtime. To calculate operational availability:

$$(A_O) = \text{MBMA}/(\text{MBTMA} + \text{MDT})$$

Where MDT = mean maintenance downtime, which includes supply time, logistics time, administrative delays, active maintenance time, and so on.

Chapter 29

E. Supplier Management

Define and describe key measures of supplier performance (quality, price, delivery, level of service, etc.) and commonly used metrics (defect rates, functional performance, timeliness, responsiveness, technical support, etc.). (Understand)

Body of Knowledge IV.E

SUPPLIER MANAGEMENT PROCESSES

Supplier management consists of:

1. Selecting a supplier
2. Monitoring the supplier
3. Correcting the performance of a deficient supplier
4. Eliminating suppliers that are incapable or unwilling to supply what the buyer needs

Suppliers should be selected on the basis of expected performance. The best indicator of future supplier performance is past performance, but auditing the supplier's manufacturing processes is an excellent way to determine if the supplier is capable of producing what the buyer needs. Buyers can determine which supplier would best meet their needs by checking historical records of previous product and service deliveries, by auditing the potential or candidate supplier, or a combination of both these methods.

Companies often establish a list of preapproved suppliers who have provided satisfactory products and services in the past and/or who have been rated in a supplier rating process. Buyers may determine which suppliers should be listed in the preapproved supplier listing by either an internally developed rating system or by using external certification models such as the ISO 9000:2000 criteria.

After a supplier has been selected and an order placed for either products or services, the buyer should continue to monitor the supplier to ensure that it is delivering what is needed. The same criteria for rating the supplier's performance that were used to select suppliers should be employed to monitor ongoing supplier performance.

Some measures of supplier performance that can be used to select a supplier or to monitor a supplier's performance are, according to Bhote:[1]

1. *No rating.* The rationale is that the purchasing and quality departments know which companies are good or bad. No formal rating system is needed.

2. *Quality rating only.* A rating based on incoming inspection statistics.

3. *Quality and delivery rating: graphic methods.* A quality rating charted against the delivery rating.

4. *Quality and delivery rating: cost index method.* Use of a rating system based on a fixed dollar penalty for nonconformances.

5. *Comprehensive method.* The measuring and rating of variables such as quality, cost, delivery, service, and so on, with applied weightings.

In those instances when a supplier is found to be furnishing products or services that are substandard, the buyer can either direct the supplier to correct their performance or the buyer can eliminate the supplier. Directing the supplier to improve performance requires cooperation between the supplier and the buyer. In many instances the relationship between supplier and buyer is strained and adversarial, and in these cases it is difficult to get the supplier to make the necessary changes to improve product and/or service quality and delivery. This is one reason quality experts have always stressed the importance of developing collaborative relations between suppliers and buyers whenever possible.[2]

Many companies establish collaborative relationships with their suppliers to remove any barriers to excellent supplier performance. Buyer support to the supplier might include training supplier personnel or providing services that the supplier could not otherwise afford. Such support services to the supplier benefits the buyers as much or more than it does the supplier since it reduces the risk of receiving poor materials and services from the supplier. Collaborative relations include:

- Involving the supplier early in product design

- Supplier participation in planning product development

- Funding supplier development efforts that lead to better supplies

- Reimbursing supplier for reasonable supply development expenses

- Giving timely quality performance feedback

KEY MEASURES OF SUPPLIER PERFORMANCE

The buyer should develop a standard set of measures for assessing supplier performance. The following factors should be included:

- Quality

- Price
- Delivery
- Service
- Availability
- Impact of problems or deficient quality

SUPPLIER ASSESSMENT METRICS

The metrics in Table 29.1 can be useful in assessing supplier performance.[3]

Table 29.1   Supplier performance assessment metrics.

Quality metrics	Percent defective lots
	ppm (parts)
	DPMO (defects)
	Lots accepted versus lots rejected
	Percent nonconforming
	Special characteristic measurements (MTBF, ohms, and so on)
Timeliness metrics	Percent of on-time deliveries
	Percent of late deliveries
Delivery metrics	Percent of over-order quantity
	Percent of under-order quantity
	Percent of early delivery
Cost metrics	Dollars rejected/dollars purchased (or vice versa)
	Dollar value in nonconforming product (scrap, returns, rework)
	Dollar value of purchases (per reporting time segment)
Compliance metrics	Percent of reported quality information (of that required)
	Percent of supplied certifications (of that required)
Subjective metrics	Responsiveness of supplier service functions
	Timeliness and/or thoroughness of supplier technical assistance
	Responsiveness of supplier in providing price quotes
	Ability to follow instructions (in a variety of formats)

Chapter 30

F. Elements of Corrective and Preventive Action

> Identify elements of the corrective action process including containment, problem identification, root cause analysis, correction, recurrence prevention, verification and validation of effectiveness, and concepts of preventive action. (Understand)
>
> **Body of Knowledge IV.F**

Corrective actions are those actions that are taken to repair a problem after it has occurred and establish and implement a plan to prevent its recurrence. Preventive actions are those taken to avoid making a mistake or creating a defective product in the first place. Preventive measures might include the following:

- Systematic process improvement techniques
- Planning
- Project management practices
- Design reviews
- Capabilities studies
- Technology and usage forecasting
- Pilot projects
- Prototype, engineering model, and first-article testing
- Equipment maintenance and repair
- Reengineering
- Kaizen and reengineering
- Vendor evaluation and selection
- Field testing
- Procedure writing and reviews

- Safety reviews
- Personnel reviews
- Job descriptions
- Applicant screening
- Training
- Quality incentives
- Mistake-proofing (poka-yoke)
- Design improvement

In the past, some companies made large quantities of products and then sorted the good from the bad. Today, modern companies try to prevent making bad products altogether. Six Sigma, ISO 9000, and other quality management programs all stress the importance of preventing mistakes over repairing them. It is nearly always more costly to fix a defect and the situation that the defect caused than it is to avoid making the mistake or creating the defect. Prevention is always the preferred method of quality programs rather than corrective action. In fact, when corrective action must be taken, it is a good idea to learn from the experience and analyze the causes of the individual defect in order to develop practices and techniques for avoiding making the same mistakes again in the future on other products.

Corrective action begins once a problem has been identified. Corrective action includes the following steps:

1. Assign responsibility for correcting the situation
2. Assess the importance of the problem
3. Contain the negative effects resulting from the defect
4. Identify root causes
5. Identify corrective action
6. Implement the controls necessary to prevent recurrence
7. Dispose of any bad products

ASSIGN RESPONSIBILITY FOR CORRECTING THE SITUATION

Senior management is responsible for instituting a corrective and preventive action program. Corrective action begins when a quality-related problem is detected and continues until the detected problem is resolved and there is reasonable confidence that the problem will not recur. Upper management often delegates the corrective action for a new problem to the individuals and teams in the organization best suited to understand the problem and how to fix it. A *material review board* (MRB) or *corrective action board* (CAB) is often designated as the party responsible to begin the corrective and preventive activities resulting from the discovery of a new quality problem.

The MRB is a multidisciplinary panel of representatives established as a material review authority. The MRB reviews, evaluates, and directs either the disposal of or rework of nonconforming materials and services. The MRB treats immediate problems when a nonconformity is detected and influences corporate practices to initiate and invest in long-term improvements to the development of future products. The MRB initiates corrective actions that prevent the recurrence of problems by identifying substandard suppliers and sources, instituting better source selection, and tracking the use of all material, parts, and components.

Immediate actions the MRB might direct when a nonconforming material is detected may include analyzing the offending material to determine why it is defective (root cause analysis), assessing the seriousness of the defect by analyzing the impact the offending material has on the product's usability, and preventing delivery of the product until it has been reworked or a waiver is negotiated. Long-term actions the MRB might take when a defect is detected include determining if other products have employed the same nonconforming material, reassessing the rating of the supplier or source, determining if alternative sources or supplies of conforming materials exist, and instigating process improvement and product enhancement projects that will eliminate the continued use of the nonconforming material in the future.

In many companies, a CAB exists to ensure that appropriate corrective and preventive action is being taken by the responsible managers. The CAB typically reviews material review results, customer complaints, returned material and products, internal test rejections, internal and external audits, cost-of-quality results, and product reliability concerns in an attempt to detect quality problems with individual products or the processes that generate them. In some companies the CAB functions are performed by an executive committee and in others the board is given upper management's authority for quality issued but staffed by managers at lower levels.

Ultimately, the manager of the function that is primarily responsible for a quality problem is responsible for implementing the necessary corrective and preventive action. For example, if products are found to be defective due to the use of poor raw materials, the procurement department manager must ultimately accept the responsibility for improving material selection to avoid future problems. If, on the other hand, product defects are found to be the result of inattentive manufacturing machine operators, the manufacturing department manager must take the necessary steps to ensure greater attention of the workforce.

ASSESS THE IMPORTANCE OF THE PROBLEM

Assessing the impact of the quality problem that has been identified is a necessary step. If the problem is minor—that is, if it has little impact on the end user's ability to use the product—then the problem may be solved in a more leisurely fashion. If, alternatively, the problem is serious—that is, if it could cause serious harm to a user—then the problem must be dealt with immediately. Understanding the impact the problem will have on corporate revenue, operating costs, customer satisfaction, and public image can help the problem's solvers obtain the appropriate support they will need to solve the problem.

There are three classifications of defects:

- *Critical* defects are any material, part, component, or product condition that could cause hazardous or unsafe conditions for individuals using or maintaining the product and would prevent the product from performing its intended function.

- *Major* defects are any material, part, component, or product condition that is likely to fail or materially reduce the usability of a product from its intended purpose.

- *Minor* defects are departures from established standards having little or no significant bearing on the effective use or operation of the product and are not likely to reduce the usability of the product for its intended purpose.

CONTAIN THE NEGATIVE EFFECTS RESULTING FROM THE DEFECT

Corrective action is often divided into two steps. First, temporary measures are taken to remedy the situation in the short term, such as 100 percent inspection, repair at the customer's site, or additional processing. This short-term remedy is maintained until a long-term corrective action can be implemented. The second step, long-term remedy, consists of implementing preventive measures that look for problems in processing and actions that prevent or reduce the production of defective products.

Containment actions are measures taken to eliminate the impact defective products have on the business and customer. Defective products usually result in the customer's inability to use the product. Sometimes defective products result in warranty claims or lawsuits, and they often result in costs for reworking and reinspecting the reworked product. Actions taken to minimize the negative effects resulting from a defect are temporary and should not be considered a management policy.

Containment activities may include:

- Defining the problem as clearly as possible

- Forming a team of appropriate personnel to analyze and solve the problem

- Identifying solutions to the problem and determining the best one

- Ascertaining the feasibility of rapidly implementing the solution

- Identifying the long-term effect that rapid implementation will have on the company and customers

- Developing an immediate action plan including how to contain, repair, and inspect the corrected product

- Assigning the appropriate personnel to perform sorting and/or repairing the defective product and implementing the containment activities quickly

- Monitoring the implementation

- Documenting the containment activity and results

- Notifying everyone affected by the problem and the containment activity

IDENTIFY ROOT CAUSES

A cross-functional, cross-disciplinary team is usually best at determining the root causes of the defect. In many organizations this team is called the *corrective action team* (CAT).

Often the situations that led to the defective product are complex, and a single, simple root cause can not be identified. To find long-term solutions to the problem in order to prevent the defect from happening again it is necessary to explore the impact and relationship of many contributing factors. Using problem-solving techniques described elsewhere in this handbook, the team should determine the leading contributors to the problem and which ones are the most likely root causes.

Tools for determining the root causes of a defective product include:

- Five whys

- PDCA

- Operator observations

- Brainstorming

- Nominal group technique

- Six hats thinking exercises

- Ishikawa (fishbone) diagrams

- Process flow analysis

- Fault tree analysis

- Failure mode and effects analysis (FMEA)

- Check sheets

- Pareto analysis

- Subgrouping of data

- Regression analysis

- Data matrix analysis

- Partitioning of variation

- Control charting

- Design of experiments

- Analytical testing

- Trial testing

IDENTIFY CORRECTIVE ACTION

Having identified the root causes of the problem, the CAT develops solutions, actions that will, when taken together, remove the root causes of the problem leading to the defective product.

Corrective actions—those steps taken to remove the causes of identified defects—are dictated by the unique circumstances and conditions.

IMPLEMENT THE CONTROLS NECESSARY TO PREVENT RECURRENCE

This step involves creating an implementation plan and persuading the decision makers to make those changes necessary to both correct the current problem and to ensure that the problem does not recur in the future. To this end, the CAT should develop a convincing case to sell management on the necessary corrections. The following factors should be considered in creating a convincing argument for the necessary changes:

- The cost of the changes must be outweighed by the reduction in cost to correct future defects

- The recommended solution must be feasible in the current corporate environment

- The solutions must be consistent with the overall corporate mission and quality policies

- Those individuals who resist the changes should be given an opportunity to express their concerns, and the issues they raise must be dealt with honestly

- A best estimate of the cost of implementing the solution should be given, and the rationale upon which the estimate is made should be understood by the decision makers

- Hard data to support the CAT's recommendations should be presented but not used to overwhelm those managers who are not data-friendly

- The impact of the solution on other departments and functions in the organization should be thoroughly understood

DISPOSE OF ANY BAD PRODUCTS

Disposing of bad products may include reworking or repairing them, scrapping them, or using them as-is. If they are used as-is, the supplier must obtain a waiver from an appropriate authority, usually the customer, to use the nonconforming product.

Chapter 31

G. Material Identification, Status, and Traceability

Describe methodologies used for material identification and conformance status. Apply various methods of identifying and segregating nonconforming materials, and describe the requirements for preserving the identity of a product and its origin. (Apply)

[Note: Product recall procedures will not be included.]

Body of Knowledge IV.G

The quality department is traditionally responsible for inspecting products to determine whether or not they conform to the requirements stated in the contract, specifications, or other approved product description. In some companies the manufacturing department or a supplier may perform this quality inspection rather than the quality department. The inspector often marks the products to indicate the result of the inspection. There are three possible markings the inspector may apply to inspected products:

- *Control marking* indicates that the inspection complies with an inspection requirement indicated by a special requirement in the contract, drawing, or product description. These markings often include a unique serial number or symbol to tie the inspection to a specific inspector. For example, the marking stamp may say, "This product was inspected by Inspector 47."

- *Status/acceptance stamp* indicating the status and acceptability of the inspected product. Examples of this stamp are "incoming acceptance" stamp, "in-process" stamp, "sample piece identification" stamp, and "nonconforming" stamp.

- *Lot or group marking* indicating that a container of several usually small items have been inspected according to an established sampling method.

In most firms, the quality department maintains the inspection stamps to provide centralized control of them so they are not used inappropriately. The quality department stocks, issues, and maintains records of all inspection stamps and their identifying numbers and symbols.

One of the most important steps in any manufacturing process is to identify, segregate, and trace nonconforming materials. "Nonconforming" means that the requirements specified in the contract, drawing, or other approved product description are not being met. The term "nonconforming product" or "nonconformity" is synonymous with "defect," since both are defined as a departure from specified requirements.

A product may still be acceptable to the customer even if the material used to construct it does not fully conform to the specifications. Clothing manufacturers often sell their "seconds" (that is, products that have a slight mistake in the sewing or that use cloth with a slight unevenness) at special outlet stores. Shoppers at outlet stores usually pay considerably less than they would if they purchased the clothing at a retail store but they must inspect the clothing to ensure that they are not dissatisfied with the defects that might be present. Often, shoppers cannot detect the defects and are perfectly satisfied with the product in the seconds outlet store and are delighted with the lower price. But, of course, the manufacturer earns less for the sale of a second than they do when they sell a conforming product in a retail store.

There are three possible courses of action when it is determined that a product contains nonconforming materials:

1. Rework the product to replace the nonconforming material with conforming material.

2. Use the product "as-is." This action should only be taken with the complete knowledge and permission of a relevant authority, usually the customer.

3. Preclude the product's intended use, usually by scrapping the product.

The formal process for using a nonconforming product as-is is known as a waiver. The supplier should request a waiver from the customer only if it is believed that the defect is minor and the customer should approve or disapprove the proposed use of the nonconforming product.

ISO 9001:2000[1] specifies that the manufacturer or supplier must maintain records of nonconformites and the disposition of nonconforming material. If the nonconformity is corrected, the reworked product must be reinspected, marked, and recorded. If the nonconformity is discovered after the product has been delivered, the supplier must take appropriate action to mitigate real or potential damage that might result.

In addition, ISO/TS 16949[2] requires that:

• Product with unidentified or suspect status be classified as nonconforming

• Instructions for rework and reinspection be accessible and utilized

• Customers be informed promptly if nonconforming product has been shipped

- A customer waiver, concession, or deviation permit be obtained whenever processing differs from that currently approved

- Any product shipped on a deviation authorization be properly identified on each container

- Records of the expiration dates and quantities for any authorized deviations be maintained

- Upon expiration of any deviation authorization, the original specification and requirement must be met

- These waiver requirements be extended equally to any product supplied to an organization before submission to the (final) customer

The quality department is not the sole and final responsible agency for identifying, segregating, and tracing nonconforming material. Many functions in the organization are responsible for performing many of the activities that are necessary to identify and segregate nonconforming products and to rework or dispose of them. These may include the manufacturing, purchasing, finance, and marketing departments in addition to the quality department.

Chapter 32

Conclusion

This handbook has been designed as a primary resource to the Certified Quality Process Analyst Body of Knowledge (BoK). The first part of the handbook, Quality Basics, describes the elements of a quality system.

The second part develops the tools needed for problem solving and improvement, including various graphical ways to display and understand data, continuous improvement models, how to apply quality management and project management tools, project management, an explanation of the Taguchi loss function, and Taguchi's signal-to-noise ratios. This section concludes with lean methodology.

The next part, Data Analysis, provides the analytical methods to interpret and compare data sets and model processes. This section also describes variables sampling and attributes sampling, statistical process control (SPC), and basic statistical decision-making tools.

Customer–supplier relations is the last section of the BoK. An understanding of customer–supplier relations is crucial to the development of processes and metrics that serve to enhance customer satisfaction.

Of course, a true professional in this discipline, as in all other professions, must continue to learn and to develop his or her skills even after being certified. While this book is a good place to start on your journey, it does not contain all the information needed by a professional quality process analyst. Further reading and study is recommended, including the following texts:

Ellis R. Ott, Edward G. Schilling, Dean V. Neubauer. *Process Quality Control: Troubleshooting and Interpretation of Data*, 4th ed. Milwaukee: ASQ Quality Press, 2005.

Russell T. Westcott. *The Certified Manager of Quality/Organizational Excellence Handbook*, 3rd ed. Milwaukee: ASQ Quality Press, 2006.

Joseph M. Juran and Frank M. Gryna, eds. *Juran's Quality Control Handbook*, 4th ed. New York: McGraw-Hill, 1988.

ASQ's Foundations in Quality Self-Directed Learning Series, CQM. Milwaukee: ASQ Quality Press, 2001.

Part V
Appendices

Appendix A

Quality Process Analyst Body of Knowledge

Included in this body of knowledge are explanations (subtext) and cognitive levels for each topic or subtopic in the test. These details will be used by the Examination Development Committee as guidelines for writing test questions and are designed to help candidates prepare for the exam by identifying specific content within each topic that can be tested. Except where specified, the subtext is not intended to limit the subject or be all-inclusive of what might be covered in an exam but is intended to clarify how topics are related to the role of the Certified Quality Process Analyst (CQPA). The descriptor in parentheses at the end of each subtext entry refers to the highest cognitive level at which the topic will be tested. A more complete description of cognitive levels is provided at the end of this document.

I. Quality Basics (24 Questions)

    A. *ASQ code of ethics.* Identify appropriate behaviors for situations requiring ethical decisions. (Apply)

    B. *Quality planning.* Define a quality plan, understand its purpose for the organization as a whole and who in the organization contributes to its development. (Understand)

    C. *Cost of quality (COQ).* Describe and distinguish the classic COQ categories (prevention, appraisal, internal failure, external failure) and apply COQ concepts. (Apply)

    D. *Quality standards, requirements, and specifications.* Define and distinguish between quality standards, requirements, and specifications. (Understand)

    E. *Documentation systems.* Identify and describe common elements and different types of documentation systems such as configuration management, quality manual, document control, etc. (Understand)

    F. *Audits.*

        1. *Audit types.* Define and describe various audit types: internal, external, system, product, and process. (Understand)

2. *Audit process.* Describe various elements, including audit preparation, performance, record keeping, and closure. (Understand) [NOTE: Corrective action is covered in IV.F.]

3. *Roles and responsibilities.* Identify and define roles and responsibilities of audit participants (lead auditor, audit team member, client, and auditee). (Understand)

G. *Teams.*

1. *Types of teams.* Distinguish between various types of teams such as process improvement, work group, self-managed, temporary/ad hoc, cellular, etc. (Analyze)

2. *Team-building techniques.* Define basic steps in team-building such as introductory meeting for team members to share information about themselves, the use of ice-breaker activities to enhance team membership, the need for developing a common vision and agreement on team objectives, etc. (Apply)

3. *Roles and responsibilities.* Explain the various team roles and responsibilities, such as sponsor, champion, facilitator, team leader, and team member, and responsibilities with regard to various group dynamics, such as recognizing hidden agendas, handling distractions and disruptive behavior, keeping on task, etc. (Understand)

H. *Training components.* Define and describe methods that can be used to train individuals on new or improved procedures and processes, and use various tools to measure the effectiveness of that training, such as feedback from training sessions, end-of-course test results, on-the-job behavior or performance changes, department or area performance improvements, etc. (Understand)

II. Problem Solving and Improvement [23 questions]

A. *Basic quality tools.* Select, apply, and interpret these tools: flow charts, Pareto charts, cause-and-effect diagrams, check sheets, scatter diagrams, and histograms. (Analyze) [NOTE: The application of control charts is covered in section III.E.]

B. *Continuous improvement models.* Define and explain elements of plan–do–check–act (PDCA), kaizen, and incremental and breakthrough improvement. (Apply)

C. *Basic quality management tools.* Select and apply affinity diagrams, tree diagrams, process decision program charts, matrix diagrams, interrelationship digraphs, prioritization matrices, and activity network diagrams. (Apply)

D. *Project management tools.* Select and interpret scheduling and monitoring tools such as Gantt charts, program evaluation and review technique (PERT), critical path method (CPM), etc. (Analyze)

E. *Taguchi loss function.* Identify and describe Taguchi concepts and techniques such as signal-to-noise ratio, controllable and uncontrollable factors, and robustness. (Understand)

F. *Lean.* Identify and apply lean tools and processes, including setup reduction (SUR), pull (including just-in-time [JIT] and kanban), 5S, continuous flow manufacturing (CFM), value stream, poka-yoke, and total preventive/predictive maintenance (TPM) to reduce waste in areas of cost, inventory, labor, and distance. (Apply)

G. *Benchmarking.* Define and describe this technique and how it can be used to support best practices. (Understand)

III. Data Analysis [35 questions]

A. *Terms and definitions.*

1. *Basic statistics.* Define, compute, and interpret mean, median, mode, standard deviation, range, and variance. (Apply)

2. *Basic distributions.* Define and explain frequency distributions (normal, binomial, Poisson, and Weibull) and the characteristics of skewed and bimodal distributions. (Understand)

3. *Probability.* Describe and apply basic terms and concepts (independence, mutual exclusivity, etc.) and perform basic probability calculations. (Apply)

4. *Measurement scales.* Define and apply nominal, ordinal, interval, and ratio measurement scales. (Apply)

B. *Data types and collection methods.*

1. *Types of data.* Identify, define, and classify continuous (variables) data and discrete (attributes) data, and identify when it is appropriate to convert attributes data to variables measures. (Apply)

2. *Methods for collecting data.* Define and apply methods for collecting data such as using data coding, automatic gaging, etc. (Apply)

C. *Sampling.*

1. *Characteristics.* Identify and define sampling characteristics such as lot size, sample size, acceptance number, operating characteristic (OC) curve, etc. (Understand)

2. *Sampling methods.* Define and distinguish between various sampling methods such as random, sequential, stratified, fixed sampling, attributes and variables sampling, etc. (Understand)
[Note: Reading sampling tables is not required.]

D. *Measurement terms.* Define and distinguish between accuracy, precision, repeatability, reproducibility, bias, and linearity. (Understand)

E. *Statistical process control (SPC).*

1. *Techniques and applications.* Select appropriate control charts for monitoring or analyzing various processes and explain their construction and use. (Apply)

2. *Control limits and specification limits.* Identify and describe different uses of control limits and specification limits. (Understand)

3. *Variables charts.* Identify, select, construct, and interpret $\bar{\bar{X}} - R$ and $\bar{\bar{X}} - s$ charts. (Analyze)

4. *Attributes charts.* Identify, select, construct, and interpret p, np, c, and u charts. (Analyze)

5. *Rational subgroups.* Define and describe the principles of rational subgroups. (Understand)

6. *Process capability measures.* Define the prerequisites for measuring capability, and calculate and interpret C_p, C_{pk}, P_p, and P_{pk} in various situations. (Analyze)

7. *Pre-control chart.* Define the concept and use of pre-control charts. (Understand)

8. *Common and special cause variation.* Interpret various control chart patterns (runs, hugging, and trends) to determine process control, and use rules to distinguish between common cause and special cause variation. (Analyze)

9. *Data plotting.* Identify the advantages and limitations of analyzing data visually instead of numerically. (Understand)

F. *Regression and correlation.* Describe how regression and correlation models are used for estimation and prediction. (Apply)

G. *Hypothesis testing.* Determine and calculate confidence intervals using t tests and the z statistic, and determine whether the result is significant. (Analyze)
[Note: The F test is covered in area III.I.]

H. *Design of experiments (DOE).* Define basic terms such as blocking, randomization, etc. (Remember)

I. *Analysis of variance (ANOVA).* Define and determine the applicability of ANOVAs. (Understand)

IV. Customer–Supplier Relations [18 questions]

A. *Internal and external customers and suppliers.* Define and distinguish between internal and external customers and suppliers and their impact on products and services, and identify strategies for working with them to improve products, services, and processes. (Apply)

B. *Customer satisfaction analysis.* Describe the different types of tools used to gather and analyze customer feedback: surveys, complaint forms, warranty analysis, quality function deployment (QFD), etc. (Understand)

C. *Product/process approval systems.* Identify and describe how validation and qualification methods (alpha/beta testing, first-article, etc.) are used in new or revised products, processes, and services. (Understand)

D. *Reliability.* Define basic concepts such as mean time to failure (MTTF), mean time between failures (MTBF), mean time between maintenance actions (MTBMA), and mean time to repair (MTTR), and identify failure models such as bathtub curve, prediction, growth, etc. (Remember)

E. *Supplier management.* Define and describe key measures of supplier performance (quality, price, delivery, level of service, etc.) and commonly used metrics (defect rates, functional performance, timeliness, responsiveness, technical support, etc.). (Understand)

F. *Elements of corrective and preventive action.* Identify elements of the corrective action process including containment, problem identification, root cause analysis, correction, recurrence prevention, verification and validation of effectiveness, and concepts of preventive action. (Analyze)

G. *Material identification, status, and traceability.* Describe methodologies used for material identification and conformance status. Apply various methods of identifying and segregating nonconforming materials, and describe the requirements for preserving the identity of a product and its origin. (Apply)
[Note: Product recall procedures will not be included.]

LEVELS OF COGNITION
BASED ON BLOOM'S TAXONOMY (REVISED)

In addition to *content* specifics, the subtext detail also indicates the intended *complexity level* of the test questions for that topic. These levels are based on the Revised "Levels of Cognition" (from Bloom's Taxonomy, 2001) and are presented below in rank order, from least complex to most complex.

Remember

(Also commonly referred to as recognition, recall, or rote knowledge.) Be able to remember or recognize terminology, definitions, facts, ideas, materials, patterns, sequences, methodologies, principles, etc.

Understand

Be able to read and understand descriptions, communications, reports, tables, diagrams, directions, regulations, etc.

Apply

Be able to apply ideas, procedures, methods, formulas, principles, theories, etc., in job-related situations.

Analyze

Be able to break down information into its constituent parts and recognize the parts' relationship to one another and how they are organized; identify sublevel factors or salient data from a complex scenario.

Evaluate

Be able to make judgments regarding the value of proposed ideas, solutions, methodologies, etc., by using appropriate criteria or standards to estimate accuracy, effectiveness, economic benefits, etc.

Create

Be able to put parts or elements together in such a way as to show a pattern or structure not clearly there before; able to identify which data or information from a complex set is appropriate to examine further or from which supported conclusions can be drawn.

Appendix B
Areas under Standard Normal Curve

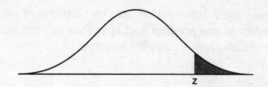

z	Area	z	Area	z	Area	z	Area	z	Area	z	Area	z	Area
0.00	0.5000	0.50	0.3085	1.00	0.1587	1.50	0.0668	2.00	0.0228	2.50	0.0062	3.00	1.35E-03
0.01	0.4960	0.51	0.3050	1.01	0.1562	1.51	0.0655	2.01	0.0222	2.51	0.0060	3.01	1.31E-03
0.02	0.4920	0.52	0.3015	1.02	0.1539	1.52	0.0643	2.02	0.0217	2.52	0.0059	3.02	1.26E-03
0.03	0.4880	0.53	0.2981	1.03	0.1515	1.53	0.0630	2.03	0.0212	2.53	0.0057	3.03	1.22E-03
0.04	0.4840	0.54	0.2946	1.04	0.1492	1.54	0.0618	2.04	0.0207	2.54	0.0055	3.04	1.18E-03
0.05	0.4801	0.55	0.2912	1.05	0.1469	1.55	0.0606	2.05	0.0202	2.55	0.0054	3.05	1.14E-03
0.06	0.4761	0.56	0.2877	1.06	0.1446	1.56	0.0594	2.06	0.0197	2.56	0.0052	3.06	1.11E-03
0.07	0.4721	0.57	0.2843	1.07	0.1423	1.57	0.0582	2.07	0.0192	2.57	0.0051	3.07	1.07E-03
0.08	0.4681	0.58	0.2810	1.08	0.1401	1.58	0.0571	2.08	0.0188	2.58	0.0049	3.08	1.04E-03
0.09	0.4641	0.59	0.2776	1.09	0.1379	1.59	0.0559	2.09	0.0183	2.59	0.0048	3.09	1.00E-03
0.10	0.4602	0.60	0.2743	1.10	0.1357	1.60	0.0548	2.10	0.0179	2.60	0.0047	3.10	9.68E-04
0.11	0.4562	0.61	0.2709	1.11	0.1335	1.61	0.0537	2.11	0.0174	2.61	0.0045	3.11	9.36E-04
0.12	0.4522	0.62	0.2676	1.12	0.1314	1.62	0.0526	2.12	0.0170	2.62	0.0044	3.12	9.04E-04
0.13	0.4483	0.63	0.2643	1.13	0.1292	1.63	0.0516	2.13	0.0166	2.63	0.0043	3.13	8.74E-04
0.14	0.4443	0.64	0.2611	1.14	0.1271	1.64	0.0505	2.14	0.0162	2.64	0.0041	3.14	8.45E-04
0.15	0.4404	0.65	0.2578	1.15	0.1251	1.65	0.0495	2.15	0.0158	2.65	0.0040	3.15	8.16E-04
0.16	0.4364	0.66	0.2546	1.16	0.1230	1.66	0.0485	2.16	0.0154	2.66	0.0039	3.16	7.89E-04
0.17	0.4325	0.67	0.2514	1.17	0.1210	1.67	0.0475	2.17	0.0150	2.67	0.0038	3.17	7.62E-04
0.18	0.4286	0.68	0.2483	1.18	0.1190	1.68	0.0465	2.18	0.0146	2.68	0.0037	3.18	7.36E-04
0.19	0.4247	0.69	0.2451	1.19	0.1170	1.69	0.0455	2.19	0.0143	2.69	0.0036	3.19	7.11E-04
0.20	0.4207	0.70	0.2420	1.20	0.1151	1.70	0.0446	2.20	0.0139	2.70	0.0035	3.20	6.87E-04
0.21	0.4168	0.71	0.2389	1.21	0.1131	1.71	0.0436	2.21	0.0136	2.71	0.0034	3.21	6.64E-04
0.22	0.4129	0.72	0.2358	1.22	0.1112	1.72	0.0427	2.22	0.0132	2.72	0.0033	3.22	6.41E-04

continued

continued

z	Area	z	Area	z	Area	z	Area	z	Area	z	Area	z	Area
0.23	0.4090	0.73	0.2327	1.23	0.1093	1.73	0.0418	2.23	0.0129	2.73	0.0032	3.23	6.19E-04
0.24	0.4052	0.74	0.2296	1.24	0.1075	1.74	0.0409	2.24	0.0125	2.74	0.0031	3.24	5.98E-04
0.25	0.4013	0.75	0.2266	1.25	0.1056	1.75	0.0401	2.25	0.0122	2.75	0.0030	3.25	5.77E-04
0.26	0.3974	0.76	0.2236	1.26	0.1038	1.76	0.0392	2.26	0.0119	2.76	0.0029	3.26	5.57E-04
0.27	0.3936	0.77	0.2206	1.27	0.1020	1.77	0.0384	2.27	0.0116	2.77	0.0028	3.27	5.38E-04
0.28	0.3897	0.78	0.2177	1.28	0.1003	1.78	0.0375	2.28	0.0113	2.78	0.0027	3.28	5.19E-04
0.29	0.3859	0.79	0.2148	1.29	0.0985	1.79	0.0367	2.29	0.0110	2.79	0.0026	3.29	5.01E-04
0.30	0.3821	0.80	0.2119	1.30	0.0968	1.80	0.0359	2.30	0.0107	2.80	0.0026	3.30	4.83E-04
0.31	0.3783	0.81	0.2090	1.31	0.0951	1.81	0.0351	2.31	0.0104	2.81	0.0025	3.31	4.67E-04
0.32	0.3745	0.82	0.2061	1.32	0.0934	1.82	0.0344	2.32	0.0102	2.82	0.0024	3.32	4.50E-04
0.33	0.3707	0.83	0.2033	1.33	0.0918	1.83	0.0336	2.33	0.0099	2.83	0.0023	3.33	4.34E-04
0.34	0.3669	0.84	0.2005	1.34	0.0901	1.84	0.0329	2.34	0.0096	2.84	0.0023	3.34	4.19E-04
0.35	0.3632	0.85	0.1977	1.35	0.0885	1.85	0.0322	2.35	0.0094	2.85	0.0022	3.35	4.04E-04
0.36	0.3594	0.86	0.1949	1.36	0.0869	1.86	0.0314	2.36	0.0091	2.86	0.0021	3.36	3.90E-04
0.37	0.3557	0.87	0.1922	1.37	0.0853	1.87	0.0307	2.37	0.0089	2.87	0.0021	3.37	3.76E-04
0.38	0.3520	0.88	0.1894	1.38	0.0838	1.88	0.0301	2.38	0.0087	2.88	0.0020	3.38	3.62E-04
0.39	0.3483	0.89	0.1867	1.39	0.0823	1.89	0.0294	2.39	0.0084	2.89	0.0019	3.39	3.50E-04
0.40	0.3446	0.90	0.1841	1.40	0.0808	1.90	0.0287	2.40	0.0082	2.90	0.0019	3.40	3.37E-04
0.41	0.3409	0.91	0.1814	1.41	0.0793	1.91	0.0281	2.41	0.0080	2.91	0.0018	3.41	3.25E-04
0.42	0.3372	0.92	0.1788	1.42	0.0778	1.92	0.0274	2.42	0.0078	2.92	0.0018	3.42	3.13E-04
0.43	0.3336	0.93	0.1762	1.43	0.0764	1.93	0.0268	2.43	0.0075	2.93	0.0017	3.43	3.02E-04
0.44	0.3300	0.94	0.1736	1.44	0.0749	1.94	0.0262	2.44	0.0073	2.94	0.0016	3.44	2.91E-04
0.45	0.3264	0.95	0.1711	1.45	0.0735	1.95	0.0256	2.45	0.0071	2.95	0.0016	3.45	2.80E-04
0.46	0.3228	0.96	0.1685	1.46	0.0721	1.96	0.0250	2.46	0.0069	2.96	0.0015	3.46	2.70E-04
0.47	0.3192	0.97	0.1660	1.47	0.0708	1.97	0.0244	2.47	0.0068	2.97	0.0015	3.47	2.60E-04
0.48	0.3156	0.98	0.1635	1.48	0.0694	1.98	0.0239	2.48	0.0066	2.98	0.0014	3.48	2.51E-04
0.49	0.3121	0.99	0.1611	1.49	0.0681	1.99	0.0233	2.49	0.0064	2.99	0.0014	3.49	2.42E-04

Appendix C
Control Limit Formulas

VARIABLES CHARTS

$\bar{x}$ and R chart: Averages chart: $\bar{\bar{x}} \pm A_2 \bar{R}$ Range chart: $LCL = D_3 \bar{R}$ $UCL = D_4 \bar{R}$

$\bar{x}$ and s chart: Averages chart: $\bar{\bar{x}} \pm A_3 \bar{s}$ Std. dev. chart: $LCL = B_3 \bar{s}$ $UCL = B_4 \bar{s}$

Individuals and moving range chart (two-value moving window):

Individuals chart: $\bar{x} \pm 2.66 \bar{R}$ Moving range: $UCL = 3.267 \bar{R}$

Moving average and moving range (two-value moving window):

Moving average: $\bar{\bar{x}} \pm 1.88 \bar{R}$ Moving range: $UCL = 3.267 \bar{R}$

ATTRIBUTE CHARTS

p chart: $\bar{p} \pm 3 \sqrt{\dfrac{\bar{p}(1-\bar{p})}{n}}$

np chart : $n\bar{p} \pm 3 \sqrt{n\bar{p}(1-\bar{p})}$

c chart: $\bar{c} \pm 3\sqrt{\bar{c}}$

u chart: $\bar{u} \pm 3 \sqrt{\dfrac{\bar{u}}{n}}$

Appendix D

Factors for Control Charts

Obs	Chart for averages			Chart for standard deviations					
	Factors for control limits			Factors for central line		Factors for control limits			
n	A	A_2	A_3	c_4	$1/c_4$	B_3	B_4	B_5	B_6
2	2.121	1.881	2.659	0.7979	1.2533	0	3.267	0	2.606
3	1.732	1.023	1.954	0.8862	1.1284	0	2.568	0	2.276
4	1.500	0.729	1.628	0.9213	1.0854	0	2.266	0	2.088
5	1.342	0.577	1.427	0.9400	1.0638	0	2.089	0	1.964
6	1.225	0.483	1.287	0.9515	1.0509	0.030	1.970	0.029	1.874
7	1.134	0.419	1.182	0.9594	1.0424	0.118	1.882	0.113	1.806
8	1.061	0.373	1.099	0.9650	1.0362	0.185	1.815	0.179	1.751
9	1.000	0.337	1.032	0.9693	1.0317	0.239	1.761	0.232	1.707
10	0.949	0.308	0.975	0.9727	1.0281	0.284	1.716	0.276	1.669
11	0.905	0.285	0.927	0.9754	1.0253	0.321	1.679	0.313	1.637
12	0.866	0.266	0.886	0.9776	1.0230	0.354	1.646	0.346	1.610
13	0.832	0.249	0.850	0.9794	1.0210	0.382	1.618	0.374	1.585
14	0.802	0.235	0.817	0.9810	1.0194	0.406	1.594	0.399	1.563
15	0.775	0.223	0.789	0.9823	1.0180	0.428	1.572	0.421	1.544
16	0.750	0.212	0.763	0.9835	1.0168	0.448	1.552	0.440	1.526
17	0.728	0.203	0.739	0.9845	1.0157	0.466	1.534	0.458	1.511
18	0.707	0.194	0.718	0.9854	1.0148	0.482	1.518	0.475	1.496
19	0.688	0.187	0.698	0.9862	1.0140	0.497	1.503	0.490	1.483
20	0.671	0.180	0.680	0.9869	1.0132	0.510	1.490	0.504	1.470
21	0.655	0.173	0.663	0.9876	1.0126	0.523	1.477	0.516	1.459
22	0.640	0.167	0.647	0.9882	1.0120	0.534	1.466	0.528	1.448
23	0.626	0.162	0.633	0.9887	1.0114	0.545	1.455	0.539	1.438
24	0.612	0.157	0.619	0.9892	1.0109	0.555	1.445	0.549	1.429
25	0.600	0.153	0.606	0.9896	1.0105	0.565	1.435	0.559	1.420

Continued

Continued

Obs	Chart for ranges						
	Factors for central line			Factors for control limits			
n	d_2	$1/d_2$	d_3	D_1	D_2	D_3	D_4
2	1.128	0.8865	0.853	0	3.686	0	3.267
3	1.693	0.5907	0.888	0	4.358	0	2.575
4	2.059	0.4857	0.880	0	4.698	0	2.282
5	2.326	0.4299	0.864	0	4.918	0	2.114
6	2.534	0.3946	0.848	0	5.079	0	2.004
7	2.704	0.3698	0.833	0.205	5.204	0.076	1.924
8	2.847	0.3512	0.820	0.388	5.307	0.136	1.864
9	2.970	0.3367	0.808	0.547	5.394	0.184	1.816
10	3.078	0.3249	0.797	0.686	5.469	0.223	1.777
11	3.173	0.3152	0.787	0.811	5.535	0.256	1.744
12	3.258	0.3069	0.778	0.923	5.594	0.283	1.717
13	3.336	0.2998	0.770				
14	3.407	0.2935	0.763				
15	3.472	0.2880	0.756				
16	3.532	0.2831	0.750				
17	3.588	0.2787	0.744				
18	3.640	0.2747	0.739				
19	3.689	0.2711	0.733				
20	3.735	0.2677	0.729				
21	3.778	0.2647	0.724				
22	3.819	0.2618	0.720				
23	3.858	0.2592	0.716				
24	3.895	0.2567	0.712				
25	3.931	0.2544	0.708				

Endnotes

Chapter 2

1. A. V. Feigenbaum, *Total Quality Control*, 3rd ed. 40th Anniversary Ed. (New York: McGraw-Hill, 1991).
2. K. Ishikawa, *What Is Total Quality Control? The Japanese Way* (Englewood Cliffs, NJ: Prentice Hall, 1985).
3. B. L. Wortman, *CQPA Primer* (Terre Haute, IN: Quality Council of Indiana, 2006).
4. J. M. Juran, *Juran on Quality by Design* (New York: Free Press, 1992).

Chapter 3

1. ASQ Quality Costs Committee, *Principles of Quality Costs: Principles, Implementation, and Use*, 3rd ed., J. Campanella, ed. (Milwaukee: ASQ Quality Press, 1999): 3–5.

Chapter 4

1. http://en.wikipedia.org/wiki/Quality_assurance.

Chapter 5

1. International Organization for Standardization, ISO 9001:2000, *Quality management systems—Requirements* (Geneva: ISO, 2000).
2. IEEE, ANSI/IEEE Standard 828-1990, *IEEE Standard for Software Configuration Management Plans* (New York: IEEE, 1990).
3. ISO, ISO 9001:2000.

Chapter 6

1. ASQ Quality Audit Division, *The ASQ Auditing Handbook*, 3rd ed. (Milwaukee: ASQ Quality Press, 2006).

Chapter 7

1. D. Okes and R. T. Westcott, eds., *The Certified Quality Manager Handbook*, 2nd ed. (Milwaukee: ASQ Quality Press, 2001): 37–41.
2. R. T. Westcott, *The Certified Manager of Quality/Organizational Excellence Handbook*, 3rd ed. (Milwaukee: ASQ Quality Press, 2006): 61–82.

3. C. C. Manz and H. P. Sims, Jr., *Business without Bosses: How Self-Managed Teams Are Building High-Performance Companies* (New York: John Wiley & Sons, 1993).
4. Westcott, *CMQ/OE Handbook*.
5. Ibid.
6. Ibid.
7. Ibid.
8. Ibid., pp. 71–73.
9. Ibid.
10. American Society for Quality, *ASQ's Foundations in Quality Self-Directed Learning Series,* Certified Quality Manager, Module 1 (Milwaukee: ASQ Quality Press, 2001).
11. Westcott, *CMQ/QE Handbook*.
12. American Society for Quality, *The Quality Improvement Handbook* (Milwaukee: ASQ Quality Press, 2006): Chapter 4.
13. Ibid.
14. P. R. Scholtes, *The Team Handbook* (Madison, WI: Joiner Associates Consulting Group, 1988): 6–37.
15. Ibid, pp. 6–24

Chapter 8

1. R. T. Westcott, *The Certified Manager of Quality/Organizational Excellence Handbook,* 3rd ed. (Milwaukee: ASQ Quality Press, 2006): Appendix B.
2. Ibid.
3. D. L. Kirkpatrick, "Techniques for Evaluating Training Programs," a four-part series beginning in the November 1959 issue of the *Training Director's Journal*.
4. D. G. Robinson and J. C. Robinson, *Training for Impact: How to Link Training to Business Needs and Measure the Results* (San Francisco: Jossey-Bass, 1990): 168.
5. R. T. Westcott, "A Quality System Needs Assessment," in *In Action: Conducting Needs Assessment (17 Case Studies),* J. J. Phillips and E. F. Holton III, eds. (Alexandria, VA: American Society for Training and Development, 1995): 235; "Applied Behavior Management Training," in *Action: Measuring Return on Investment,* vol. 1, J. Phillips, ed. (Alexandria, VA: American Society for Training and Development, 1994): 85–104; "Behavior Management Training," *Human Resources Management and Development Handbook,* 2nd ed. (New York: AMACOM, 1994): 897–911; and "ROQI: Overlooked Quality Tool," *The Total Quality Review* (November/December, 1994): 37–44.
5. J. J. Phillips, *The ASTD Training and Development Handbook,* 1st ed. (New York: McGraw-Hill, 1996).
6. Ibid.
7. J. M. Juran and F. M. Gryna, eds., *Juran's Quality Control Handbook,* 4th ed. (New York: McGraw-Hill, 1988): 11.9–11.10.

Chapter 9

1. N. R. Tague, *The Quality Toolbox,* 2nd ed. (Milwaukee: ASQ Quality Press, 2004): 292–99.

Chapter 10

1. http://www.asq.org/learn-about-quality/project-planning-tools/links-resources/audiocasts.html.
2. V. E. Sower and F. K. Fair, "There Is More to Quality than Continuous Improvement: Listening to Plato," *Quality Management Journal* 12, no. 1 (January 2005): 8–20.

3. D. W. Benbow and T. M. Kubiak, *The Certified Six Sigma Black Belt Handbook* (Milwaukee: ASQ Quality Press, 2005).
4. Ibid.

Chapter 11

1. Adapted from N. R. Tague, *The Quality Toolbox,* 2nd ed. (Milwaukee: ASQ Quality Press, 2004).
2. Ibid, p. 444–46.
3. Ibid, p. 338–44.

Chapter 12

1. Project Management Institute, *Guide to the Project Management Body of Knowledge,* 3rd ed. (Newton Square, PA: PMI Publications, 2004).
2. Ibid.
3. Ibid.

Chapter 13

1. G. Taguchi, *Introduction to Quality Engineering* (White Plains, NY: UNIPUB, Kraus International Publications, Asian Productivity Organization, 1986).

Chapter 14

1. R. T. Westcott, *The Certified Manager of Quality/Organizational Excellence Handbook,* 3rd ed. (Milwaukee: ASQ Quality Press, 2006).
2. G. Alukal, "Create a Lean, Mean Machine," *ASQ Quality Progress* 36, no. 4 (April, 2003): 29–35.
3. The seven basic steps are an excerpt from http://en.wikipedia.org/wiki/SMED.
4. S. Shingo, *A Revolution in Manufacturing: The SMED System* (Portland, OR: Productivity Press, 1985): 26–52.
5. T. J. Shuker, "The Leap to Lean" in *54th Annual Quality Congress Proceedings* (Indianapolis, IN: 2000): 105–12.
6. Ibid.
7. R. T. Westcott, *The Certified Manager of Quality/Organizational Excellence Handbook,* 3rd ed. (Milwaukee: ASQ Quality Press, 2006): Appendix B.
8. Ibid.
9. Ibid.
10. Ibid., p. 390.
11. Ibid, Appendix B.
12. P. Bayers, "Apply Poka-Yoke Devices Now to Eliminate Defects," in *51st Annual Quality Congress Proceedings* (May 1997): 451–55.
13. A. Douglas, "Improving Manufacturing Performance," *56th Annual Quality Congress Proceedings* (Denver, CO: 2002): 725–32.

Chapter 15

1. D. L. Goetsch and S. B. Davis, *Quality Management: Introduction to Total Quality Management for Production, Processing, and Services,* 3rd ed. (Upper Saddle River, NJ: Prentice Hall, 2000).

Chapter 16

1. http://en.wikipedia.org/wiki/Weibull_distribution.
2. J. Mitchell, "Measurement Scales and Statistics: A Clash of Paradigms," *Psychological Bulletin* 3 (1986): 398–407.
3. P. F. Velleman and L. Wilkinson, "Nominal, Ordinal, Interval, and Ratio Typologies Are Misleading," *The American Statistician* 47, no. 1 (1993): 65–72. http://www.spss.com/research/wilkinson/Publications/Stevens.pdf.

Chapter 18

1. W. E. Deming, *Out of the Crisis* (Cambridge, MA: M. I. T. Center for Advanced Engineering, 1986).
2. ASQC Statistics Division, *Glossary and Tables for Statistical Quality Control*, 2nd ed. (Milwaukee: ASQC Quality Press, 1983).
3. Ibid.
4. Deming, *Out of the Crisis*.
5. http://en.wikipedia.org/wiki/Sampling_%28statistics%29.

Chapter 19

1. J. A. Simpson, "Foundations of Metrology," *Journal of Research of the National Bureau of Standards* 86, no. 3 (May/June 1981): 36–42.
2. W. J. Darmody, "Elements of a Generalized Measuring System," in *Handbook of Industrial Metrology* (Englewood Cliffs, NJ: Prentice Hall, 1967).
3. Ibid.
4. A. F. Rashed and A. M. Hamouda, *Technology for Real Quality* (Alexandria, Egypt: Egyptian University House, 1974).
5. ASQC Statistics Division, *Glossary and Tables for Statistical Quality Control*, 2nd ed. (Milwaukee: ASQC Quality Press, 1983).
6. American Society for Testing and Materials, *ASTM Standards on Precision and Accuracy for Various Applications* (Philadelphia: ASTM, 1977).
7. A. McNish, "The Nature of Measurement," in *Handbook of Industrial Metrology* (Englewood Cliffs, NJ: Prentice Hall, 1967).
8. D. W. Benbow, A. K. Elshennawy, and H. F. Walker, *The Certified Quality Technician Handbook* (Milwaukee: ASQ Quality Press, 2003): 70.
9. Ibid, pp. 121–23.
10. E. R. Ott, et al., *Process Quality Control*, 4th ed. (Milwaukee: ASQ Quality Press, 2005): 526.

Chapter 20

1. W. E. Deming, *Quality, Productivity, and Competitive Position* (Cambridge, MA: Massachusetts Institute of Technology, 1982).
2. D. J. Wheeler, "Normality and the Process Behavior Chart," www.spcpress.com.
3. http://en.wikipedia.org/wiki/Cp_index.
4. http://en.wikipedia.org/wiki/Ppk_index.
5. http://en.wikipedia.org/wiki/Cpk_index.
6. D. C. Montgomery, *Introduction to Statistical Quality Control*, 4th ed. (New York: John Wiley & Sons, 2001).
7. E. R. Ott, et. al., *Process Quality Control*, 4th ed. (Milwaukee: ASQ Quality Press, 2005).

Chapter 21

1. http://en.wikipedia.org/wiki/Regression_analysis.
2. J. Cohen, P. Cohen, S. G. West, and L. S. Aiken, *Applied Multiple Regression/Correlation Analysis for the Behavioral Sciences*, 2nd ed. (Hillsdale, NJ: Lawrence Erlbaum Associates, 2003).
3. http://en.wikipedia.org/wiki/Correlated.

Chapter 22

1. J. H. Zar, *Biostatistical Analysis* (Englewood Cliffs, NJ: Prentice Hall International, 1984): 43–45.
2. R. L. Ott, *An Introduction to Statistical Methods and Data Analysis*, 4th ed. (Belmont, CA: Duxbury Press, 1993): 86.
3. http://en.wikipedia.org/wiki/Student%27s_t-distribution.
4. http://en.wikipedia.org/wiki/Hypothesis_testing.
5. http://en.wikipedia.org/wiki/Null_hypothesis.
6. http://en.wikipedia.org/wiki/Statistical_significance.
7. http://en.wikipedia.org/wiki/P-value.

Chapter 23

1. http://en.wikipedia.org/wiki/Design_of_experiments.
2. G. E. P. Box and R. D. Meyer, "An Analysis for Unreplicated Fractional Factorials," *Technometrics* 28, no. 1 (February 1986): 11–18.

Chapter 24

1. H. R. Lindman, *Analysis of Variance in Complex Experimental Designs* (San Francisco: W. H. Freeman & Co, 1974).
2. E. R. Ott, E. G. Schilling, and D. V. Neubauer, *Process Quality Control: Troubleshooting and Interpretation of Data*, 4th ed. (Milwaukee: ASQ Quality Press, 2005): 440.

Chapter 25

1. J. M. Juran, *Juran's Quality Handbook*, 5th ed. (New York: McGraw-Hill, 1999): 2.3.
2. R. T. Westcott, *The Certified Manager of Quality/Organizational Excellence Handbook*, 3rd ed. (Milwaukee: ASQ Quality Press, 2006): Appendix B.
3. J. E. Bauer, G. L. Duffy, and R. T. Westcott, *The Quality Improvement Handbook*, 2nd ed. (Milwaukee: ASQ Quality Press, 2007): Chapter 10.
4. *ASQ's Foundations in Quality Self-Directed Learning Series*, Certified Quality Manager, Module 4 (Milwaukee: ASQ Quality Press, 2001).
5. A. V. Feigenbaum and D. S. Feigenbaum, "The Power of Management Capital," *Quality Progress* 37, no. 11 (November 2004): 24–29.
6. Bauer, et al., *The Quality Improvement Handbook*.
7. R. T. Westcott, "Quality Level Agreements for Clarity of Expectations," *The Informed Outlook* (December 1999).
8. Westcott, *The Certified Manager of Quality*, pp. 456–457.
9. Ibid, p. 457.
10. R. Lawton, "8 Dimensions of Excellence," *Quality Progress* (April 2006).
11. Bauer, et al., *The Quality Improvement Handbook*.
12. Lawton, "8 Dimensions."

Chapter 26

1. M. LeBoeuf, *How to Win Customers and Keep Them for Life* (New York: G. P. Putnam & Sons, 1987).
2. Ibid.
3. N. Scriabina and S. Fomichov, "Six Ways to Benefit from Customer Complaints," *Quality Progress* 38, no. 9 (September 2005): 49–54.
4. *ASQ's Foundations in Quality Self-Directed Learning Series*, Certified Quality Manager, Module 4 (Milwaukee: ASQ Quality Press, 2001): 4-47.
5. Ibid, p. 4-86.

Chapter 27

1. J. M. Juran, *Juran on Planning for Quality* (New York: The Free Press, 1988).
2. International Organization for Standardization, ISO 9001:2000, *Quality management systems—Requirements* (Geneva: ISO, 2000).
3. Ibid.
4. Ibid.
5. http://www.fda.gov/cder/guidance/pv.htm.
6. Ibid.
7. ISO, ISO 9001:2000.

Chapter 28

1. MIL-STD-721, *Definitions of Terms for Reliability and Maintainability*, Revision C (Washington, DC: Department of Defense, June 12, 1981).
2. For additional information, see W. G. Ireson, C. F. Coombs, and R. Y. Moss, *Handbook of Reliability Engineering and Management*, 2nd ed. (New York: McGraw-Hill, 1996): 25.32–25.33.

Chapter 29

1. K. R. Bhote, *Strategic Supply Management* (New York: AMACOM, 1989).
2. J. B. Ayers, *Supply Chain Project Management: A Structured Collaborative and Measurable Approach* (Boca Raton, FL: St. Lucie Press, 2003).
3. B. L. Wortman, *CQPA Primer* (Terre Haute, IN: Quality Council of Indiana, 2006).

Chapter 31

1. American Society for Quality, ANSI/ISO/ASQ Q9001-2000, *Quality management systems—Requirements* (Milwaukee: ASQ Quality Press, 2000).
2. International Organization for Standardization, ISO/TS 16949:2002, *Quality management systems—Particular requirements for the application of ISO 9001:2000 for automotive production and relevant service part organizations*, 2nd ed. (Geneva: International Organization for Standardization, 2002).

Glossary

1-α—See *confidence level*.

1-β—The power of testing a hypothesis is 1-β. It is the probability of correctly rejecting the null hypothesis, H_0.

2^k factorial design—A factorial design in which k factors are studied, each at exactly two levels.

100 percent inspection—An inspection of selected characteristics of every item under consideration.

A

α (alpha)—1: The maximum probability, or risk, of making a *Type I error* when dealing with the *significance level* of a test. 2: The *probability* or risk of incorrectly deciding that a shift in the process mean has occurred when the process is unchanged when referring to α in general or as the *p-value* obtained in a test). 3: α_i is usually designated as producer's risk.

Ac—See *acceptance number*.

acceptable process level (APL)—The process level that forms the outer boundary of the zone of acceptable processes. (A process located at the APL will have only a *probability of rejection* designated a when the plotted statistical measure is compared to the acceptance control limits.)

Note: In the case of two-sided tolerances, upper and lower acceptable process levels will be designated UAPL and LAPL. (These need not be symmetrical around the standard level.)

acceptance (control chart or acceptance control chart usage)—A decision that the *process* is operating in a satisfactory manner with respect to the *statistical measure* being plotted.

acceptance control chart—A *control chart* intended primarily to evaluate whether or not the plotted measure can be expected to satisfy specified *tolerances*.

acceptance control limit (ACL)—*Control limits* for an *acceptance control chart* that permit some assignable shift in process level based on specified requirements, provided within-*subgroup* variability is in a *state of statistical control*.

acceptance number (Ac)—The largest number of *nonconformities* or nonconforming items found in the *sample* by *acceptance sampling inspection by attributes* that permits the acceptance of the lot as given in the *acceptance sampling plan*.

acceptance quality limit (AQL)—The AQL is the *quality* level that is the worst tolerable product average when a continuing series of lots is submitted for *acceptance sampling*.

Note 1: This concept only applies when an *acceptance sampling scheme* with rules for switching and for discontinuation is used.

Note 2: Although individual lots with quality as bad as the acceptance quality limit can be accepted with fairly high probability, the designation of an acceptance quality limit does not suggest that this is a desirable quality level.

Note 3: *Acceptance sampling schemes* found in standards, with their rules for switching and for discontinuation of sampling inspection, are designed to encourage suppliers to have process averages consistently better than the acceptance quality limit. If suppliers fail to do so, there is a high probability of being switched from *normal inspection* to *tightened inspection*, where *lot* acceptance becomes more difficult. Once on tightened inspection, unless corrective action is taken to improve product quality, it is very likely that the rule requiring discontinuance of sampling inspection will be invoked.

Note 4: The use of the abbreviation AQL to mean acceptable quality level is no longer recommended since modern thinking is that no fraction defective is really acceptable. Using "acceptance quality limit" rather than "acceptable quality level" indicates a technical value where acceptance occurs.

acceptance sampling—A sampling after which decisions are made to accept or not to accept a *lot,* or other grouping of products, materials, or services, based on sample results.

acceptance sampling inspection—An acceptance *inspection* where the acceptability is determined by sampling *inspection*.

acceptance sampling inspection by attributes—An *acceptance sampling inspection* whereby the presence or absence of specified *characteristics* of each item in a sample is observed to statistically establish the acceptability of a *lot* or *process*.

acceptance sampling inspection by variables—An *acceptance sampling inspection* in which the acceptability of a *process* is determined statistically from measurements on specified quality *characteristics* of each item in a *sample* from a *lot*.

Note: Lots taken from an acceptable process are assumed to be acceptable.

acceptance sampling inspection system—A collection of *acceptance sampling plans* or *acceptance sampling schemes* together with criteria by which appropriate plans or schemes may be chosen.

acceptance sampling plan—A plan that states the *sample size(s)* to be used and the associated criteria for *lot* acceptance.

acceptance sampling procedure—The operational requirements and/or instructions related to the use of a particular *acceptance sampling plan*.

Note: This covers the planned method of selection, withdrawal, and preparation of *sample(s)* from a *lot* to yield knowledge of the *characteristic(s)* of the lot.

acceptance sampling scheme—The combination of *acceptance sampling plans* with *switching rules* for changing from one plan to another.

acceptance sampling system—A collection of sampling schemes.

accessibility—A measure of the relative ease of admission to the various areas of an item for the purpose of operation or maintenance.

accreditation—Certification, by a duly recognized body, of the facilities, capability, objectivity, competence, and integrity of an agency, service, or operational group or individual to provide the specific service or operation needed. For example, the Registrar Accreditation Board (U.S.) accredits those organizations that register companies to the ISO 9000 series standards.

accuracy—The closeness of agreement between a test result or measurement result and the true value.

ACL—See *acceptance control limit*.

ACSI—The American Customer Satisfaction Index is an economic indicator, a cross-industry measure of the satisfaction of U.S. customers with the quality of the goods and services available to them—both those goods and services produced within the United States and those provided as imports from foreign firms that have substantial market shares or dollar sales.

action limits—The *control chart control limits* (for a process *in a state of statistical control*) beyond which there is a very high probability that a value is not due to chance. When a measured value lies beyond an action limit, appropriate corrective action should be taken on the process.

Example: Typical action limits for an $\bar{x}$ *chart* are $\pm 3\hat{\sigma}$ (three *standard deviations*).

action plan—The detailed plan to implement the actions needed to achieve strategic goals and objectives (similar to, but not as comprehensive as a project plan).

active listening—Paying attention solely to what others are saying (for example, rather than what you think of what they're saying or what you want to say back to them).

activity network diagram (AND)—See *arrow diagram*.

ad hoc team—See *temporary team*.

ADDIE—An instructional design model (analysis, design, development, implementation, and evaluation).

adult learning principles—Key principles about how adults learn that impact how education and training of adults should be designed.

affinity diagram—A management and planning tool used to organize ideas into natural groupings in a way that stimulates new, creative ideas. Also known as the KJ method.

agile approach—See *lean approach*.

AIAG—Automotive Industry Action Group.

alias—An *effect* that is completely confounded with another effect due to the nature of the designed experiment. Aliases are the results of *confounding*, which may or may not be deliberate.

alpha testing—In-house tests conducted on prototypes to ensure that the product is meeting requirements under controlled conditions. Usually done in the development stage.

alternative hypothesis, H_1—A hypothesis that is accepted if the *null hypothesis* (H_0) is disproved.

Example: Consider the null hypothesis that the statistical model for a *population* is a *normal distribution*. The alternative hypothesis to this null hypothesis is that the statistical model of the population is not a normal distribution.

Note 1: The alternative hypothesis is a statement that contradicts the null hypothesis. The corresponding test statistic is used to decide between the null and alternative hypotheses.

Note 2: The alternative hypothesis can be denoted H_1 or H_A with no clear preference as long as the symbolism parallels the null hypothesis notation.

alpha testing—Usually done in the development stage, alpha testing consists of in-house tests conducted on prototypes to ensure that the product is meeting requirements under controlled conditions.

analogies—A technique used to generate new ideas by translating concepts from one application to another.

analysis of covariance (ANCOVA)—A technique for estimating and testing the effects of treatments when one or more *concomitant variables* influence the *response variable*.

Note: Analysis of covariance can be viewed as a combination of *regression analysis* and *analysis of variance*.

analysis of variance (ANOVA)—A technique to determine if there are statistically significant differences between group means by analyzing group *variances*. An ANOVA is an analysis technique that evaluates the importance of several factors of a set of data by subdividing the variation into component parts. An analysis of variance table generally contains columns for:

- Source

- *Degrees of freedom*

- Sum of squares

- Mean square

- *F*-ratio or *F-test* statistic

- *p-value*

- Expected mean square

The basic assumptions are that the *effects* from the sources of *variation* are additive and that the e*xperimental errors* are independent, normally distributed, and have equal *variances*. ANOVA tests the *hypothesis* that the within-group variation is homogeneous and does not vary from group to group. The *null hypothesis* is that the group means are equal to each other. The *alternative hypothesis* is that at least one of the group means is different from the others.

analytical thinking—Breaking down a problem or situation into discrete parts to understand how each part contributes to the whole.

ANCOVA—See *analysis of covariance.*

andon board—A visual device (usually lights) displaying status alerts that can easily be seen by those who should respond.

ANOVA—See *analysis of variance.*

AOQ—See *average outgoing quality.*

AOQL—See *average outgoing quality limit.*

APL—See *acceptable process level.*

AQL—See *acceptance quality limit.*

arithmetic average—See *arithmetic mean.*

arithmetic mean—A calculation or estimation of the center of a set of values. It is the sum of the values divided by the number in the sum.

$$\bar{x} = \frac{1}{n}\sum_{i=1}^{n} x_i$$

where $\bar{x}$ is the arithmetic mean.

The average of a set of n observed values is the sum of the observed values divided by n:

$$\bar{x} = \frac{x_1 + x_2 + \ldots + x_n}{n}$$

See also *sample mean* and *population mean.*

ARL (average run length)—See *average run length.*

arrow diagram—A management and planning tool used to develop the best possible schedule and appropriate controls to accomplish the schedule; the critical path method (CPM) and the program evaluation review technique (PERT) make use of arrow diagrams.

AS9100—An international quality management standard for the aeronautics industry embracing the ISO 9001 standard.

ASME—American Society of Mechanical Engineers.

ASN—See *average sample number.*

ASQ—American Society for Quality, a society of individual and organizational members dedicated to the ongoing development, advancement, and promotion of quality concepts, principles, and technologies.

assessment—An estimate or determination of the significance, importance, or value of something.

assignable cause—A specifically identified factor that contributes to *variation* and is feasible to detect and identify. Eliminating assignable causes so that the points plotted on a *control chart* remain within the *control limits* helps achieve a *state of statistical control.*

Note: Although assignable cause is sometimes considered synonymous with *special cause*, a special cause is assignable only when it is specifically identified. See also *special cause.*

ASTD—American Society for Training and Development.

ASTM—American Society for Testing and Materials.

ATI—See *average total inspection.*

attribute—A countable or categorized quality *characteristic* that is *qualitative* rather than *quantitative* in nature. Attribute data come from *discrete, nominal,* or *ordinal scales.* Examples of attribute data are irregularities or flaws in a sample and results of pass/fail tests.

attributes control chart—A *Shewhart control chart* where the measure plotted represents countable or categorized data.

attributes data—Does/does not exist data. The control charts based on attribute data include fraction defective chart, number of affected units chart, count chart, count-per-unit chart, quality score chart, and demerit chart.

attributes, inspection by—See *inspection by attributes.*

audit—A planned, independent, and documented assessment to determine whether agreed-upon requirements are being met.

audit team—The group of trained individuals conducting an audit under the direction of a team leader, relevant to a particular product, process, service, contract, or project.

auditee—The individual or organization being audited.

auditor—An individual or organization carrying out an audit.

autocorrelation—The internal *correlation* between members of a series of observations ordered in time.

Note: Autocorrelation can lead to misinterpretation of *runs* and trends in *control charts*.

autonomation—Use of specially equipped automated machines capable of detecting a defect in a single part, stopping the process, and signaling for assistance. See *jidoka*.

availability—1. The ability of a product to be in a state to perform its designated function under stated conditions at a given time. Availability can be expressed by the ratio:

$$\frac{uptime}{total\ time}$$

2. A measure of the degree to which an item is in the operable and commitable state at the start of the mission when the mission is called for at an unknown (random) time.

average—The *arithmetic mean.*

average outgoing quality (AOQ)—The expected average quality level of outgoing product for a given value of incoming product quality.

Note: Unless otherwise specified, the average outgoing quality (AOQ) is calculated over all accepted *lots* plus all nonaccepted lots after the latter have been 100% inspected and the nonconforming items replaced by conforming items. An approximation often used is AOQ equals incoming *process* quality multiplied by the *probability of acceptance.* This formula is exact for accept-zero plans and overestimates other plans.

average outgoing quality limit (AOQL)—The maximum *average outgoing quality* over all possible values of incoming product quality level for a given *acceptance sampling plan* and rectification of all nonaccepted *lots* unless specified otherwise.

average run length (ARL)—The expected number of samples (or *subgroups*) plotted on a *control chart* up to and including the decision point that a *special cause* is present. The choice of ARL is a compromise between taking action when the process has not changed (ARL too small) or not taking action when a special cause is present (ARL too large).

average sample number (ASN)—The average number of sample *units* per *lot* used for making decisions (acceptance or nonacceptance).

average total inspection (ATI)—The average number of items inspected per lot including *100% inspection* of items in nonaccepted *lots.*

Note: Applicable when the procedure calls for 100% inspection of nonaccepted lots.

B

β (beta)—1. The maximum *probability*, or risk, of making a *Type II error*. See comment on *α (alpha)*. 2. The probability or risk of incorrectly deciding that a shift in the *process mean* has not occurred when the process has changed. 3. β is usually designated as *consumer's risk*. See *power curve*.

beta testing—The final test before the product is released to the commercial market. Beta testers are existing customers.

balanced design—A design where all *treatment* combinations have the same number of observations. If *replication* in a design exists, it would be balanced only if the replication was consistent across all the treatment combinations. In other words, the number of replicates of each treatment combination is the same.

balanced incomplete block design—An *incomplete block design* in which each *block* contains the same number (k) of different *levels* from the (l) *levels* of the principal *factor* arranged so that every pair of *levels* occurs in the same number (l) of *blocks* from the b *blocks*.

Note: This design implies that every *level* of the principal *factor* appears the same number of times in the experiment.

balanced scorecard—Translates an organization's mission and strategy into a comprehensive set of performance measures to provide a basis for strategic measurement and management, typically using four balanced views: financial, customers, internal business processes, and learning and growth.

batch—A definite quantity of product accumulated under conditions considered uniform or accumulated from a common source. This term is sometimes synonymous with *lot*.

batch processing—Running large batches of a single product through the process at one time, resulting in queues awaiting next steps in the process.

BATF—Bureau of Alcohol, Tobacco, and Firearms.

bathtub curve—Also called life history curve or Weibull curve. A graphic demonstration of the relationship of failures over the life of a product versus the probable failure rate. Includes three phases: early or infant failure (break-in), a stable rate during normal use, and wearout.

behavioral theories—Motivational theories, notably those of Abraham Maslow, Frederick Herzberg, Douglas McGregor, and William Ouchi.

benchmarking—An improvement process in which a company measures its performance against that of best-in-class companies (or others who are good performers), determines how those companies achieved their performance levels, and uses the information to improve its own performance. The areas that can be benchmarked include strategies, operations, processes, and procedures.

beta testing—The final test before a product is released to the commercial market. Beta testers are existing customers.

bias—Inaccuracy in a *measurement system* that occurs when the *mean* of the measurement result is consistently or systematically different than its *true value*.

bimodal—A probability distribution having two distinct statistical modes.

binomial confidence interval (one sample)—To construct $100(1 - \alpha)\%$ *confidence intervals* on the parameter p of a binomial distribution, if n is large and $p \leq 0.1$. For instance:

$$\hat{p} - z_{\alpha/2}\sqrt{\frac{\hat{p}(1-\hat{p})}{n}} \leq p \leq \hat{p} + z_{\alpha/2}\sqrt{\frac{\hat{p}(1-\hat{p})}{n}}$$

where the unbiased point estimator of p is

$$\hat{p} = x / n$$

If n is small, then tables of the *binomial distribution* should be used to establish the confidence interval on p. If n is large but p is small, then the Poisson approximation to the binomial is useful in constructing confidence intervals.

binomial confidence interval (two sample)—If there are two binomial parameters of interest, $p1$ and $p2$, then an approximate $100(1 - \alpha)\%$ *confidence interval on the difference is*

$$\hat{p}_1 - \hat{p}_2 - z_{\alpha/2}\sqrt{\frac{\hat{p}_1(1-\hat{p}_1)}{n_1} + \frac{\hat{p}_2(1-\hat{p}_2)}{n_2}} \leq p_1 - p_2 \leq \hat{p}_1 - \hat{p}_2 + z_{\alpha/2}\sqrt{\frac{\hat{p}_1(1-\hat{p}_1)}{n_1} + \frac{\hat{p}_2(1-\hat{p}_2)}{n_2}}$$

where
$$\hat{p}_1 = x_1 / n_1 \text{ and } \hat{p}_2 = x_2 / n_2$$

binomial distribution—A two-parameter discrete distribution involving the *mean* (μ) and the *variance* (σ^2) of the variable x with *probability* p where p is a constant, $0 \leq p \leq 1$, and sample size n. Mean = np and variance = $np(1 - p)$.

blemish—An *imperfection* that causes awareness but does not impair function or usage.

block—A collection of *experimental units* more homogeneous than the full set of experimental units. Blocks are usually selected to allow for *special causes*, in addition to those introduced as *factors* to be studied. These special causes may be avoidable within blocks, thus providing a more homogeneous experimental subspace.

block diagram—A diagram that shows the operation, interrelationships, and interdependencies of components in a system. Boxes, or blocks (hence the name), represent the components; connecting lines between the blocks represent interfaces. There are two types of block diagrams: a functional block diagram, which shows a system's subsystems and lower-level products, their interrelationships, and interfaces with other systems, and a reliability block diagram, which is similar to the functional block diagram except that it is modified to emphasize those aspects influencing reliability.

block effect—An *effect* resulting from a *block* in an *experimental design*. Existence of a block effect generally means that the method of blocking was appropriate.

blocking—The method of including *blocks* in an experiment in order to broaden the applicability of the conclusions or to minimize the impact of selected *assignable causes*. The randomization of the experiment is restricted and occurs within blocks.

body language—The expression of thoughts and emotions through movement or positioning of the body.

Box-Behnken design—A type of *response surface design* constructed by judicious combination of 2^k *factorial designs* with *balanced incomplete block designs*. It is most useful when it is difficult to have more than three *levels* of a *factor* in an experiment or when trying to avoid extreme combinations of *factor levels*.

box plot—Box plots, which are also called box-and-whisker plots, are particularly useful for showing the distributional characteristics of data. A box plot consists of a box, whiskers, and *outliers*. A line is drawn across the box at the *median*. By default, the bottom of the box is at the *first quartile* (Q1) and the top is at the *third quartile* (Q3) value. The whiskers are the lines that extend from the top and bottom of the box to the adjacent values. The adjacent values are the lowest and highest observations that are still inside the region defined by the following limits:

$$\text{Lower limit: } Q1 - 1.5 (Q3 - Q1)$$

$$\text{Upper limit: } Q3 + 1.5 (Q3 - Q1)$$

Outliers are points outside of the lower and upper limits and usually are plotted with asterisks (*).

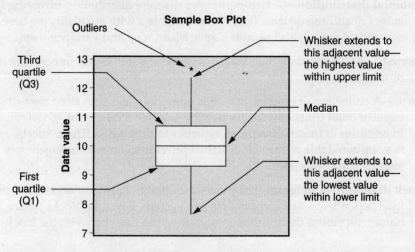

Sample Box Plot

BPR—Business process reengineering. See *reengineering*.

brainstorming—A problem-solving tool that teams use to generate as many ideas as possible related to a particular subject. Team members begin by offering all their ideas; the ideas are not discussed or reviewed until after the brainstorming session.

breakthrough—A method of solving chronic problems that results from the effective execution of a strategy designed to reach the next level of quality. Such change often requires a paradigm shift within the organization.

burn-in—The operation of an item under stress to stabilize its characteristics.

business partnering—The creation of cooperative business alliances between constituencies within an organization or between an organization and its customers or suppliers. Partnering occurs through a pooling of resources in a trusting atmosphere focused on continuous, mutual improvement. See also *customer–supplier partnership*.

business processes—Processes that focus on what the organization does as a business and how it goes about doing it. A business has functional processes (generating output within a single department) and cross-functional processes (generating output across several functions or departments).

C

c (count)—The number of *events* (often *nonconformities*) of a given classification occurring in a *sample* of fixed size. A *nonconforming unit* may have more than one nonconformity.

Example: *Blemishes* per 100 meters of rubber hose. Counts are *attribute* data. See *c chart*.

c chart (count chart)—An *attribute control chart* that uses *c (count)* or number of *events* as the plotted values where the opportunity for occurrence is fixed. Events, often *nonconformities*, of a particular type form the count. The fixed opportunity relates to samples of constant size or a fixed amount of material. Examples include flaws in each 100 square meters of fabric; errors in each 100 invoices; number of absentees per month.

$$\text{Central line: } \bar{c}$$

$$\text{Control limits: } \bar{c} \pm 3\sqrt{\bar{c}}$$

where: $\bar{c}$ is the *arithmetic mean* of the number of events.

Note: If the *lower control limit* calculates ≤ 0, there is no lower control limit. See also *u chart*, a count chart where the opportunity for occurrence is variable.

CAB—Corrective Action Board

capability—1. The performance of a *process* demonstrated to be in a *state of statistical control*. See *process capability* and *process performance*. 2. A measure of the ability of an item to achieve mission objectives given the conditions during the mission.

capability index—See *process capability index.*

capability ratio (C_p)—Is equal to the specification tolerance width divided by the process capability.

cascading training—Training implemented in an organization from the top down, where each level acts as trainers to those below.

case study—A prepared scenario (story) that, when studied and discussed, serves to illuminate the learning points of a course of study.

cause—A cause is an identified reason for the presence of a symptom, *defect,* or problem. See *effect* and *cause-and-effect diagram.*

cause-and-effect diagram—A tool for analyzing process variables. It is also referred to as the Ishikawa diagram, because Kaoru Ishikawa developed it, and the fishbone diagram, because the complete diagram resembles a fish skeleton. The diagram illustrates the main causes and subcauses leading to an effect (symptom). The cause-and-effect diagram is one of the seven basic tools of quality.

CBT—Computer-based training. Training delivered via computer software.

cell—A layout of workstations and/or various machines for different operations (usually in a U shape) in which multitasking operators proceed with a part from machine to machine to perform a series of sequential steps to produce a whole product or major subassembly.

cellular team—The cross-trained individuals who work within a cell.

center line—See *central line.*

central line—A line on a *control chart* representing the long-term *average* or a standard value of the *statistical measure* plotted. The central line calculation for each type of *control chart* is given under the specific control chart term; that is, for the *individuals control chart,* the central line is $\bar{x}$.

certification—The receipt of a document from an authorized source stating that a device, process, or operator has been certified to a known standard.

chain acceptance sampling inspection—An *acceptance sampling inspection* in which the criteria for acceptance of the current *lot* are governed by the sampling results of that *lot* and those of a specified number of the preceding consecutive lots.

chain reaction—A series of interacting events described by W. Edwards Deming: improve quality > decrease costs > improve productivity > increase market share with better quality and lower price > stay in business, provide jobs, and provide more jobs.

chaku–chaku—(Japanese) Meaning load–load in a cell layout where a part is taken from one machine and loaded into the next.

champion—An individual who has accountability and responsibility for many processes or who is involved in making strategic-level decisions for the organization. The champion ensures ongoing dedication of project resources and monitors strategic alignment (also referred to as a sponsor).

chance cause—See *random cause.*

chance variation—See *random cause.*

change agent—The person who takes the lead in transforming a company into a quality organization by providing guidance during the planning phase, facilitating implementation, and supporting those who pioneer the changes.

change management—The strategies, processes, and practices involved in creating and managing change.

changeover—Changing a machine or process from one type of product or operation to another.

characteristic—A distinguishing feature or inherent property of a product, *process,* or system related to a requirement. A property that helps to differentiate among items of a given *sample* or *population.* The differentiation may be either *quantitative* (by *variable*) or *qualitative* (by *attribute*).

charter—A documented statement officially initiating the formation of a committee, team, project, or other effort in which a clearly stated purpose and approval is conferred.

check sheet—A simple data-recording device. The check sheet is custom-designed for the particular use, allowing ease in interpreting the results. The check sheet is one of the seven basic tools of quality. Check sheets should not be confused with data sheets and checklists.

checklist—A tool for organizing and ensuring that all important steps or actions in an operation have been taken. Checklists contain items that are important or relevant to an issue or situation. Checklists should not be confused with check sheets and data sheets.

checkout—Tests or observations of an item to determine its condition or status.

χ^2 **(chi-square) distribution**—A positively skewed distribution that varies with the *degrees of freedom* with a minimum value of zero.

χ^2 **(chi-square) statistic**—A value obtained from the χ^2 *distribution* at a given percentage point and a given *degree of freedom.*

χ^2 **(chi-square) test**—A *statistic* used in testing a *hypothesis* concerning the discrepancy between *observed* and expected results.

chronic problem—A long-standing adverse situation that can be remedied by changing the status quo. For example, actions such as revising an unrealistic manufacturing process or addressing customer defections can change the status quo and remedy the situation.

clearance number—As associated with a *continuous sampling plan,* the number of successively inspected units of product that must be found acceptable during the *100 percent inspection* sequence before action to change the amount of inspection can be taken. The clearance number is often designated as *i.*

coaching—A continual improvement technique by which people receive one-to-one learning through demonstration and practice and that is characterized by immediate feedback and correction.

code of conduct—The expected behavior that has been mutually developed and agreed upon by a team.

code of ethics—A guide for ethical behavior of the members of ASQ and those individuals who hold certification in its various divisions.

coefficient of determination (R^2)—A measure of the part of the *variance* for one *variable* that can be explained by its linear relationship with another variable (or variables). The coefficient of determination is the square of the correlation between the observed y values and the fitted y values, and is also the fraction of the variation in y that is explained by the fitted equation. It is a percentage between zero and 100 with higher values indicating a stronger degree of the combined linear relationship of several predictor variables $x_1, x_2, \ldots x_p$ to the *response variable Y. See regression analysis.*

coefficient of variation (CV)—This measures relative dispersion. It is the *standard deviation* divided by the *mean* and is commonly reported as a percentage.

comment cards—Printed cards or slips of paper used to solicit and collect comments from users of a service or product.

commercial/industrial market—Refers to business market customers who are described by variables such as location, NAICS code, buyer industry, technological sophistication, purchasing process, size, ownership, and financial strength.

common cause—See *random cause.*

common causes of variation—Causes that are inherent in any process all the time. A process that has only common causes of variation is said to be stable or predictable or in control. Also called chance causes.

competence—Refers to a person's ability to learn and perform a particular activity. Competence consists of knowledge, experience, skills, aptitude, and attitude components (KESAA factors).

competency-based training—A training methodology that focuses on building mastery of a predetermined segment or module before moving on to the next.

complaint handling—The process and practices involved in receiving and resolving complaints from customers.

complete block—A *block* that accommodates a complete set of *treatment* combinations.

completely randomized design—A design in which the *treatments* are assigned at random to the full set of *experimental units*. No *blocks* are involved in a *completely randomized design*.

completely randomized factorial design—A *factorial design* in which all the *treatments* are assigned at random to the full set of *experimental units*. See *completely randomized design*.

compliance—An affirmative indication or judgment that the supplier of a product or service has met the requirements of the relevant specifications, contract, or regulation; also the state of meeting the requirements.

computer-based instruction—Any instruction delivered via a computer.

concomitant variable—A *variable* or *factor* that cannot be accounted for in the data analysis or design of an experiment but whose *effect* on the results should be accounted for. For example, the *experimental units* may differ in the amount of some chemical constituent present in each unit, which can be measured, but not adjusted. See *analysis of covariance*.

confidence coefficient (1 − α)—See *confidence level*.

confidence interval—A confidence interval is an estimate of the interval between two *statistics* that includes the *true value* of the *parameter* with some probability. This probability is called the *confidence level* of the estimate. Confidence levels typically used are 90 percent, 95 percent, and 99 percent. The interval either contains the parameter or it does not. See *z-confidence interval, t-confidence interval*.

confidence level (confidence coefficient [1 − α])—1. The *probability* that the *confidence interval* described by a set of *confidence limits* actually includes the population parameter. 2. The probability that an interval about a sample *statistic* actually includes the population parameter.

confidence limits—The endpoints of the interval about the sample *statistic* that is believed, with a specified *confidence level*, to include the population parameter. See *confidence interval*.

configuration management (CM)—A discipline and a system to control changes that are made to hardware, software, firmware, and documentation throughout a system's lifecycle.

conflict resolution—A process for resolving disagreements in a manner acceptable to all parties.

conformance—An affirmative indication or judgment that a product or service has met the requirements of a relevant specification, contract, or regulation.

confounding—Indistinguishably combining an *effect* with other effects or *blocks*. When done deliberately, higher-order effects are systematically *aliased* so as to allow estimation of lower-order effects. Sometimes, confounding results from inadvertent changes to a design during the running of an experiment or poor planning of the design. This can diminish or even invalidate the effectiveness of the experiment.

consensus—Finding a proposal acceptable enough that all team members can support the decision and no member opposes it.

consistency—See *precision*.

consultative—A decision-making approach in which a person talks to others and considers their input before making a decision.

consumer market customers—End users of a product or service.

consumer's risk (β)—The *probability of acceptance* when the quality level has a value stated by the *acceptance sampling plan* as unsatisfactory.

Note 1: Such acceptance is a *Type II error*.

Note 2: Consumer's risk is usually designated as *β (beta)*.

continual process improvement—Includes the actions taken throughout an organization to increase the effectiveness and efficiency of activities and processes in order to provide added benefits to the customer and organization. It is considered a subset of total quality management and operates according to the premise that organizations can always make improvements. Continual improvement can also be equated with reducing process variation.

continuous distribution—A distribution where data is from a *continuous scale*. Examples of continuous scales include the *normal, t*, and *F* distributions.

continous quality improvement (CQI)—A philosophy and actions for repeatedly improving an organization's capabilities and processes with the objective of customer satisfaction.

continuous sampling plan—An *acceptance sampling inspection* applicable to a continuous flow process, which involves acceptance or nonacceptance on an item-by-item basis and uses alternative periods of *100 percent inspection* and sampling, depending on the quality of the observed process output.

continuous scale—A scale with a continuum of possible values.

Note: A continuous scale can be transformed into a *discrete scale* by grouping values, but this leads to some loss of information.

contrast—Linear function of the *response values* for which the sum of the coefficients is zero with not all coefficients equal to zero. Questions of logical interest from an experiment may be expressed as contrasts with carefully selected coefficients.

contrast analysis—A technique for estimating the *parameters* of a *model* and making *hypothesis* tests on preselected linear combinations of the *treatments* or *contrasts*.

control chart—A chart that plots a *statistical measure* of a series of *samples* in a particular order to steer the *process* regarding that measure and to control and reduce variation.

Note 1: The order is usually time- or sample number order–based.

Note 2: The control chart operates most effectively when the measure is a process *characteristic* correlated with an ultimate product or service characteristic.

control chart, acceptance—See *acceptance control chart*.

control chart, standard given—A *control chart* whose *control limits* are based on adopted standard values applicable to the *statistical measures* plotted on the chart.

Note 1: Standard values are adopted at any time for computing the control limits to be used as criteria for action in the immediate future or until additional evidence indicates need for revision.

Note 2: This type of control chart is used to discover whether *observed values* differ from standard values by an amount greater than chance. An example of this use is to establish and verify an *internal reference material*.

Note 3: The subscript zero is used for standard chart parameters, that is, $\bar{x}_0, \bar{y}_0, R_0$.

control chart factor—A factor, usually varying with *sample size,* that converts specified statistics or parameters into *control limits* or a *central line* value.

Note: Common control chart factors for *Shewhart control charts* are $d_2, A_2, D_3,$ and D_4. See $\bar{x}$ *chart* and *range*.

control factor—In *robust parameter design*, a *control factor* is a *predictor variable* that is controlled as part of the standard experimental conditions. In general, inference is made on control factors while *noise factors* are allowed to vary to broaden the conclusions.

control limit—A line on a *control chart* used for judging the stability of a *process*.

Note 1: Control limits provide statistically determined boundaries for the *deviations* from the *central line* of the statistic plotted on a *Shewhart control chart* due to *random causes* alone.

Note 2: Control limits (with the exception of the *acceptance control chart*) are based on actual process data, not on *specification limits*.

Note 3: Other than points outside of *control limits*, out-of-control criteria can include *runs,* trends, cycles, periodicity, and unusual patterns within the control limits.

Note 4: The control limit calculation for each type of control chart is given under the specific control chart term; that is, for the *individuals control chart* under calculation of its control limits.

control plan—1. A document describing the system elements to be applied to control *variation* of *processes*, products, and services in order to minimize deviation from their preferred values. 2. A document, or documents, that may include the characteristics for quality of a product or service, measurements, and methods of control.

corrective action—Action taken to eliminate the root cause(s) and symptom(s) of an existing deviation or nonconformity to prevent recurrence.

correlation—Correlation measures the linear association between two variables. It is commonly measured by the *correlation coefficient r*. See also *regression*.

correlation coefficient (*r*)—A number between –1 and 1 that indicates the degree of linear relationship between two sets of numbers.

$$r = \frac{s_{xy}}{s_x s_y} = \frac{n\Sigma xy - \Sigma y}{\sqrt{\left[n\Sigma x^2 - (\Sigma x)^2 - (\Sigma y)^2 \right]}}$$

where s_x is the *standard deviation* of x, s_y is *standard deviation* of y, and s_{xy} is the *covariance* of x and y.

Correlation coefficients of –1 and +1 represent perfect linear agreement between two variables; $r = 0$ implies no linear relationship at all. If r is positive, y increases as x increases. In other words, if a *linear regression equation* were fit, the *linear regression coefficient* would be positive. If r is negative, y decreases as x increases. In other words, if a linear regression equation were fit, the linear regression coefficient would be negative.

cost of quality (COQ)—The total costs incurred relating to the quality of a product or service. There are four categories of quality costs: internal failure costs (costs associated with defects found before delivery of the product or service), external failure costs (costs associated with defects found during or after product or service delivery), appraisal costs (costs incurred to determine the degree of conformance to quality requirements), and prevention costs (costs incurred to keep failure and appraisal costs to a minimum).

covariance—This measures the relationship between pairs of observations from two variables. It is the sum of products of *deviations* of pairs of variables in a random *sample* from their sample means divided by the number in the sum minus one.

$$\text{Sample covariance} = \frac{1}{n-1} \sum_{i=1}^{n} (x_i - \bar{x})(y_i - \bar{y})$$

where $\bar{x}$ is the sample *mean* for the first sample and $\bar{y}$ is the sample mean for the second sample.

Note: Using $n - 1$ provides an unbiased estimator of the *population covariance*. When the covariance is standardized to cover a range of values from –1 to 1, it is more commonly known as the *correlation coefficient*.

C_p **(process capability index)**—An index describing *process capability* in relation to specified *tolerance* of a *characteristic* divided by a measure of the length of the *reference interval* for a process in a *state of statistical control*.

$$C_p = \frac{U - L}{6\sigma}$$

where U is upper specification limit, L is lower *specification limit*, and 6σ is the process capability.

See also P_p, a similar term, except that the *process* may not be in a *state of statistical control*.

C_{pk} **(minimum process capability index)**—An index that represents the smaller of C_{pk_U} *(upper process capability index)* and C_{pk_L} *(lower process capability index)*.

C_{pk_L} **(lower process capability index; CPL)**—An index describing *process capability* in relation to the lower *specification limit*.

$$C_{pk_L} = \frac{\mu - L}{3\sigma}$$

where μ = process average, L = lower specification limit, and 3σ = half of the *process capability*.

C_{pk_U} **(upper process capability index; CPU)**—An index describing *process capability* in relation to the upper *specification limit*.

$$C_{pk_U} = \frac{U - \mu}{3\sigma}$$

where μ = process average, U = upper specification limit, and 3σ = half of the *process capability*.

CPM—See *critical path method*

C_{pm} **(process capability index of the mean)**—An index that takes into account the location of the mean, defined as

$$C_{pm} = \frac{(U - L)/2}{3\sqrt{\sigma^2 + (\mu - T)^2}}$$

where U = upper *specification limit*, L = lower specification limit, σ = *standard deviation*, μ = expected value, and T = *target value*.

CQPA—Certified Quality Process Analyst

critical path—Refers to the sequence of tasks that takes the longest time and determines a project's completion date.

critical path method (CPM)—An activity-oriented project management tool that uses arrow-diagramming techniques to demonstrate both the time and cost required to complete a project. It provides one time estimate—normal time—and allows for computing the critical path.

critical thinking—The careful analysis, evaluation, reasoning (both deductive and inductive), clear thinking, and systems thinking leading to effective decisions.

crossed design—A design where all the *factors* are crossed with each other. See *crossed factor*.

crossed factor—Two *factors* are crossed if every *level* of one factor occurs with every level of the other in an experiment.

cross-functional team—A group consisting of members from more than one department or work unit that is organized to accomplish a project.

CSR—Customer service representative.

cumulative frequency distribution—The sum of the frequencies accumulated up to the upper boundary of a class in a distribution.

cumulative sum chart (CUSUM chart)—The CUSUM *control chart* calculates the cumulative sum of *deviations* from *target* to detect shifts in the level of the measurement. The CUSUM chart can be graphical or numerical. In the graphical form, a V-mask indicates action: whenever the CUSUM trace moves outside the V-mask, take control action. The numerical (or electronic) form computes CUSUM from a selected bias away from target value and has a fixed *action limit*. Additional tests for abnormal patterns are not required.

curtailed inspection—An *acceptance sampling procedure* that contains a provision for stopping inspection when it becomes apparent that adequate data have been collected for a decision.

curvature—The departure from a straight line relationship between the *response variable* and *predictor variable*. Curvature has meaning with *quantitative* predictor variables, but not with *categorical* (nominal) or *qualitative* (ordinal) predictor variables. Detection of curvature requires more than two *levels* of the *factors* and is often done using *center points*.

customer—Recipient of a product or service provided by a supplier. See *external customer* and *internal customer*.

customer delight—The result achieved when customer requirements are exceeded in unexpected ways the customer finds valuable.

customer loyalty/retention—The result of an organization's plans, processes, practices, and efforts designed to deliver their services or products in ways that create retained and committed customers.

customer relationship management (CRM)—Refers to an organization's knowledge of its customers' unique requirements and expectations, and using that knowledge to develop a closer and more profitable link to business processes and strategies.

customer satisfaction—The result of delivering a product or service that meets customer requirements, needs, and expectations.

customer segmentation—Refers to the process of differentiating customers based on one or more dimensions for the purpose of developing a marketing strategy to address specific segments.

customer service—The activities of dealing with customer questions; also sometimes the department that takes customer orders or provides post-delivery services.

customer value—The market-perceived quality adjusted for the relative price of a product.

customer-oriented organization—An organization whose mission, purpose, and actions are dedicated to serving and satisfying its customers.

customer–supplier partnership—A long-term relationship between a buyer and supplier characterized by teamwork and mutual confidence. The supplier is considered an extension of the buyer's organization. The partnership is based on several commitments. The buyer provides long-term contracts and uses fewer suppliers. The supplier implements quality assurance processes so that incoming inspection can be minimized. The supplier also helps the buyer reduce costs and improve product and process designs.

CUSUM chart—See *cumulative sum chart*.

CV—See *coefficient of variation*.

CWQC—Companywide quality control.

cycle time—Refers to the time that it takes to complete a process from beginning to end.

cycle time reduction—To reduce the time that it takes, from start to finish, to complete a particular process.

D

δ **(delta lowercase)**—The smallest shift in *process* level that it is desired to detect, stated in terms of the number of *standard deviations* from the *average*. It is given by

$$(\delta = D / \sigma)$$

where D is the size of the shift it is desired to detect and σ is the standard deviation.

D—The smallest shift in *process* level that it is desired to detect, stated in terms of original units.

debugging—A process to detect and remedy inadequacies.

decision matrix—A matrix used by teams to evaluate problems or possible solutions. For example, after a matrix is drawn to evaluate possible solutions, the team lists them in the far left vertical column. Next, the team selects criteria to rate the possible solutions, writing them across the top row. Then, each possible solution is rated on a scale of 1 to 5 for each criterion and the rating recorded in the corresponding grid. Finally, the ratings of all the criteria for each possible solution are added to determine its total score. The total score is then used to help decide which solution deserves the most attention.

defect—The nonfulfillment of a requirement related to an intended or specified use.

Note: The distinction between the concepts *defect* and *nonconformity* is important as it has legal connotations, particularly those associated with product liability issues. Consequently, the term *defect* should be used with extreme caution.

defective (defective unit)—A *unit* with one or more *defects*.

defining relation—Used for *fractional factorial designs*. It consists of *generators* that can be used to determine which *effects* are *aliased* with each other.

degrees of freedom (v, df)—In general, the number of independent comparisons available to estimate a specific *parameter* that allows entry to certain statistical tables.

Deming cycle—See *plan–do–check–act cycle*.

Deming Prize—Award given annually to organizations that, according to the award guidelines, have successfully applied companywide quality control based on statistical quality control and will keep up with it in the future. Although the award is named in honor of W. Edwards Deming, its criteria are not specifically related to Deming's teachings. There are three separate divisions for the award: the Deming Application Prize, the Deming Prize for Individuals, and the Deming Prize for Overseas Companies. The award process is overseen by the Deming Prize Committee of the Union of Japanese Scientists and Engineers in Tokyo.

demographics—Variables among buyers in the consumer market, which include geographic location, age, sex, marital status, family size, social class, education, nationality, occupation, and income.

demonstrated—That which has been measured by the use of objective evidence gathered under specified conditions.

dependability—A measure of the degree to which an item is operable and capable of performing its required function at any (random) time during a specified mission profile, given item availability at the start of the mission.

dependent variable—See *response variable*.

deployment—Used in strategic planning to describe the process of cascading goals, objectives, and plans throughout an organization.

derating—(a) using an item in such a way that applied stresses are below rated values or (b) the lowering of the rating of an item in one stress field to allow an increase in rating in another stress field.

design for manufacturing (DFM)—The design of a product for ease in manufacturing. Also, design for assembly (DFA) is used.

design for Six Sigma (DFSS)—The aim is for a robust design that is consistent with applicable manufacturing processes and assures a fully capable process that will produce quality products.

design generator—See *generator*.

design of experiments (DOE; DOX)—The arrangement in which an experimental program is to be conducted, including the selection of *factor* combinations and their *levels*.

Note: The purpose of designing an experiment is to provide the most efficient and economical methods of reaching valid and relevant conclusions from the experiment. The selection of the design is a function of many considerations, such as the type of questions to be answered, the applicability of the conclusions, the homogeneity of *experimental units*, the *randomization* scheme, and the cost to run the experiment. A properly designed experiment will permit simple interpretation of valid results.

design resolution—See *resolution*.

design review—Documented, comprehensive, and systematic examination of a design to evaluate its capability to fulfill the requirements for quality.

design space—The multidimensional region of possible *treatment* combinations formed by the selected *factors* and their *levels*.

desired quality—Refers to the additional features and benefits a customer discovers when using a product or service that lead to increased customer satisfaction. If missing, a customer may become dissatisfied.

deviation—The difference between a measurement and (measurement usage) its stated value or intended level.

deviation control—Deviation control involves *process* control where the primary interest is detecting small shifts in the *mean*. Typical *control charts* used for this purpose are *CUSUM*, *EWMA*, and *Shewhart control charts* with *runs* rules. This type of control is appropriate in advanced stages of a quality improvement process. See *threshold control*.

discrete distribution—A *probability distribution* where data are from a *discrete scale*. Examples of *discrete distributions* are *binomial* and *Poisson* distributions. *Attribute* data involve discrete distributions.

discrete scale—A scale with only a set or sequence of distinct values.

Examples: Defects per unit, events in a given time period, types of defects, number of orders on a truck.

discrimination—See *resolution*.

dispersion—A term synonymous with *variation*.

dispersion effect—The influence of a single *factor* on the *variance* of the *response variable*.

dissatisfiers—Those features or functions that the customer or employee has come to expect and, if they were no longer present, would result in dissatisfaction.

distance learning—Learning where student(s) and instructor(s) are not collocated, often carried out through electronic means.

distribution—Describes the amount of potential variation in outputs of a process; it is usually described in terms of its shape, average, and standard deviation.

DMAIC—Pertains to a methodology used in the Six Sigma approach: define, measure, analyze, improve, control.

documentation control (DC)—a system used to track and store electronic documents and/or paper documents

DOE—See *design of experiments*.

dot plot—A plot of a *frequency distribution* where the values are plotted on the *x*-axis. The *y*-axis is a count. Each time a value occurs, the point is plotted according to the count for the value.

double sampling—A multiple *acceptance sampling inspection* in which, at most, two *samples* are taken.

Note: The decisions are made according to defined rules.

downsizing—The planned reduction in workforce due to economics, competition, merger, sale, restructuring, or reengineering.

duplication—See *repeated measures*.

durability—A measure of useful life (a special case of reliability).

E

effect—1. The result of taking an action; the expected or predicted impact when an action is to be taken or is proposed. An effect is the symptom, *defect*, or problem. See *cause* and *cause-and effect diagram*. 2. (Design of experiments usage) A relationship between a *factor*(s) and a *response variable*(s). Specific types include *main effect*, *dispersion effect*, or *interaction effect*.

eighty–twenty (80–20) rule—A term referring to the Pareto principle, which suggests that most effects come from relatively few causes; that is, 80 percent of the effects come from 20 percent of the possible causes.

element—See *unit*.

employee involvement—A practice within an organization whereby employees regularly participate in making decisions on how their work areas operate, including making suggestions for improvement, planning, objectives setting, and monitoring performance.

empowerment—A condition whereby employees have the authority to make decisions and take action in their work areas, within stated bounds, without prior approval. For example, an operator can stop a production process upon detecting a problem, or a customer service representative can send out a replacement product if a customer calls with a problem.

end users—External customers who purchase products/services for their own use.

environmental analysis/scanning—Relates to identifying and monitoring factors from both inside and outside the organization that may impact the long-term viability of the organization.

EPA—Environmental Protection Agency.

event—An occurrence of some *attribute* or outcome. In the quality field, events are often *nonconformities*.

evolutionary operation (EVOP)—A sequential form of experimentation conducted in production facilities during regular production. The range of variation of the *factor* levels is usually quite small in order to avoid extreme changes, so it often requires considerable *replication* and time. It is most useful when it would be difficult to do standard experimentation techniques like *factorial design*.

EVOP—See *evolutionary operation*.

EWMA—See *exponentially weighted moving average*.

EWMA chart—See *exponentially weighted moving average chart*.

excited quality—The additional benefit a customer receives when a product or service goes beyond basic expectations. Excited quality wows the customer and distinguishes the provider from the competition. If missing, the customer will still be satisfied.

executive education—Usually refers to the education (and training) provided to top management.

expected quality—Also known as basic quality, the minimum benefit or value a customer expects to receive from a product or service.

experiment space—See *design space*.

experimental design—See *design of experiments*.

experimental error—The variation in the *response variable* beyond that accounted for by the *factors*, *blocks*, or other assignable sources in the conduct of an experiment.

experimental plan—The assignment of *treatments* to an *experimental unit* and the time order in which the *treatments* are to be applied.

experimental run—A single performance of an experiment for a specific set of *treatment* combinations.

experimental unit—The smallest entity receiving a particular *treatment*, subsequently yielding a value of the *response variable*.

experimentation—See *design of experiments*.

explanatory variable—See *predictor variable*.

exploratory data analysis (EDA)—Exploratory data analysis isolates patterns and features of the data and reveals these forcefully to the analyst.

exponential distribution—A continuous distribution where data are more likely to occur below the average than above it. Typically used to describe the break-in portion of the bathtub curve.

exponentially weighted moving average (EWMA)—A *moving average* that is not greatly influenced when a small or large value enters the calculation.

$$w_t = \lambda x_i + (1 - \lambda) w_{t-1}$$

where w_t is the exponentially weighted moving average (EWMA) at the present time t, λ *(lambda)* is a defined or experimentally determined weighting factor, x_i is the observed value at time t, and w_{t-1} is the EWMA at the immediately preceding time interval.

exponentially weighted moving average chart (EWMA chart)—A *control chart* where each new result is averaged with the previous *average* value using an experimentally determined weighting factor, λ. Usually only the averages are plotted and the *range* is omitted. The action *signal* is a single point out of limits. The advantage of the EWMA chart compared to the *MA chart* is that the average does not jump when an extreme value leaves the moving average. The weighting factor λ can be determined by an effective time period. Additional tests are not appropriate because of the high correlation between successive averages.

Control limits: $\bar{X}_0 \pm 3\sigma_{w_t}$

Center line: $\bar{X}_0$

external audit—An audit performed by independent and impartial auditors who are not members of the audit team.

external customer—A person or organization who receives a product, a service, or information, but is not part of the organization supplying it. See also *internal customer*.

F

$F_{v1, v2}$—*F-test* statistic. See *F-test*.

$F_{v1, v2,}$—Critical value for *F-test*. See *F-test*.

facilitator—An individual who is responsible for creating favorable conditions that will enable a team to reach its purpose or achieve its goals by bringing together the necessary tools, information, and resources to get the job done. A facilitator addresses the processes a team uses to achieve its purpose.

factor—A *predictor variable* that is varied with the intent of assessing its *effect* on a *response variable*.

factor level—See *level*.

factorial design—An *experimental design* consisting of all possible *treatments* formed from two or more *factors*, each studied at two or more *levels*. When all combinations are run, the *interaction effects* as well as *main effects* can be estimated. See also *2^k factorial design*.

F distribution—A *continuous distribution* that is a useful reference for assessing the ratio of independent *variances*. See *variance, tests for*.

failure—The event, or inoperable state, in which an item or part of an item does not, or would not, perform as previously specified.

failure, dependent—Failure that is caused by the failure of an associated item(s). Not independent.

failure, independent—Failure that occurs without being caused by the failure of any other item. Not dependent.

failure, random—Failure whose occurrence is predictable only in a probabilistic or statistical sense. This applies to all distributions.

failure analysis—Subsequent to a failure, the logical systematic examination of an item, its construction, application, and documentation to identify the failure mode and determine the failure mechanism and its basic course.

failure mechanism—The physical, chemical, electrical, thermal, or other process that results in failure.

failure mode—The consequence of the mechanism through which the failure occurs, such as short, open, fracture, excessive wear.

failure mode analysis (FMA)—A procedure to determine which malfunction symptoms appear immediately before or after a failure of a critical parameter in a system. After all the possible causes are listed for each symptom, the product is designed to eliminate the problems.

failure mode and effects analysis (FMEA)—A procedure in which each potential failure mode in every subitem of an item is analyzed to determine its effect on other subitems and on the required function of the item.

Typically two types of FMEAs are used: DFMEA (design) and PFMEA (process).

failure rate—The total number of failures within an item population divided by the total number of life units expended by that population during a particular measurement interval under stated conditions.

FDA—Food and Drug Administration.

feedback—The response to information received in interpersonal communication (written or oral); it may be based on fact or feeling and helps the party who is receiving the information judge how well he/she is being understood by the other party. Generally, feedback is information about a process or performance and is used to make decisions that are directed toward improving or adjusting the process or performance as necessary.

feedback loops—Pertains to open-loop and closed-loop feedback. Open-loop feedback focuses on how to detect or measure problems in the inputs and how to plan for contingencies. Closed-loop feedback focuses on how to measure the outputs and how to determine the control points where adjustment can be made.

filters—Relative to human-to-human communication, those perceptions (based on culture, language, demographics, experience, and so on) that affect how a message is transmitted by the sender and how a message is interpreted by the receiver.

first-article testing—Tests a production process to demonstrate process capability and that tooling meets quality standard.

fishbone diagram—See *cause-and-effect diagram*.

fitness for use—A term used to indicate that a product or service fits the customer's defined purpose for that product or service.

5S—Five practices for maintaining a clean and efficient workplace (Japanese).

five whys—A repetitive questioning technique to probe deeper to surface the root cause of a problem.

fixed factor—A *factor* that only has a limited number of *levels* that are of interest. In general, inference is not made to other levels of fixed factors not included in the experiment. For example, gender when used as a *factor* only has two possible *levels* of interest. See also *random factor*.

fixed model—A *model* that contains only *fixed factors*.

flowchart—A graphical representation of the steps in a process. Flowcharts are drawn to better understand processes. The flowchart is one of the seven basic tools of quality.

focus group—A qualitative discussion group consisting of eight to 10 participants, invited from a segment of the customer base to discuss an existing or planned product or service, led by a facilitator working from predetermined questions (focus groups may also be used to gather information in a context other than customers).

fourteen (14) points—W. Edwards Deming's 14 management practices to help organizations increase their quality and productivity. They are:

1. Create constancy of purpose for improving products and services.

2. Adopt a new philosophy.

3. Cease dependence on inspection to achieve quality.

4. End the practice of awarding business on price alone; instead, minimize total cost by working with a single supplier.

5. Improve constantly and forever every process for planning, production, and service.

6. Institute training on the job.

7. Adopt and institute leadership.

8. Drive out fear.

9. Break down barriers between staff areas.

10. Eliminate slogans, exhortations, and targets for the workforce.

11. Eliminate numerical quotas for the workforce and numerical goals for management.

12. Remove barriers that rob people of pride of workmanship and eliminate the annual rating or merit system.

13. Institute a vigorous program of education and self-improvement for everyone.

14. Put everybody in the company to work to accomplish the transformation.

fractional factorial design—An *experimental design* consisting of a subset (fraction) of the *factorial design*. Typically, the fraction is a simple proportion of the full set of possible *treatment* combinations. For example, half-fractions, quarter-fractions, and so forth are common. While fractional factorial designs require fewer runs, some degree of *confounding* occurs.

frequency—The number of occurrences or *observed values* in a specified class, *sample*, or *population*.

frequency distribution—A set of all the various values that individual observations may have and the *frequency* of their occurrence in the *sample* or *population*. See *cumulative frequency distribution*.

F-test—A statistical test that uses the *F distribution*. It is most often used when dealing with a *hypothesis* related to the ratio of independent *variances*.

$$F = \frac{s_L^2}{s_S^2}$$

where: s_L^2 is the larger variance and s_S^2 is the smaller variance. See *analysis of variance.*

fully nested design—See *nested design.*

G

gage R&R study—A type of *measurement system analysis* done to evaluate the performance of a test method or *measurement system.* Such a study quantifies the capabilities and limitations of a measurement instrument, often estimating its *repeatability* and *reproducibility.* It typically involves multiple operators measuring a series of measurement items multiple times.

Gantt chart—A type of bar chart used in process/project planning and control to display planned work and finished work in relation to time. Also called a milestone chart when interim checkpoints are added.

gap analysis—A technique that compares a company's existing state to its desired state (as expressed by its long-term plans) to help determine what needs to be done to remove or minimize the gap.

gatekeeping—The role of an individual (often a facilitator) in a group meeting in helping ensure effective interpersonal interactions (for example, someone's ideas are not ignored due to the team moving on to the next topic too quickly).

Gaussian distribution—See *normal distribution.*

g chart—An *attributes control chart* based on a shifted *geometric distribution* for the number of occurrences of a given *event* per *unit* of *process* output. The *g* chart is for the total number of events.

Center line (CL): $\overline{t}$

Upper control limit (UCL): $\overline{t} + k\sqrt{n\left(\dfrac{\overline{t}}{n} - \alpha\right)\left(\dfrac{\overline{t}}{n} - \alpha + 1\right)}$

Lower control limit (LCL): $\overline{t} - k\sqrt{n\left(\dfrac{\overline{t}}{n} - \alpha\right)\left(\dfrac{\overline{t}}{n} - \alpha + 1\right)}$

where *a* is the known minimum possible number of events, *n* is the *sample size*, *k* is the σ multiplier (for example, 3σ), and $\overline{t}$ is the average total number of events per *subgroup.* Calculation for $\overline{t}$:

$$\overline{t} = \frac{t_1 + t_2 + \ldots + t_r}{r}$$

where $t_1 + t_2 + \ldots + t_r$ is the total number of events occurring in each subgroup and *r* is the number of subgroups. (Note that *r* is not the correlation coefficient *r* used elsewhere in this glossary.) The *g* chart is commonly used in healthcare and hospital settings. See *h chart* for the average number of events chart.

geometric distribution—A case of the *negative binomial distribution* where $c = 1$ (c is integer parameter). The geometric distribution is a *discrete distribution*.

generator—In *design of experiments*, a generator is used to determine the level of *confounding* and the pattern of *aliases* in a *fractional factorial design*.

goal—A statement of general intent, aim, or desire; it is the point toward which management directs its efforts and resources; goals are usually nonquantitative.

Graeco-Latin square design—An extension of a *Latin square design* that involves four *factors* in which the combination of the *level* of any one of them with the *levels* of the other three appears once and only once.

groupthink—Occurs when most or all team members coalesce in supporting an idea or decision that hasn't been fully explored, or when some members secretly disagree but go along with the other members in apparent support.

H

H_0—See *null hypothesis*.

H_1—See *alternative hypothesis*.

H_A—See *alternative hypothesis*.

h chart—An *attributes control chart* based on a shifted *geometric distribution* for the number of occurrences of a given *event* per unit of *process* output. The h chart is for the *average* number of events.

Center line (CL): $\dfrac{\bar{t}}{n}$

Upper control limit (UCL): $\dfrac{\bar{t}}{n} + \dfrac{k}{\sqrt{n}}\sqrt{n\left(\dfrac{\bar{t}}{n} - \alpha\right)\left(\dfrac{\bar{t}}{n} - \alpha + 1\right)}$

Lower control limit (LCL): $\dfrac{\bar{t}}{n} - \dfrac{k}{\sqrt{n}}\sqrt{n\left(\dfrac{\bar{t}}{n} - \alpha\right)\left(\dfrac{\bar{t}}{n} - \alpha + 1\right)}$

where a is the known minimum possible number of events, n is the *sample size*, k is the σ multiplier (for example, 3σ), and $\bar{t}$ is the average total number of events per *subgroup*. Calculation for $\bar{t}$:

$$\bar{t} = \frac{t_1 + t_2 + \ldots + t_r}{r}$$

where $t_1 + t_2 + \ldots + t_r$ is the total number of events occurring in each subgroup and r is the number of subgroups. (Note that r is not the correlation coefficient r used elsewhere in this glossary.) The h chart is commonly used in healthcare and hospital settings. See *g chart* for the total number of events chart.

hierarchical design—See *nested design.*

histogram—A plot of a *frequency distribution* in the form of rectangles (cells) whose bases are equal to the class interval and whose areas are proportional to the frequencies. The pictorial nature of the histogram lets people see patterns that are difficult to see in a simple table of numbers. The histogram is one of the seven basic tools of quality.

hoshin kanri, hoshin planning—Japanese-based strategic planning/policy deployment process that involves consensus at all levels as plans are cascaded throughout the organization, resulting in actionable plans and continual monitoring and measurement.

Hotelling's T^2—See T^2.

house of quality—A diagram (named for its house-shaped appearance) that clarifies the relationship between customer needs and product features. It helps correlate market or customer requirements and analysis of competitive products with higher-level technical and product characteristics and makes it possible to bring several factors into a single figure. Also known as quality function deployment (QFD).

hypothesis—A statement about a *population* to be tested. See *null hypothesis, alternative hypothesis,* and *hypothesis testing.*

hypothesis testing—A statistical *hypothesis* is a conjecture about a *population parameter.* There are two statistical hypotheses for each situation—the *null hypothesis (H_0)* and the *alternative hypothesis (H_1).* The null hypothesis proposes that there is no difference between the population of the sample and the specified population; the alternative hypothesis proposes that there is a difference between the sample and the specified population. See *moving average.*

I

i—See *moving average.*

I **chart (individuals control chart)**—See *individuals control chart.*

imprecision—A measure of *precision* computed as a *standard deviation* of the test or measurement results. Less precision is reflected by a larger standard deviation.

incomplete block—A *block* that accommodates only a subset of *treatment* combinations. See *balanced incomplete block design.*

incomplete block design—A design in which the *design space* is subdivided into *blocks* in which there are insufficient *experimental units* available to run a complete *replicate* of the experiment.

in-control process—A condition where the existence of *special causes* is no longer indicated by a *Shewhart control chart*. It indicates (within limits) a predictable and *stable process*, but it does not indicate that only *random causes* remain; nor does it imply that the distribution of the remaining values is normal Gaussian.

independent variable—See *predictor variable*.

indicators—Predetermined measures used to measure how well an organization is meeting its customers' needs and its operational and financial performance objectives. Such indicators can be either leading or lagging indicators. Indicators are also devices used to measure lengths or flow.

indifference quality level—The quality level that, in an *acceptance sampling plan*, corresponds to a *probability of acceptance* of 0.5 when a continuing series of *lots* is considered.

indifference zone—The region containing quality levels between the *acceptance quality limit* and the *limiting quality level*.

indirect customers—Customers who do not receive process output directly, but are affected if the process output is incorrect or late.

individual development—A process that may include education and training, but also includes many additional interventions and experiences to enable an individual to grow and mature both intellectually as well as emotionally.

individuals control chart (*x* chart; *I* chart)—A *variables control chart* with individual values plotted in the form of a *Shewhart control chart* for individuals and *subgroup* size $n = 1$.

Note 1: The x chart is widely used in the chemical industry because of cost of testing, test turnaround time, and the time interval between independent samples.

Note 2: This chart is usually accompanied by a *moving range chart*, commonly with $n = 2$.

Note 3: An individuals chart sacrifices the advantages of averaging subgroups (and the assumptions of the *normal distribution* central limit theorem) to minimize *random variation*.

Central line: $\bar{x}$ (x_0 if standard given)

Control limits: $\bar{x} \pm E_2 \overline{MR}$ or $\bar{x} \pm E_3 \bar{s}$ (if standard given, $x_0 \pm 3s_0$)

where $\bar{x}$ is the average of the individual variables

$$\bar{x} = \frac{x_1 + x_2 + \ldots + x_n}{n}$$

where R is the range, $\bar{s}$ is average sample standard deviation, x_0 is the standard value for the average, and σ_0 is the standard value for the population standard deviation. (Use the formula with $\overline{MR}$ when the sample size is small; the formula with $\overline{s}$ when the sample is larger, generally > 10 to 12.)

inherent process variation—The variation in a *process* when the process is operating in a *state of statistical control*.

inherent R and M value—A measure of reliability or maintainability that includes only the effects of an item design and its application, and assumes an ideal operation and support environment.

input—Material, product, or service that is obtained from an upstream internal provider or an external supplier, and is used to produce an output.

input variable—A variable that can contribute to the variation in a *process*.

inspection—Measuring, examining, testing, and gauging one or more characteristics of a product or service and comparing the results with specified requirements to determine whether conformity is achieved for each characteristic.

inspection, 100 percent—See *100 percent inspection*.

inspection, attributes—See *inspection by attributes*.

inspection, curtailed—See *curtailed inspection*.

inspection, normal—See *normal inspection*.

inspection, rectifying—See *rectifying inspection*.

inspection, reduced—See *reduced inspection*.

inspection, tightened—See *tightened inspection*.

inspection, variables—See *inspection by variables*.

inspection by attributes—An *inspection* noting the presence, or absence, of the *characteristic*(s) in each of the items in the group under consideration and counting how many items do, or do not, possess the *characteristic*(s), or how many such events occur in the *item*, group, or opportunity space.

Note: When *inspection* is performed by simply noting whether the item is nonconforming or not, the inspection is termed inspection for *nonconforming* items. When inspection is performed by noting the number of nonconformities on each *unit*, the inspection is termed inspection for number of nonconformities.

inspection by variables—An *inspection* measuring the magnitude(s) of the *characteristic*(s) of an item.

inspection level—An index of the relative amount of *inspection* of an *acceptance sampling scheme* chosen in advance and relating the *sample size* to the *lot* size.

Note 1: A lower/higher inspection level can be selected if experience shows that a less or more discriminating *operating characteristic curve* will be appropriate.

Note 2: The term should not be confused with severity of sampling, which concerns *switching rules* that operate automatically.

inspection lot—A collection of similar units or a specific quantity of similar material offered for inspection and subject to a decision with respect to acceptance.

interaction effect—The *effect* for which the apparent influence of one *factor* on the *response variable* depends on one or more other factors. Existence of an *interaction effect* means that the factors cannot be changed independently of each other.

interaction plot—The plot providing the *average* responses at the combinations of *levels* of two distinct *factors*.

interactive multimedia—A term encompassing technology that allows the presentation of facts and images with physical interaction by the viewers; for example, taking a simulated certification exam on a computer or training embedded in transaction processing software.

intercept—See *regression analysis*.

intermediate customers—Distributors, dealers, or brokers who make products and services available to the end user by repairing, repackaging, reselling, or creating finished goods from components or subassemblies.

internal audit—An audit performed by impartial members of the organization being audited

internal customer—The recipient (person or department) of another person's or department's output (product, service, or information) within an organization. See also *external customer*.

internal reference material—A reference material that is not traceable to a national body, but is stable (within stated time and storage conditions), homogeneous, and appropriately characterized.

interrelationship digraph—A management and planning tool that displays the relationship between factors in a complex situation. It identifies meaningful categories from a mass of ideas and is useful when relationships are difficult to determine.

intervention—An action taken by a leader or a facilitator to support the effective functioning of a team or work group.

inventory, active—The group of items assigned an operational status.

inventory, inactive—The group of items being held in reserve for possible future assignment to an operational status.

Ishikawa diagram—See *cause-and-effect diagram*.

isolated lot—A unique *lot* or one separated from the sequence of *lots* in which it was produced or collected.

isolated sequence of lots—A group of *lots* in succession but not forming part of a large sequence or produced by a continuing *process*.

item—A nonspecific term used to denote any product, including systems, materials, parts, subassemblies, sets, accessories, and so on (*Source:* MIL-STD-280).

J

jidoka—Japanese method of autonomous control involving the adding of intelligent features to machines to start or stop operations as control parameters are reached, and to signal operators when necessary.

job aid—Any device, document, or other media that can be provided to a worker to aid in correctly performing tasks (for example, laminated setup instruction card hanging on machine, photos of product at different stages of assembly, or a metric conversion table).

job description—A narrative explanation of the work, responsibilities, and basic requirements of a job.

job enrichment—Increasing the worker's responsibilities and authority in work to be done.

job specification—A list of the important functional and quality attributes (knowledge, skills, aptitudes, and personal characteristics) needed to succeed in the job.

joint planning meeting—A meeting involving representatives of a key customer and the sales and service team for that account to determine how better to meet the customer's requirements and expectations.

Juran's trilogy—See *quality trilogy*.

just-in-time (JIT) manufacturing—An optimal material requirement planning system for a manufacturing process in which there is little or no manufacturing material inventory on hand at the manufacturing site and little or no incoming inspection.

just-in-time training—Providing job training coincidental with, or immediately prior to, an employee's assignment to a new or expanded job.

K

kaikaku—A Japanese term that means a breakthrough improvement in eliminating waste.

kaizen—A Japanese term that means gradual unending improvement by doing little things better and setting and achieving increasingly higher standards.

The term was made famous by Masaaki Imai in his book *Kaizen: The Key to Japan's Competitive Success.*

kaizen blitz/event—An intense, short time frame (typically three to five consecutive days), team approach to employ the concepts and techniques of continual improvement (for example, to reduce cycle time or increase throughput).

kanban—A system inspired by Taiichi Ohno's (Toyota) visit to a U.S. supermarket. The system signals the need to replenish stock or materials or to produce more of an item (also called a pull approach).

Kano model—A representation of the three levels of customer satisfaction defined as dissatisfaction, neutrality, and delight. Named after Noriaki Kano.

kurtosis—A measure of peakedness or flattening of a distribution near its center in comparison to the *normal distribution.*

L

λ **(lambda)**—The weighting factor in *EWMA* calculations where $0 < \lambda \le 1$.

L—See *lower specification limit.*

lateral thinking—A process that includes recognizing patterns, becoming unencumbered with old ideas, and creating new ones.

Latin square design—A design involving three *factors* in which the combination of the *levels* of any one of them with the levels of the other two appears once and only once. It is often used to reduce the impact of two blocking factors by balancing out their contributions. A basic assumption is that these block factors do not interact with the factor of interest or with each other. This design is particularly useful when the assumptions are valid for minimizing the amount of experimentation. See *Graeco-Latin square design.*

LCALI—A process for operating a listening-post system for capturing and using formerly unavailable customer data. Acronym stands for listen, capture, analyze, learn, and improve.

LCL—See *lower control limit.*

leader—An individual recognized by others as the person to lead an effort. One cannot be a leader without one or more followers. The term is often used interchangeably with manager. A leader may or may not hold an officially designated management-type position.

leadership—An essential part of a quality improvement effort. Organization leaders must establish a vision, communicate that vision to those in the organization, and provide the tools, knowledge, and motivation necessary to accomplish the vision.

lean approach/lean thinking—A focus on reducing cycle time and waste using a number of different techniques and tools, for example, value stream mapping, and identifying and eliminating monuments and non-value-added steps. Lean and agile are often used interchangeably.

lean manufacturing—Applying the lean approach to improving manufacturing operations.

learner-controlled instruction—When a learner works without an instructor, at an individual pace, building mastery of a task. Computer-based training is a form of LCI. Also called self-directed learning.

learning curve—The time it takes to achieve mastery of a task or body of knowledge.

learning organization—An organization that has as a policy to continue to learn and improve its products, services, processes, and outcomes; "an organization that is continually expanding its capacity to create its future" (Senge).

least squares, method of—A technique of estimating a *parameter* that minimizes the sum of the difference squared, where the difference is between the *observed value* and the predicted value (*residual*) derived from the *model*.

Note: *Experimental errors* associated with individual observations are assumed to be independent. The usual *analysis of variance, regression analysis,* and *analysis of covariance* are all based on the method of least squares.

lesson plan—A detailed plan created to guide an instructor in delivering training and/or education.

level—A potential setting, value, or assignment of a *factor* or the value of the *predictor variable*.

level of significance—See *significance level.*

life units—A measure of duration applicable to the item, for example, operating hours, cycles, distance, rounds fired, attempts to operate.

lifecycle—A product lifecycle is the total time frame from product concept to the end of its intended use; a project lifecycle is typically divided into five stages: concept, planning, design, implementation, evaluation, and closeout.

limiting quality level (LQL)—The quality *level* that, for the purposes of *acceptance sampling inspection,* is the limit of an unsatisfactory process average when a continuing series of *lots* is considered.

Note: The limiting quality level is sometimes referred to as the *rejectable quality level* (RQL), *unacceptable quality level* (UQL), or limiting quality (LQ), but limiting quality level (LQL) is the preferred term. When the percentage of *nonconforming units* is expressed as a percent defective, this may be referred to as the *lot tolerance percent defective* (LTPD).

line balancing—A method of proportionately distributing workloads within the value stream to meet takt time.

line graph budget—A graphic representation of the planned and/or actual expenditures of a project's budget throughout the performance of the project lifecycle

linear regression—The mathematical application of the concept of a scatter diagram where the correlation is actually a cause-and-effect relationship.

linear regression coefficients—The values associated with each *predictor variable* in a *linear regression equation*. They tell how the *response variable* changes with each unit increase in the *predictor variable*. See *regression analysis*.

linear regression equation—A function that indicates the linear relationship between a set of *predictor variables* and a *response variable*. See *regression analysis*.

linearity (general sense)—The degree to which a pair of variables follow a straight-line relationship. Linearity can be measured by the *correlation coefficient*.

linearity (measurement system sense)—The difference in *bias* through the range of measurement. A measurement system that has good *linearity* will have a constant bias no matter the magnitude of measurement. If one views the relation between the observed measurement result on the *y*-axis and the *true value* on the *x*-axis, an ideal *measurement system* would have a line of slope equal to one.

lognormal distribution—If log *x* is normally distributed, it is a lognormal distribution. See *normal distribution*.

lot—A definite part of a population constituted under essentially the same conditions as the *population* with respect to the sampling purpose.

Note: The sampling purpose may, for example, be to determine lot acceptability or to estimate the mean value of a particular *characteristic*.

lot quality—A *statistical measure* of quality of the product from a given *lot*. These measures may relate to the occurrence of *events* or to physical measurements. For the purposes of this glossary, the most commonly used measure of lot quality is likely to be the percentage or proportion of *nonconforming units* in a *lot*.

lot size (N)—The number of items in a *lot*.

lot tolerance percent defective (LTPD)—See note under *limiting quality level*.

lot-by-lot—The *inspection* of a product submitted in a series of *lots*.

lower control limit (LCL)—The *control chart* control limit that defines the lower control boundary.

lower specification limit (lower tolerance limit, L)—The *specification* or *tolerance limit* that defines the lower limiting value.

LQL—See *limiting quality level*.

LRPL—See *rejectable process level*.

LTPD—See note under *limiting quality level*.

M

μ (mu)—See *population mean*.

MA chart—See *moving average chart*.

main effect—The influence of a single *factor* on the *mean* of a *response variable*.

main effects plot—A plot giving the *average* responses at the various *levels* of individual *factors*.

maintainability—The measure of the ability of an item to be retained or restored to specified condition when maintenance is performed by personnel having specified skill levels, using prescribed procedures and resources, at each prescribed level of maintenance and repair.

maintenance—All actions necessary for retaining an item in or restoring it to a specified condition.

maintenance, corrective—All actions performed, as a result of failure, to restore an item to a specified condition. Corrective maintenance can include any or all of the following steps: localization, isolation, disassembly, interchange, reassembly, alignment, and checkout.

maintenance, preventive—All actions performed in an attempt to retain an item in a specified condition by providing systematic inspection, detection, and prevention of incipient failures.

maintenance action rate—The reciprocal of the mean time between maintenance actions = 1/MTBMA.

maintenance time constraint (*t*)—The maintenance time that is usually prescribed as a requirement of the mission.

Malcolm Baldrige National Quality Award (MBNQA)—An award established by Congress in 1987 to raise awareness of quality management and to recognize U.S. companies that have implemented successful quality management systems. Criteria for Performance Excellence are published each year. Three awards may be given annually in each of five categories: manufacturing businesses, service businesses, small businesses, education institutions, and healthcare organizations. The award is named after the late Secretary of Commerce Malcolm Baldrige, a proponent of quality management. The U.S. Commerce Department's National Institute of Standards and Technology manages the award, and ASQ administers it. The major emphasis in determining success is achieving results.

market-perceived quality—The customer's opinion of your products or services as compared to those of your competitors.

materials review board (MRB)—A quality control committee or team, usually employed in manufacturing or other materials-processing installations, that has the responsibility and authority to deal with items or materials that do not conform to fitness-for-use specifications. An equivalent, the error review board, is sometimes used in software development.

matrix chart/diagram—A management and planning tool that shows the relationships among various groups of data; it yields information about the relationships and the importance of task/method elements of the subjects.

mean—See *arithmetic mean.*

mean maintenance downtime (MDT)—Includes supply time, logistics time, administrative delays, active maintenance time, and so on.

mean maintenance time—The measure of item maintainability taking into account maintenance policy. The sum of preventive and corrective maintenance times, divided by the sum of scheduled and unscheduled maintenance events, during a stated period of time.

mean time between failure (MTBF)—A basic measure of reliability for repairable items: the mean number of life units during which all parts of the item perform within their specified limits during a particular measurement interval under stated conditions.

mean time between maintenance (MTBM)—A measure of reliability taking into account maintenance policy. The total number of life units expended by a given time divided by the total number of maintenance events (scheduled and unscheduled) due to that item.

mean time between maintenance actions (MTBMA)—A measure of the system reliability parameter related to demand for maintenance manpower: the total number of system life units divided by the total number of maintenance actions (preventive and corrective) during a stated period of time.

mean time to failure (MTTF)—A basic measure of system reliability for nonrepairable items: the total number of life units of an item divided by the total number of failures within that population during a particular measurement interval under stated conditions.

mean time to repair (MTTR)—A basic measure of maintainability: the sum of corrective maintenance times at any specific level of repair divided by the total number of failures within an item repaired at that level during a particular interval under stated conditions.

means, tests for—Testing for *means* includes computing a *confidence interval* and hypothesis testing by comparing means to a *population* mean (known or unknown) or to other *sample* means. Which test to use is determined by whether s is known, whether the test involves one or two samples, and other factors.

If *s* is known, the *z-confidence interval* and *z-test* apply. If *s* is unknown, the *t-confidence interval* and *t-test* apply.

measurement system—Everything that can introduce variability into the measurement *process,* such as equipment, operator, sampling methods, and accuracy.

measurement systems analysis (MSA)—A statistical analysis of a *measurement system* to determine the variability, *bias, precision,* and *accuracy* of the measurement system(s). Such studies may also include the limit of detection, selectivity, *linearity,* and other *characteristics* of the system in order to determine the suitability of the measurement system for its intended purpose.

median—The value for which half the *data* is larger and half is smaller. The median provides an estimator that is insensitive to very extreme values in a data set, whereas the *average* is affected by extreme values.

Note: For an odd number of units, the median is the middle measurement; for an even number of units, the median is the average of the two middle measurements.

mentor—A person who voluntarily assumes a role of trusted advisor and teacher to another person. The mentor may or may not be the mentored person's organizational superior or even in the same organization. Usually the only reward the mentor receives is self-gratification in having helped someone else.

meta-analysis—The use of statistical methods to combine the results of multiple studies into a single conclusion.

method of least squares—See *least squares, method of.*

M_i—See *moving average.*

midrange—(Highest value + lowest value)/2.

mind mapping—A technique for creating a visual representation of a multitude of issues or concerns by forming a map of the interrelated ideas.

mission profile—A time-phased description of the events and environments an item experiences from initiation to completion of a specified mission, to include the criteria of mission success of critical failures.

mission statement—An explanation of purpose or reasons for existing as an organization; it provides the focus for the organization and defines its scope of business.

mistake-proofing—The use of *process* or design features to prevent manufacture of nonconforming product. See *poka-yoke.*

mixed model—A model where some *factors* are fixed, but other *factors* are random. See *fixed factors* and *random factors.*

mixed variables and attribute sampling—The *inspection* of a sample by *attributes*, in addition to *inspection by variables* already made of a previous *sample* from the *lot*, before a decision as to acceptability or rejectability of the lot can be made.

mixture design—A design constructed to handle the situation in which the *predictor variables* are constrained to sum to a fixed quantity, such as proportions of ingredients that make up a formulation or blend.

mode—The most frequent value of a *variable*.

model—The description relating a *response variable* to *predictor variable*(s) and including attendant assumptions.

model I analysis of variance (fixed model)—An analysis of variance in which the *levels* of all *factors* are fixed rather than random selections over the range of versions to be studied for those factors.

model II analysis of variance (random model)—An analysis of variance in which the *levels* of all *factors* are assumed to be selected at random over the range of versions to be studied for those factors.

moving average—Let $x_1, x_2, \ldots$ denote individual observations. The moving average of span w at time i is

$$M_i = \frac{x_i + x_{i-1} + \ldots + x_{i-w+1}}{w}$$

At time period i, the oldest observation in the moving average set is dropped and the newest one added to the set.

The variance, V, of the moving average M_i:

$$V(M_i) = \frac{1}{w^2} \sum_{j=1-w+1}^{i} V(x_j) = \frac{1}{w^2} \sum_{J=i-w+1}^{i} \sigma^2 = \frac{\sigma^2}{w}$$

moving average chart (MA chart)—A *variables control chart* where the simple unweighted *moving average* of the last w observations is plotted and provides a convenient substitute for the $\bar{x}$ *chart*. A fixed number of samples, w, is required for averaging, but the choice of the interval for w may be difficult to optimize. The control *signal* is a single point out of limits. Additional tests are not appropriate because of the high correlation introduced between successive averages that share several common values. The MA chart is sensitive to trends and is sometimes used in conjunction with *individuals charts* and/or *MR charts*. It is, however, probably less effective than either the *CUSUM* or *exponentially weighted moving average* (EWMA) chart for this purpose.

Note 1: This chart is particularly useful when only one observation per *subgroup* is available. Examples are process characteristics such as temperature, pressure, and time.

Note 2: This chart has the disadvantage of an unweighted carryover effect lasting w points.

$$\text{Each plotted point: } M_i = \frac{x_i + x_{i-1} + \ldots + x_{i-w+1}}{w}$$

where M_i is the moving range of span w at time i. At time period i, the oldest observation in the moving average set is dropped and the newest one added to the set. For time periods where $i < w$, the average of the observations for periods $1, 2 \ldots i$ is plotted. Example (where $w = 4$):

Observation, i	x_i	M_i
1	9.45	9.45
2	7.99	8.72
3	9.29	8.91
4	11.66	9.5975
5	12.16	10.275

$$\text{Central line: } \bar{x}$$

$$\text{Control limits: } \bar{x} \pm \frac{3\sigma}{\sqrt{w}}$$

where $\bar{x}$ is the average of the individual variables

$$\bar{x} = \frac{x_1 + x_2 + \ldots + x_n}{n}$$

where $\bar{n}$ is the total number of samples, σ the standard deviation, and w is the moving range span.

moving range chart (MR chart)—A *control chart* that plots the absolute difference between the current and previous value. It often accompanies an *individuals chart* or *moving average chart*.

Note: The current observation replaces the oldest of the latest $n + 1$ observations.

$$\text{Central line: } \bar{R}$$

$$\text{Control limits: } UCL = D_4\bar{R}$$

$$LCL = D_3\bar{R}$$

where D_4 and D_3 are *control chart factors*.

MR chart—See *moving range chart*.

MRB—See *material review board*.

MSA—See *measurement systems analysis.*

muda—(Japanese) An activity that consumes resources, but creates no value; seven categories are correction, processing, inventory, waiting, overproduction, internal transport, and motion (waste).

multilevel continuous sampling—A continuous *acceptance sampling inspection* of consecutively produced items where two or more sampling inspection rates are either alternated with *100 percent inspection* or with each other depending on the quality of the observed *process* output.

multimodal—A *probability distribution* having more than one mode. See *bimodal* and *trimodal.*

multiple acceptance sampling inspection—An *acceptance sampling inspection* in which, after each *sample* has been inspected, a decision is made based on defined decision rules to accept, not accept, or take another *sample.*

Note: For most multiple sampling plans, the largest number of samples that can be taken is specified with an accept or not accept decision being forced at that point.

multiple linear regression—See *regression analysis.*

multivariate control chart—A variables *control chart* that allows plotting of more than one variable. These charts make use of the T^2 *statistic* to combine information from the *dispersion* and *mean* of several variables.

multivoting—A decision-making tool that enables a group to sort through a long list of ideas to identify priorities.

mystery shopper—A person who pretends to be a regular shopper in order to get an unencumbered view of how a company's service process works.

N

n—See *sample size.*

N—See *lot size.*

natural process limits—See *reference interval.*

natural team—A work group having responsibility for a particular process.

negative binomial distribution—A two-parameter, *discrete distribution.*

negotiation—A process for individuals or groups to work together to achieve common goals under an agreement that assures that each party's values and priorities be addressed to reach a 'win–win' result.

nested design—A design in which the second *factor* is nested within *levels* of the first factor and each succeeding factor is nested within levels of the previous factor.

nested factor—A *factor* (A) is nested within another factor (B) if the *levels* of A are different for every level of B. See *crossed factor.*

next operation as customer—Concept that the organization has service/product providers and service/product receivers or internal customers.

noise factor—In *robust parameter design,* a noise factor is a *predictor variable* that is hard to control or is not desired to control as part of the standard experimental conditions. In general, it is not desired to make inference on noise factors, but they are included in an experiment to broaden the conclusions regarding *control factors.*

nominal scale—A scale with unordered, labeled categories, or a scale ordered by convention. Examples: Type of defect, breed of dog, complaint category.

Note: It is possible to count by category, but not order or measure.

nonconforming unit—A *unit* with one or more *nonconformities.*

nonconformity—The result of nonfulfillment of a specified requirement. See also *blemish* and *defect.*

nondestructive testing and evaluation (NDT)—Testing and evaluation methods that do not damage or destroy the product being tested.

non-value-added—Refers to tasks or activities that can be eliminated with no deterioration in product or service functionality, performance, or quality in the eyes of the customer.

normal distribution (Gaussian distribution)—A *continuous,* symmetrical *frequency distribution* that produces a bell-shaped curve. The location *parameter* (*x*-axis) is the *mean,* μ. The scale parameter, σ, is the *standard deviation* of the normal distribution. When measurements have a normal distribution, 68.26% of the values lie within plus or minus one standard deviation of the mean ($\mu \pm 1\sigma$); 95.44% lie within plus or minus two standard deviations of the mean ($\mu \pm 2\sigma$); while 99.73% lie within plus or minus three standard deviations of the mean ($\mu \pm 3\sigma$). The probability function is

$$f(x) = \frac{1}{\sigma\sqrt{2\pi}} \exp\left[-\frac{(x-\mu)^2}{2\sigma^2} \right]$$

where

$$-\infty < x < \infty \text{ and with parameters } -\infty < \mu < \infty \text{ and } \sigma > 0.$$

normal inspection—The inspection that is used when there is no reason to think that the quality level achieved by the *process* differs from a specified level. See *tightened inspection* and *reduced inspection.*

normal probability plot—See *probability plot.*

norms—Behavioral expectations, mutually agreed-upon rules of conduct, protocols to be followed, social practice.

np **(number of affected or categorized units)**—The total number of *units* in a *sample* in which an *event* of a given classification occurs. A unit (area of opportunity) is counted only once, even if several events of the same classification are encountered.

Note: In the quality field, the classification generally is number of *nonconforming units.*

np **chart (number of categorized units control chart)**—An *attributes control chart* for number of *events* per unit where the opportunity is variable. (The *np* chart is for the number nonconforming whereas the *p chart* is for the proportion nonconforming.)

Note: Events of a particular type, for example, number of absentees or number of sales leads, form the count. In the quality field, events are often expressed as *nonconformities* and the variable opportunity relates to *subgroups* of variable size or variable amounts of material. If a standard is given:

$$\text{Central line: } np$$

$$\text{Control limits: } np \pm 3\sqrt{np(1-p)}$$

where *np* is the standard value of the events per unit and *p* is the standard value of the fraction nonconforming.

If no standard is given:

$$\text{Central line: } \overline{np}$$

$$\text{Control limits: } \overline{np} \pm 3\sqrt{\overline{np}(1-\overline{p})}$$

where $\overline{np}$ is the average value of the events per unit and $\overline{p}$ is the average value of the fraction nonconforming.

null hypothesis, H_0—The *hypothesis* in *tests of significance* that there is no difference (null) between the *population* of the *sample* and the specified population (or between the populations associated with each sample). The null hypothesis can never be proved true, but it can be shown (with specified risks or error) to be untrue; that is, a difference exists between the populations. If it is not disproved, one assumes there is no adequate reason to doubt it is true, and the null hypothesis is accepted. If the null hypothesis is shown to be untrue, then the *alternative hypothesis* is accepted. Example: In a random sample of independent random variables with the same *normal distribution* with unknown *mean* and unknown *standard deviation,* a typical null hypothesis for the mean, μ, is that the mean is less than or equal to a given value, μ_0. The hypothesis is written as: $H_0 = \mu \leq \mu_0$.

O

objective—A quantitative statement of future expectations and an indication of when the expectations should be achieved; it flows from goals and clarifies what people must accomplish.

objective evidence—Verifiable qualitative or quantitative observations, information, records, or statements of fact pertaining to the quality of an item or service or to the existence and implementation of a quality system element.

objective setting—See *S.M.A.R.T. W.A.Y.*

observational unit—The smallest entity on which a *response variable* is measured. It may be or not be the same as the *experimental unit.*

observed value—The particular value of a *characteristic* determined as a result of a test or measurement.

OC curve—See *operating characteristic curve.*

off-the-job training—Training that takes place away from the actual work site.

Ogive—A type of graph that represents the cumulative frequencies for the classes in a *frequency distribution.*

on-the-job training (OJT)—Training conducted usually at the workstation, typically done one-on-one.

one-tailed test—A *hypothesis* test that involves only one of the tails of a distribution. Example: We wish to reject the *null hypothesis* H_0 only if the true *mean* is larger than μ_0.

$$H_0: \mu = \mu_0$$

$$H_1: \mu < \mu_0$$

A one-tailed test is either right-tailed or left-tailed, depending on the direction of the inequality of the *alternative hypothesis.*

one-to-one marketing—The concept of knowing customers' unique requirements and expectations and marketing to these. See also *customer relationship management.*

operable—The state of being able to perform the intended function.

operating characteristic curve (OC curve)—A curve showing the relationship between the *probability* of acceptance of product and the incoming quality level for a given *acceptance sampling plan.*

Note: Several terms in this glossary need to be referenced for better understanding of operating characteristic curves: α, β, *acceptance quality level, consumer's risk, indifference quality level, indifference zone, limiting quality level, probability of acceptance, probability of nonacceptance,* and *producer's risk.*

ordinal scale—A scale with ordered, labeled categories.

Note 1: There is sometimes a blurred borderline between ordinal and *discrete scales.* When subjective opinion ratings such as excellent, very good, neutral, poor, and very poor are coded (as numbers 1–5), the apparent effect is conversion from an ordinal to a discrete scale. Such numbers should not be treated as ordinary numbers, however, because the distance between 1 and 2 may not be the same as between 2 and 3, or 3 and 4, and so forth. On the other hand, some categories that are ordered objectively according to magnitude, such as the Richter scale, which ranges from zero to 8 according to the amount of energy release, could be related to a discrete scale as well.

Note 2: Sometimes *nominal scales* are ordered by convention. An example is the blood groups A, B, and O, which are always stated in this order. The same is the case if different categories are denoted by single letters; they are then ordered by convention, according to the alphabet.

organization culture—Refers to the collective beliefs, values, attitudes, manners, customs, behaviors, and artifacts unique to an organization.

original inspection—The *inspection* of a *lot*, or other amount, not previously inspected.

Note: This is in contrast, for example, to inspection of a *lot* that has previously been designated as not acceptable and is submitted again for inspection after having been further sorted, reprocessed, and so on.

orthogonal contrasts—A set of *contrasts* whose coefficients satisfy the condition that, if multiplied in corresponding pairs, the sum of the products equals zero. See *contrast analysis*.

orthogonal design—A design in which all pairs of *factors* at particular *levels* appear together an equal number of times. Examples include a wide variety of special designs such as a *Latin square, completely randomized factorial design,* and *fractional factorial design*. Statistical analysis of the results for *orthogonal designs* is relatively simple because each *main effect* and *interaction effect* may be evaluated independently.

OSHA—Occupational Safety and Health Administration.

out of spec—A term used to indicate that a unit does not meet a given specification.

outlier—An extremely high or an extremely low data value compared to the rest of the data values. Great caution must be used when trying to identify an outlier.

out-of-control process—A *process* operating with the presence of *special causes*. See *in-control process*.

output—The deliverables resulting from a process, project, a quality initiative, an improvement, and so on. Outputs include data, information, documents, decisions, and tangible products. Outputs are generated both from the planning and management of the activity and the delivered product, service, or program. Output is also the item, document, or material delivered by an internal provider (supplier) to an internal receiver (customer).

output variable—The variable representing the outcome of the *process*.

outsourcing—A strategy and an action to relieve an organization of processes and tasks in order to reduce costs, improve quality, reduce cycle time (for example, by parallel processing), reduce the need for specialized skills, and increase efficiency.

P

p—The ratio of the number of *units* in which at least one *event* of a given classification occurs to the total number of units. A unit is counted only once even if several events of the same classification are encountered within it. *p* can also be expressed as a percent.

p chart—See *proportion chart*.

P—See *probability*.

$p_{.95}$, $p_{.50}$, $p_{.10}$, $p_{.05}$—The submitted quality in terms of the proportion of variant units for which the *probability of acceptance* is 0.95, 0.50, 0.10, 0.05 for a given *sampling plan*.

p_1—The percent of *nonconforming* individual items occurring when the *process* is located at the *acceptable process level (APL)*.

p_2—The percent of *nonconforming* individual items occurring when the *process* is located at the *rejectable process level (RPL)*.

P_a—See *probability of acceptance*.

P_r—See *probability of rejection*.

panels—Groups of customers recruited by an organization to provide ad hoc feedback on performance or product development ideas.

paradigm—The standards, rules, attitudes, mores, and so on, that influence the way by which an organization lives and behaves.

parameter—A constant or coefficient describing some *characteristic* of a *population*. Examples: *standard deviation* and *mean*.

Pareto chart—A basic tool used to graphically rank causes from most significant to least significant. It utilizes a vertical bar graph in which the bar height reflects the frequency or impact of causes.

partially nested design—A *nested design* in which several *factors* may be *crossed* as in *factorial experiments* and other factors nested within the crossed combinations.

participative management—A style of managing whereby the manager tends to work from theory Y assumptions about people, involving the workers in decisions made. See *theory Y*.

partnership/alliance—A strategy leading to a relationship with suppliers or customers aimed at reducing costs of ownership, maintenance of minimum stocks, just-in-time deliveries, joint participation in design, exchange of information on materials and technologies, new production methods, quality improvement strategies, and the exploitation of market synergy.

parts per million (PPM or ppm)—A measurement that is expressed by dividing the data set into 1,000,000 or 106 equal groups.

part-to-part variation—The variability of the data due to measurement items rather than the *measurement system*. This *variation* is typically estimated from the measurement items used in a study, but could be estimated from a *representative sample* of product.

Pearson's correlation coefficient—See *correlation coefficient*.

percentile—The division of the data set into 100 equal groups.

performance appraisal/evaluation—A formal method of measuring employees' progress against performance standards and providing feedback to them.

performance plan—A performance management tool that describes desired performance and provides a way to assess the performance objectively.

PERT—See *program evaluation and review technique*.

pilot run—A trial production run consisting of various production tests conducted to ensure that the product is being built to specification.

Plackett-Burman design—An *experimental design* used where there are many *factors* to study. It only studies the *main effects* and is primarily used as a *screening design* before applying other types of designs. In general, Plackett-Burman designs are two-level, but three-, five-, and seven-level designs are available. They allow for efficient estimation of the main effects, but assume that interactions can initially be ignored.

plan–do–check–act (PDCA) cycle—A four-step process for quality improvement. In the first step (plan), a plan to effect improvement is developed. In the second step (do), the plan is carried out, preferably on a small scale. In the third step (check), the effects of the plan are observed. As part of the last step (act), the results are studied to determine what was learned and what can be predicted. The plan–do–check–act cycle is sometimes referred to as the Shewhart cycle, because Walter A. Shewhart discussed the concept in his book *Statistical Method from the Viewpoint of Quality Control,* and as the Deming cycle because W. Edwards Deming introduced the concept in Japan. The Japanese subsequently called it the Deming cycle. Sometimes referred to as plan–do–study–act (PDSA).

plan–do–study–act (PDSA) cycle—A variation of PDCA.

PMI—Project Management Institute (*www.pmi.org*)

Poisson distribution—The Poisson distribution describes occurrences of isolated events in a continuum of time or space. It is a one-parameter, *discrete distribution* depending only on the *mean*.

poka-yoke—(Japanese) A term that means to mistake-proof a process by building safeguards into the system that avoid or immediately find errors. It comes from poka, which means *error,* and yokeru, which means *to avoid.*

pooled standard deviation—A *standard deviation* value resulting from some combination of individual standard deviation values. It is most often used when individual standard deviation values are similar in magnitude and can be denoted by *sp*. See *t-test (two sample)*.

population—1. The entire set (totality) of *units*, quantity of material, or observations under consideration. A population may be real and finite, real and infinite, or completely hypothetical. See *sample*. 2. A group of people, objects, observations, or measurements about which one wishes to draw conclusions.

population covariance—See *covariance*.

population mean (μ)—The true *mean* of the *population*, represented by μ *(mu)*. The *sample mean*, $\bar{x}$, is a common estimator of the population mean.

population standard deviation—See *standard deviation*.

population variance—See *variance*.

power—The equivalent to one minus the *probability* of a *Type II error* $(1 - \beta)$. A higher power is associated with a higher probability of finding a statistically significant difference. Lack of power usually occurs with smaller sample sizes.

power curve—The curve showing the relationship between the *probability* $(1 - \beta)$ of rejecting the *hypothesis* that a sample belongs to a given *population* with a given *characteristic(s)* and the actual population value of that characteristic(s).

Note: If β, the probability of accepting the hypothesis, is used instead of $(1 - \beta)$, the curve is called an *operating characteristic (OC) curve*.

P_p (process performance index)—An index describing *process performance* in relation to specified *tolerance*

$$P_p = \frac{U - L}{6s}$$

s is used for *standard deviation* instead of σ since both *random* and *special causes* may be present.

Note: A *state of statistical control* is not required.

P_{pk} (minimum process performance index)—The smaller of *upper process performance index* and *lower process performance index*.

P_{pk_L} (lower process performance index or PPL)—An index describing *process performance* in relation to the *lower specification limit*. For a symmetrical normal distribution:

$$P_{pk_L} = \frac{\bar{x} - L}{3s}$$

where *s* is defined under P_p.

P_{pk_U} **(upper process performance index or PPU)**—An index describing *process performance* in relation to the *upper specification limit*. For a symmetrical *normal distribution:*

$$P_{pk_U} = \frac{U - \bar{x}}{3s}$$

where *s* is defined under P_p.

PPL—See P_{pk_L}.

PPU—See P_{pk_U}.

precision—The closeness of agreement between randomly selected individual measurements or test results. See *repeatability* and *reproducibility*.

precision to tolerance ratio (PTR)—A measure of the *capability* of the *measurement system*. It can be calculated by

$$PTR = \frac{5.15 \times \hat{\sigma}_{ms}}{USL - LSL}$$

where $\hat{\sigma}_{ms}$ is the estimated *standard deviation* of the total measurement system variability. In general, reducing the PTR will yield an improved measurement system.

predicted—That which is expected at some future date, postulated on analysis of past experience and tests.

predicted value—The prediction of future observations based on the formulated *model.*

prediction interval—Similar to a *confidence interval,* it is an interval based on the *predicted value* that is likely to contain the values of future observations. It will be wider than the confidence interval because it contains bounds on individual observations rather than a bound on the *mean* of a group of observations.

predictor variable—A *variable* that can contribute to the explanation of the outcome of an experiment.

prevention versus detection—A term used to contrast two types of quality activities. Prevention refers to those activities designed to prevent nonconformances in products and services. Detection refers to those activities designed to detect nonconformances already in products and services. Another phrase used to describe this distinction is designing-in quality versus inspecting-in quality.

preventive action—Action taken to eliminate the potential causes of a nonconformity, defect, or other undesirable situation in order to prevent occurrence.

primary customer—The individual or group who directly receives the output of a process.

priorities matrix—A tool used to choose between several options that have many useful benefits, but where not all of them are of equal value.

probability (*P*)—The chance of an *event* occurring.

probability distribution—A function that completely describes the probabilities with which specific values occur. The values may be from a *discrete scale* or a *continuous scale*.

probability of acceptance (*P_a*)—The *probability* that when using a given *acceptance sampling plan*, a *lot* will be accepted when the *lot* or *process* is of a specific quality level.

probability of nonacceptance—See *probability of rejection*.

probability of rejection (*P_r*)—The probability that when using a given *acceptance sampling plan*, a *lot* will not be accepted when *the lot* or *process* is of a specified quality level.

probability plot—The plot of ranked data versus the *sample* cumulative frequency on a special vertical scale. The special scale is chosen (that is, normal, lognormal, and so on) so that the *cumulative distribution* is a straight line.

problem solving—A rational process for identifying, describing, analyzing, and resolving situations in which something has gone wrong without explanation.

process—An activity or group of activities that takes an input, adds value to it, and provides an output to an internal or external customer; a planned and repetitive sequence of steps by which a defined product or service is delivered. A process can be graphically represented using a *flowchart*.

process analysis—Defining and quantifying the process capability from data derived from mapping and measurement of the work performed by the process.

process audit—An audit of a process.

process average—See *sample mean* or *population mean*.

process capability—The calculated inherent variability of a *characteristic* of a product. It represents the best performance of the *process* over a period of stable operations. Process capability is expressed as $6\hat{\sigma}$, where $\hat{\sigma}$ is the *sample standard deviation* (short-term component of variation) of the process under a *state of statistical control*.

process capability index—A single-number assessment of ability to meet *specification limits* on the quality *characteristic(s)* of interest. The indices compare the variability of the characteristic to the specification limits. Three basic process capability indices are C_p, C_{pk}, and C_{pm}.

Note: Since there are many different types and variations of process capability indices, details are given under the symbol for the specific type of index.

process control—*Process* management that is focused on fulfilling process requirements. Process control is the methodology for keeping a process within boundaries and minimizing the variation of a process.

process decision program chart (PDPC)—A management and planning tool that identifies all events that can go wrong and the appropriate countermeasures for these events. It graphically represents all sequences that lead to a desirable effect.

process improvement—The act of changing a process to reduce variability and cycle time and make the process more effective, efficient, and productive.

process improvement team (PIT)—A natural work group or cross-functional team whose responsibility is to achieve needed improvements in existing processes. The lifespan of the team is based on the completion of the team purpose and specific goals.

process management—The collection of practices used to implement and improve process effectiveness; it focuses on holding the gains achieved through process improvement and assuring process integrity.

process mapping—The flowcharting of a work process in detail, including key measurements.

process owner—The manager or leader who is responsible for ensuring that the total process is effective and efficient.

process performance—The statistical measure of the outcome of a *characteristic* from a *process* that may *not* have been demonstrated to be in a *state of statistical control*.

Note: Use this measure cautiously since it may contain a component of variability from *special causes* of unpredictable value. It differs from *process capability* because a state of statistical control is not required.

process performance index—A single-number assessment of ability to meet *specification limits* on the quality *characteristic*(s) of interest. The indices compare the variability of the characteristic to the specification limits. Three basic process capability indices are P_p, P_{pk}, and P_{pm}.

Note: Since there are many different types and variations of process capability indices, details are given under the symbol for the specific type of index.

process quality—A statistical measure of the quality of product from a given *process*. The measure may be an *attribute (qualitative)* or a *variable (quantitative)*. A common measure of process quality is the fraction or proportion of nonconforming units in the process.

process to tolerance ratio—See *precision to tolerance ratio*.

process variable—See *variable*.

producer's risk (α)—The *probability* of nonacceptance when the quality level has a value stated by the *acceptance sampling plan* as acceptable.

Note 1: Such nonacceptance is a *Type I error.*

Note 2: Producer's risk is usually designated as α.

Note 3: Quality level could relate to fraction nonconforming and acceptable to *AQL.*

Note 4: Interpretation of the producer's risk requires knowledge of the stated quality level.

product audit—Audit of a product.

product/service liability—The obligation of a company to make restitution for loss related to personal injury, property damage, or other harm caused by its product or service.

product warranty—The organization's stated policy that it will replace, repair, or reimburse a customer for a defective product providing the product defect occurs under certain conditions and within a stated period of time.

production part approval process (PPAP)—In the automotive industry, PPAP is used to show that suppliers have met all the engineering requirements based on a sample of parts produced from a production run made with production tooling, processes, and cycle times.

professional development plan—An individual development tool for an employee. Working together, the employee and his/her supervisor create a plan that matches the individual's career needs and aspirations with organizational demands.

program evaluation and review technique (PERT)—An event-oriented project management planning and measurement technique that utilizes an arrow diagram to identify all major project events and demonstrates the amount of time (critical path) needed to complete a project. It provides three time estimates: optimistic, most likely, and pessimistic.

project—A temporarily endeavor undertaken to create a unique project, service, or result.

project constraints—Projects are constrained by schedule, budget, and project scope.

project lifecycle—Refers to five sequential phases of project management: concept, planning, design, implementation, and evaluation.

project management (PM)—The application of knowledge, skills, tools, and techniques to project activities to meet project requirements throughout a project's lifecycle.

project management body of knowledge (PMBoK)—An inclusive term that describes the sum of knowledge within the profession of project management. The complete project management body of knowledge includes proven traditional practices that are widely applied and innovative practices that are emerging in the project management profession. A guide to the PMBoK is maintained and periodically updated by the Project Management Institute (PMI).

project plan—All the documents that comprise the details of why the project is to be initiated, what the project is to accomplish, when and where it is to be implemented, who will have responsibility, how the implementation will be carried out, how much it will cost, what resources are required, and how the project's progress and results will be measured.

project team—A designated group of people working together to produce a planned project's outputs and outcome.

proportion chart (*p* chart or percent categorized units control chart)—An *attributes control chart* for number of *units* of a given classification per total number of units in the *sample* expressed as either a proportion or percent.

Note 1: The classification often takes the form of *nonconforming units*.

Note 2: The *p* chart applies particularly when the sample size is variable.

Note 3: If the *upper control limit* (UCL) calculates ≥ 1 there is no UCL; or if the *lower control limit* (LCL) calculates ≤ 0, there is no LCL.

a. When the fraction nonconforming is known, or is a specified standard value:

Central line: p

Control limits: $p \pm 3\sqrt{\dfrac{p(1-p)}{n}}$

where p is known fraction nonconforming (or standard value).

Plotted value: $\hat{p}$

where $\hat{p}$ is the *sample* fraction nonconforming:

$$\hat{p} = D/n$$

where D is the number of product units that are nonconforming and n is the sample size.

b. When the fraction nonconforming is not known:

Central line: $\bar{p}$

Control limits: $\bar{p} \pm 3\sqrt{\dfrac{\bar{p}(1-\bar{p})}{n}}$

where $\bar{p}$ is the *average* value of the fraction of the classification (often average percent nonconforming), and n is the total number of units. These limits are considered trial limits.

c. For variable sample size:

$$\text{Central line: } \bar{p}$$

$$\text{Control limits: } \bar{p} \pm 3\sqrt{\frac{\bar{p}(1-\bar{p})}{n_i}}$$

where n_i is varying sample size.

$$\text{Plotted value: } \hat{p}$$

where $\hat{p}$ is the *sample* fraction nonconforming:

$$\hat{p} = D_i/n$$

where D_i is the number of nonconforming units in sample i, and n is the sample size.

proportions, tests for—Tests for proportions include the *binomial distribution*. See *binomial confidence interval,* and *z-test (two proportions)*. The *standard deviation* for proportions is given by

$$s = \sqrt{\frac{p(1-p)}{n}}$$

where p is the *population* proportion and n is the *sample* size.

psychographic customer characteristics—Variables among buyers in the consumer market that address lifestyle issues and include consumer interests, activities, and opinions.

PTR—See *precision to tolerance ratio.*

pull system—See *kanban.*

p-value—The *probability* of observing the *test statistic* value or any other value at least as unfavorable to the *null hypothesis.*

Q

QPA—Quality Process Analyst.

qualitative data—See *attributes data.*

quality—A subjective term for which each person has his or her own definition. In technical usage, quality can have two meanings: (1) the characteristics of a product or service that bear on its ability to satisfy stated or implied needs and (2) a product or service free of deficiencies.

quality assessment—The process of identifying business practices, attitudes, and activities that are enhancing or inhibiting the achievement of quality improvement in an organization.

quality assurance/quality control (QA/QC)—Two terms that have many interpretations because of the multiple definitions for the words assurance and control. For example, assurance can mean the act of giving confidence, the state of being certain, or the act of making certain; control can mean an evaluation to indicate needed corrective responses, the act of guiding, or the state of a process in which the variability is attributable to a constant system of chance causes. (For a detailed discussion on the multiple definitions, see ANSI/ISO/ASQC A35342, *Statistics—Vocabulary and symbols—Statistical quality control*.) One definition of quality assurance is: all the planned and systematic activities implemented within the quality system that can be demonstrated to provide confidence that a product or service will fulfill requirements for quality. One definition for quality control is: the operational techniques and activities used to fulfill requirements for quality. Often, however, quality assurance and quality control are used interchangeably, referring to the actions performed to ensure the quality of a product, service, or process.

quality characteristics—The unique characteristics of products and of services by which customers evaluate their perception of quality.

quality circles—Quality improvement or self-improvement study groups composed of a small number of employees—10 or fewer—and their supervisor, who meet regularly with an aim to improve a process.

quality council—The group driving the quality improvement effort and usually having oversight responsibility for the implementation and maintenance of the quality management system; operates in parallel with the normal operation of the business. Sometimes called *quality steering committee*.

quality engineering—The analysis of a manufacturing system at all stages to maximize the quality of the process itself and the products it produces.

quality function—The entire spectrum of activities through which an organization achieves its quality goals and objectives, no matter where these activities are performed.

quality function deployment (QFD)—A multifaceted matrix in which customer requirements are translated into appropriate technical requirements for each stage of product development and production. The QFD process is often referred to as listening to the voice of the customer. See also *house of quality*.

quality improvement—Actions taken throughout the organization to increase the effectiveness and efficiency of activities and processes in order to provide added benefits to both the organization and its customers.

quality level agreement (QLA)—Internal service/product providers assist their internal customers in clearly delineating the level of service/ product required in quantitatively measurable terms. A QLA may contain specifications for accuracy, timeliness, quality/usability, product life, service availability, or responsiveness to needs.

quality management—Coordinated activities to direct and control an organization with regard to *quality*. Such activities generally include establishment of the quality policy, quality objectives, quality planning, *quality control*, *quality assurance*, and quality improvement.

quality management system—The organizational structure, processes, procedures, and resources needed to implement, maintain, and continually improve quality management.

quality planning—The activity of establishing quality objectives and quality requirements.

quality trilogy—A three-pronged approach to managing for quality. The three legs are quality planning (developing the products and processes required to meet customer needs), quality control (meeting product and process goals), and quality improvement (achieving unprecedented levels of performance). Attributed to Joseph M. Juran.

quantitative data—See *variable (control chart usage)*.

questionnaires—See *survey*.

queue processing—Processing in batches (contrast with continuous flow processing).

queue time—Wait time of product awaiting next step in process.

R

r—See *correlation coefficient*.

R—See *range*.

R chart—See *range chart*.

$\bar{R}$ **(pronounced r-bar)**—The *average* range calculated from the set of *subgroup* ranges under consideration. See *range*.

R^2—See *coefficient of determination*.

R^2_{adj}—R^2 *(coefficient of determination)* adjusted for *degrees of freedom*.

$$R^2_{adj} = 1 - \frac{\text{Sum of squares error}/(n-p)}{\text{Sum of squares total}/(n-1)}$$

where *p* is the number of coefficients fit in the *regression* equation.[25]

RACI—A structure that ensures that every component of the project's scope of work is assigned to a person/team and that each person/team is *responsible, accountable, consulting,* or *informing* for the assigned task.

RAM—See *responsibility assignment matrix*.

random cause—The source of *process* variation that is inherent in a process over time. Also called *common cause* or *chance cause*.

Note: In a process subject only to random cause variation, the variation is predictable within statistically established limits.

random factor—A *factor* that uses *levels* that are selected at random from a large or infinite number of possibilities. In general, inference is made to other levels of the same factor. See also *fixed factor*.

random model—A *model* that contains only random *factors*.

random sampling—A sampling where a *sample* of n sampling *units* is taken from a *population* in such a way that each of the possible combinations of n sampling units has a particular probability of being taken.

random variation—Variation from *random causes*.

randomization—The process used to assign *treatments* to *experimental units* so that each experimental unit has an equal chance of being assigned a particular *treatment*. Randomization validates the assumptions made in statistical analysis and prevents unknown biases from impacting the conclusions.

randomized block design—An *experimental design* consisting of b *blocks* with t *treatments* assigned via *randomization* to the *experimental units* within each block. This is a method for controlling the variability of experimental units. For the *completely randomized design*, no stratification of the experimental units is made. In the randomized block design, the treatments are randomly allotted within each block; that is, the randomization is restricted.

randomized block factorial design—A *factorial design* run in a *randomized block design* where each *block* includes a complete set of factorial combinations.

range (R)—A measure of *dispersion*, which is the absolute difference between the highest and lowest *value* in a given *subgroup*:

$$R = \text{highest observed value} - \text{lowest observed value}.$$

range chart (R chart)—A *variables control chart* that plots the range of a *subgroup* to detect shifts in the subgroup range. See *range (R)*.

$$\text{Central line: } \bar{R}$$

$$\text{Upper control limit: } D_4 \bar{R}$$

$$\text{Lower control limit: } D_3 \bar{R}$$

where $\bar{R}$ is the average range; D_3 and D_4 are factors from Appendix D.

Note: A range chart is used when the sample size is small; if the sample is larger (generally >10 to 12), the s chart should be used.

rational subgroup—A *subgroup* wherein the variation is presumed to be only from *random causes*.

Re—See *rejection number*.

rectifying inspection—An inspection of all, or a specified number of, items in a *lot* or other amount previously rejected on *acceptance sampling* inspection, as a consequence of which all *nonconforming items* are removed or replaced.

reduced inspection—An *inspection* less severe than *normal inspection,* to which the latter is switched when inspection results of a predetermined number of *lots* indicate that the quality level achieved by the *process* is better than that specified.

redundancy—The existence of more than one means for accomplishing a given function. Each means of accomplishing the function need not necessarily be identical.

redundancy, active—That redundancy wherein all redundant items are operating simultaneously.

redundancy, standby—That redundancy wherein an alternative means of performing the function is not operating until it is activated upon failure of the primary means of performing the function.

reengineering—Completely redesigning or restructuring a whole organization, an organizational component, or a complete process. It's a "start all over again from the beginning" approach, sometimes called a breakthrough. In terms of improvement approaches, reengineering is contrasted with incremental improvement (kaizen).

reference interval—The interval bounded by the 99.865% distribution fractile, $X_{99.865\%}$, and the 0.135% distribution fractile, $X_{0.135\%}$, expressed by the difference $X_{99.865\%} - X_{0.135\%}$.

Note 1: This term is used only as an arbitrary, but standardized, basis for defining the *process performance index*, P_p, and *process capability index*, C_p.

Note 2: For a *normal distribution,* the reference interval may be expressed in terms of *standard* deviations as 6σ or $6s$ when estimated from a sample.

Note 3: For a nonnormal distribution, the reference interval may be estimated by appropriate probability scales or from the sample kurtosis and sample skewness.

Note 4: The conditions prevailing must always be stated: *process capability* conditions (a *state of statistical control* required) or *process performance* conditions (a state of statistical control not required).

regression—See *regression analysis.*

regression analysis—A technique that uses *predictor variable(s)* to predict the *variation* in a *response variable.* Regression analysis uses the method of *least squares* to determine the values of the *linear regression coefficients* and the corresponding *model.* It is particularly pertinent when the predictor variables are *continuous* and emphasis is on creating a predictive model. When some of the predictor variables are *discrete, analysis of variance* or *analysis of covariance* is likely a more appropriate method. This resulting model can then test the resulting predictions for statistical significance against an appropriate *null Hypothesis* model. The model also gives some sense of the degree of *linearity* present in the data. When only one predictor variable is used, regression analysis is often referred to as *simple linear regression*. A simple linear regression model commonly uses a *linear regression equation* expressed as $Y = \beta_0 + \beta_1 x + \varepsilon$, where Y is the response, x is the value of the predictor variable, β_0 and β_1 are the linear regression coefficients, and ε is the random error term. β_0 is often called the *intercept* and β_1 is often called the *slope*.

When multiple predictor variables are used, regression is referred to as *multiple linear regression*. For example, a multiple linear regression model with three predictor variables commonly uses a linear regression equation expressed as $Y = \beta_0 + \beta_1 x_1 + \beta_2 x_2 + \beta_3 x_3 + \varepsilon$, where Y is the response, x_1, x_2, x_3 are the values of the predictor variables, $\beta_0, \beta_1, \beta_2,$ and β_3 are the linear regression coefficients, and ε is the random error term. The random error terms in regression analysis are often assumed to be normally distributed with a constant *variance*. These assumptions can be readily checked through *residual analysis* or *residual plots*. See *regression coefficients, tests for*.

regression coefficients, tests for—A test of the individual *linear regression coefficients* to determine their significance in the *model*. These tests assume that the *response variable* is normally distributed for a fixed level of the *predictor variable*, the variability of the response variable is the same regardless of the value of the predictor variable, and that the predictor variable can be measured without error. See *regression analysis*.

reject—(acceptance sampling usage) To decide that a *batch, lot,* or quantity of product, material, or service has not been shown to satisfy the requirement criteria based on the information obtained from the *sample*(s).

Note: In *acceptance sampling*, the words "to reject" generally are used to mean to not accept, without direct implication of product usability. Lots that are rejected may be scrapped, sorted (with or without *nonconforming units* being replaced), reworked, reevaluated against more specific usability criteria, held for additional information, and so on. Because the common language usage "reject" often results in an inference of unsafe or unusable product, it is recommended that the words "not accept" be used instead.

rejectable process level (RPL)—The *process* level that forms the inner boundary of the rejectable process zone.

Note: In the case of two-sided *tolerances,* upper and lower rejectable process levels will be designated URPL and LRPL. (These need not be symmetrical around the standard level.)

rejectable process zone—See *rejectable process level*.

rejectable quality level (RQL)—See *limiting quality level (LQL)*.

rejection number (Re)—The smallest number of *nonconformities* or *nonconforming units* found in the *sample* by *acceptance sampling by attributes* that requires the *lot* to be not accepted, as given in the *acceptance sampling plan*.

relative frequency—The number of occurrences or *observed values* in a specified class divided by the total number of occurrences or observed values.

reliability—In measurement systems analysis, refers to the ability of an instrument to produce the same results over repeated administration—to measure consistently. In reliability engineering, it is the probability of a product performing its intended function under stated conditions for a given period of time. For redundant items this is equivalent to definition of mission reliability. See also *mean time between failures*.

reliability, mission—The ability of an item to perform its required functions for the duration of a specified *mission profile*.

remedy—Something that eliminates or counteracts a problem cause; a solution.

repair—Action taken on a nonconforming product so that it will fulfill the intended usage requirements, although it may not conform to the originally specified requirements. See *maintenance, corrective*.

repeatability—*Precision* under conditions where independent measurement results are obtained with the same method on identical measurement items by the same operator using the same equipment within a short period of time.

repeated measures—The measurement of a *response variable* more than once under similar conditions. Repeated measures allow one to determine the inherent variability in the *measurement system*. Also known as *duplication* or repetition.

replicate—A single repetition of the experiment. See also *replication*.

replication—Performance of an experiment more than once for a given set of *predictor variables*. Each of the repetitions of the experiment is called a *replicate*. Replication differs from *repeated measures* in that it is a repeat of the entire experiment for a given set of *predictor variables*, not just a repeat of measurements on the same experiment.

Note: Replication increases the precision of the estimates of the *effects* in an experiment. It is more effective when all elements contributing to the *experimental error* are included. In some cases replication may be limited to *repeated measures* under essentially the same conditions. In other cases, replication may be deliberately different, though similar, in order to make the results more general.

representative sample—A *sample* that by itself or as part of a sampling system or protocol exhibits *characteristics* and properties of the *population* sampled.

reproducibility—*Precision* under conditions where independent measurement results are obtained with the same method on identical measurement items with different operators using different equipment.

resample—An analysis of an additional *sample* from a *lot*. Resampling not included in the design of the *control chart* delays *corrective action* and changes the risk levels on which control charts are calculated. Resampling should not be used to replace sample results that indicate a *nonconformity*, but it is appropriate if the sample was known to be defective or invalid or to obtain additional information on the material sampled.

residual analysis—The method of using *residuals* to determine appropriateness of assumptions made by a statistical method.

residual plot—A plot used in *residual analysis* to determine appropriateness of assumptions made by a statistical method. Common forms include a plot of the *residuals* versus the *observed values* or a plot of the residuals versus the *predicted values* from the fitted model.

residuals—The difference between the observed result and the *predicted value* (estimated treatment response) based on an empirically determined model.

resistance to change—Unwillingness to change beliefs, habits, and ways of doing things.

resistant line—A line derived from using *medians* in fitting lines to data.

resolution—1: The smallest measurement increment that can be detected by the *measurement system*. 2: In the context of *experimental design*, resolution refers to the level of *confounding* in a *fractional factorial design*. For example in a resolution III design, the *main effects* are confounded with the two-way *interaction effects*.

response surface design—A design intended to investigate the functional relationship between a *response variable* and a set of *predictor variables*. It is generally most useful when the predictor variables are continuous. See *Box-Behnken design*.

response surface methodology (RSM)—A methodology that uses *design of experiments, regression analysis,* and optimization techniques to determine the best relationship between a *response variable* and a set of *predictor variables*.

response variable—A variable representing the outcome of an experiment.

responsibility assignment matrix (RAM)—A structure that relates the project organizational breakdown structure to the work breakdown structure to help ensure that each component of the project's scope of work is assigned to a responsible person/team.

resubmitted lot—A *lot* that previously has been designated as not acceptable and that is submitted again for acceptance inspection after having been further tested, sorted, reprocessed, and so on.

Note: Any *nonconforming units* discovered during the interim action must be removed, replaced, or reworked.

RETAD—Rapid exchange of tooling and dies.

rework—Action taken on a nonconforming product so that it will fulfill the specified requirements (may also pertain to a service).

right the first time—A term used to convey the concept that it is beneficial and more cost-effective to take the necessary steps up front to ensure that a product or service meets its requirements than to provide a product or service that will need rework or not meet customers' needs. In other words, an organization should engage in defect prevention rather than defect detection.

risk, consumer's (β)—See *consumer's risk (β)*.

risk, producer's (α)—See *producer's risk (α)*.

risk assessment/management—The process of determining what present or potential risks are possible in a situation (for example, project plan) and what actions might be taken to eliminate or mitigate them.

robust—A *characteristic* of a *statistic* or statistical method. A robust statistical method still gives reasonable results even though the standard assumptions are not met. A robust statistic is unchanged by the presence of unusual data points or *outliers*.

robust parameter design—A design that aims to reduce the performance variation of a product or *process* by choosing the setting of its *control factors* to make it less sensitive to variability from *noise factors*.

robustness—The condition of a product or process design that remains relatively stable with a minimum of variation even though factors that influence operations or usage, such as environment and wear, are constantly changing.

ROI—Return on investment. Umbrella term for a variety of ratios measuring an organization's business performance and calculated by dividing some measure of return by a measure of investment and then multiplying by 100 to provide a percentage. In its most basic form, ROI indicates what remains from all money taken in after all expenses are paid. ROI is always expressed as a percentage.

root cause analysis—A quality tool used to distinguish the source of defects or problems. It is a structured approach that focuses on the decisive or original cause of a problem or condition.

ROTI—Return on training investment. A measure of the return generated by the benefits from the organization's investment in training.

RPL—See *rejectable process level*.

RQL—See *rejectable quality level*.

RSM—See *response surface methodology*.

run—See *experimental run*.

run—(control chart usage) An uninterrupted sequence of occurrences of the same *attribute* or *event* in a series of observations, or a consecutive set of successively increasing (run up) or successively decreasing (run down) values in a series of *variable* measurements.

Note: In some *control chart* applications, a run might be considered a series of a specified number of points consecutively plotting above or below the *center line,* or five consecutive points, three of which fall outside warning limits.

running weighted average—An equation used to smooth data sequences by replacing each data value with the *average* of the data values around it and with weights multiplied in each averaging operation. The weights sum to 1.

S

σ **(sigma)**—See *standard deviation.*

σ^2 **(sigma square)**—See *variance.*

σ_{wt}—The *standard deviation* of the *exponentially weighted moving average:*

$$\sigma_{w_t} = \sigma\sqrt{\lambda/(2-\lambda)} \quad \text{[Asymptotic value]}$$

$$\sigma_{w_t} = \sigma\sqrt{\lambda/(2-\lambda)\left[1-(1-\lambda)^{2i}\right]} \quad \text{[Initial values]}$$

See *exponentially weighted moving average chart.*

$\sigma_{\bar{x}}$ **(sigma x-bar)**—The *standard deviation* (or *standard error*) of $\bar{x}$.

$\hat{\sigma}$ **(sigma-hat)**—In general, any estimate of the *population standard deviation.* There are various ways to get this estimate depending on the particular application.

s—See *standard deviation.*

s **chart**—See *standard deviation chart.*

s^2—See *variance.*

s_p—See *pooled standard deviation, t-test (two sample),* and *proportion, tests for.*

sales leveling—A strategy of establishing a long-term relationship with customers to lead to contracts for fixed amounts and scheduled deliveries in order to smooth the flow and eliminate surges.

sample—A group of *units,* portions or material, or observations taken from a larger collection of units, quantity of material, or observations that serves to provide information that may be used for making a decision concerning the larger quantity (the *population*).

Note 1: The sample may be the actual units or material or the observations collected from them. The decision may or may not involve taking action on the units or material, or on the *process.*

Note 2: *Sampling plans* are schemes set up statistically in order to provide a sampling system with minimum bias.

Note 3: There are many different ways, random and nonrandom, to select a sample. In survey sampling, sampling units are often selected with a probability proportional to the size of a known variable, giving a biased sample.

sample covariance—See *covariance*.

sample mean—The sample mean (or *average*) is the sum of random variables in a *random sample* divided by the number in the sum. It is generally designated by the symbol $\bar{x}$.

Note: The sample mean considered as a *statistic* is an estimator for the *population mean*. A common synonym is the *arithmetic mean*.

sample size (*n*)—The number of sampling *units* in a *sample*.

Note: In a multistage sample, the sample size is the total number of sampling *units* at the conclusion of the final stage of sampling.

sample standard deviation—See *standard deviation*.

sample variance—See *variance*.

sampling interval—In systematic sampling, the sampling interval is the fixed interval of time, output, running hours, and so on, between samples.

sampling plan (acceptance sampling usage)—A specific plan that states the *sample size*(s) to be used and the associated criteria for accepting the *lot*.

Note: The sampling plan does not contain the rules on how to take the sample.

sampling scheme (acceptance sampling usage)—A combination of *sampling plans* with rules (acceptance for changing from one plan to another).

saturated design—A design where the number of *factors* studied is nearly equal to the number of *experimental runs*. Saturated designs do not allow for a check of model adequacy or estimation of higher-order *effects*. This design should only be used if the cost of doing more experimental runs is prohibitive.

scatter diagram (or scatter plot)—A graphical technique to analyze the relationship between two variables. Two sets of data are plotted on a graph, with the *y*-axis being used for the variable to be predicted and the *x*-axis for the variable being used to make the prediction. The graph will show possible relationships (although two variables might appear to be related, they might not be: those who know most about the variables must make that evaluation). The scatter diagram is one of the seven basic tools of quality.

screening design—An experiment intended to identify a subset of the collection of *factors* for subsequent study. Examples include *fractional factorial designs* or *Plackett-Burman designs*.

secondary customer—Individuals or groups from outside the process boundaries who receive process output but who are not the reason for the process's existence.

segmentation—See *customer segmentation.*

self-directed learning—See *learner-controlled instruction.*

self-managed team—A team that requires little supervision and manages itself and the day-to-day work it does; self-directed teams are responsible for whole work processes with each individual performing multiple tasks.

sequential sampling—An *acceptance sampling inspection* in which, after each item has been inspected, the decision to accept the *lot,* not accept the lot, or to inspect another item is taken based on the cumulative sampling evidence.

Note 1: The decision is made according to defined rules.

Note 2: The total number of items to be inspected is not fixed in advance but a maximum number is often agreed upon.

servicing—The performance of any act to keep an item in operating condition (that is, lubrication, fueling, oiling, cleaning, and so on), but not including other preventive maintenance of parts or corrective maintenance.

setup time—The time taken to change over a process to run a different product or service.

seven basic quality tools—Tools that help organizations understand their processes in order to improve them. The tools are the cause-and-effect diagram, check sheet, control chart, flowchart, histogram, Pareto chart, and scatter diagram. See individual entries.

seven basic tools of quality management—The tools used primarily for planning and managing are the activity network diagram (AND) or arrow diagram, affinity diagram (KJ method), interrelationship digraph, matrix diagram, priorities matrix, process decision program chart (PDPC), and tree diagram.

Shewhart control chart—A *control chart* with *Shewhart control limits* intended primarily to distinguish between variation due to *random causes* and variation due to *special causes.*

Shewhart control limits—*Control limits* based on empirical evidence and economic considerations, placed about the *center line* at a distance of ± 3 *standard deviations* of the *statistic* under consideration and used to evaluate whether or not it is a *stable process.*

Shewhart cycle—See *plan–do–check–act (PDCA) cycle.*

sigma—Greek letter (σ) that stands for the standard deviation of a process.

signal—An indication on a *control chart* that a *process* is not *stable* or that a shift has occurred. Typical indicators are points outside *control limits, runs, trends,* cycles, patterns, and so on. See also *average run length.*

signal-to-noise ratio (S/N ratio)—A mathematical equation that indicates the magnitude of an experimental effect above the effect of experimental error due to chance fluctuations.

significance level—The maximum *probability* of rejecting the *null hypothesis* when in fact it is true.

Note: The significance level is usually designated by α and should be set before beginning the test.

significance tests—Significance tests are a method of deciding, with certain predetermined risks of error, (1) whether the *population* associated with a *sample* differs from the one specified, (2) whether the population associated with each of two samples differ, or (3) whether the populations associated with each of more than two samples differ. Significance testing is equivalent to the testing of *hypotheses*. Therefore, a clear statement of the *null hypothesis, alternative hypotheses,* and predetermined selection of a *confidence level* are required.

Note: The level of significance is the maximum *probability* of committing a *Type I error*. This probability is symbolized by α; that is, P (Type I error) = α. See *confidence interval; means, tests for;* and *proportions, tests for.*

simple linear regression—See *regression analysis.*

single sampling (acceptance sampling usage)—An *acceptance sampling inspection* in which the decision to accept, according to a defined rule, is based on the inspection results obtained from a single *sample* of predetermined size, *n*.

single-level continuous sampling—A continuous *acceptance sampling inspection* of consecutively produced items where a single fixed sampling rate is alternated with *100 percent inspection* depending on the quality of the observed *process* output.

single-minute exchange of dies (SMED)—A goal to be achieved in reducing the setup time required for a changeover to a new process; the methodologies employed in devising and implementing ways to reduce setup.

single-piece flow—A method whereby the product proceeds through the process one piece at a time, rather than in large batches, eliminating queues and costly waste.

SIPOC—A macro-level analysis of the suppliers, inputs, processes, outputs, and customers.

Six Sigma approach—A quality philosophy; a collection of techniques and tools for use in reducing variation; a program of improvement.

six sigma quality—a term used generally to indicate that a process is well controlled, that is, within process limits $\pm 3\sigma$ from the center line in a control chart, and requirements/tolerance limits $\pm 6\sigma$ from the center line. The term was initiated by Motorola.

skewness—A measure of symmetry about the *mean.* For a *normal distribution*, skewness is zero because the distribution is symmetric.

skip-lot sampling—An *acceptance sampling inspection* in which some *lots* in a series are accepted without *inspection* when the sampling results for a stated number of immediately preceding lots meet stated criteria.

slope—See *regression analysis.*

S.M.A.R.T goals—A template for setting goals that are specific, measurable, achievable, realistic, and traceable.

S.M.A.R.T. W.A.Y.—A template for setting objectives: specific, measurement, achievable, realistic, time, worth, assign, yield.

spaghetti chart—A before-improvement chart of existing steps in a process with lines showing the many back and forth interrelationships (can resemble a bowl of spaghetti). It is used to see the redundancies and other wasted movements of people and material.

SPC—See *statistical process control.*

special cause—A source of process variation other than *inherent process variation.*

Note 1: Sometimes special cause is considered synonymous with *assignable cause,* but a special cause is assignable only when it is specifically identified.

Note 2: A special cause arises because of specific circumstances that are not always present. Therefore, in a process subject to special causes, the magnitude of the variation over time is unpredictable.

specification—The engineering requirement used for judging the acceptability of a particular product/service based on product characteristics such as appearance, performance, and size. In statistical analysis, specifications refer to the document that prescribes the requirements to which the product or service has to conform.

specification limit(s)—The limiting value(s) stated for a *characteristic.* See *tolerance.*

split-plot design—A design where there is a hierarchical structure in the *experimental units.* One or more principal *factors* are studied using the largest experimental unit, often called a whole plot. The whole plots are subdivided into smaller experimental units, often called split-plots. Additional factors are studied using each of the split-plots. This type of design is frequently used with a principal factor whose levels are not easily changed, and the other factors can be varied readily within the runs assigned to the principal factor.

sponsor—The person who supports a team's plans, activities, and outcomes; the team's "backer." The sponsor provides resources and helps define the mission and scope to set limits. The sponsor may be the same individual as the "champion."

spread—A term sometimes synonymous with *variation* or *dispersion.*

stable process—A *process* that is predictable within limits; a process that is subject only to *random causes*. (This is also known as a *state of statistical control*.)

Note 1: A stable process will generally behave as though the results are simple random samples from the same *population*.

Note 2: This state does not imply that the *random variation* is large or small, within or outside of *specification limits,* but rather that the variation is predictable using statistical techniques.

Note 3: The *process capability* of a stable process is usually improved by fundamental changes that reduce or remove some of the *random causes* present and/or adjusting the *mean* toward the *target* value.

staggered nested design—A *nested design* in which the *nested factors* are run within only a subset of the *levels* of the first or succeeding *factors.*

stakeholder—People, departments, and organizations that have an investment or interest in the success or actions taken by the organization.

stakeholder analysis—The identification of stakeholders and delineation of their needs.

standard deviation—A measure of the *spread* of the *process* output or the spread of a sampling *statistic* from the process. When working with the *population,* the standard deviation is usually denoted by σ (sigma). When working with a *sample,* the standard deviation is usually denoted by s. They are calculated as

$$\sigma = \sqrt{\frac{1}{n}\sum(x-\mu)^2}$$

$$s = \sqrt{\frac{1}{n-1}\sum(x-\bar{x})^2}$$

where *n* is the number of data points in the sample or population, μ is the *population mean*, *x* is the observed value of the quality characteristic, and $\bar{x}$ is the sample mean. See *standard error.*

Note: Standard deviation can also be calculated by taking the square root of the *population variance* or *sample variance.*

standard deviation chart (*s* chart)—A *variables control chart* of the *standard deviation* of the results within a *subgroup*. It replaces *range chart* for large subgroup samples (rule of thumb is subgroup size 10 to 12).

Central line: $\bar{s}$

Upper control limit: $B_4\bar{s}$

Lower control limit: $B_3\bar{s}$

where $\bar{s}$ is the average value of the standard deviation of the subgroups and B_3 and B_4 are factors from Appendix D. See *standard deviation*.

standard deviation of proportions—See *proportions, tests for*.

standard deviation of the intercept—See *regression coefficients, tests for*.

standard deviation of the slope—See *regression coefficients, tests for*.

standard deviation of u or c/n—See *u chart*.

standard error—The *standard deviation* of a *sample statistic* or estimator. When dealing with sample statistics, we either refer to the standard deviation of the sample statistic or to its standard error.

standard error of predicted values—A measure of the variation of individual predicted values of the *dependent variable* about the *population* value for a given value of the *predictor variable*. This includes the variability of individuals about the sample regression line and the sample line about the population line. It measures the variability of individual observations and can be used to calculate a *prediction interval*.

standardized work—Documented and agreed-upon work instructions that express the best known methods and work sequence for each manufacturing or assembly process.

state of statistical control—See *stable process*.

statistic—A quantity calculated from a sample of observations, most often to form an estimate of some *population parameter*.

statistical measure—A *statistic* or mathematical function of a statistic.

statistical process control (SPC)—The use of statistical techniques such as *control charts* to reduce variation, increase knowledge about the *process,* and to steer the process in the desired way.

Note 1: SPC operates most efficiently by controlling variation of the process or in-process *characteristics* that correlate with a final product characteristic and/or by increasing the *robustness* of the process against this variation.

Note 2: A supplier's final product characteristic can be a process characteristic of the next downstream supplier's process.

statistical thinking—A philosophy of learning and action based on the following fundamental principles:

- All work occurs in a system of interconnected processes.

- *Variation* exists in all *processes*.

- Understanding and reducing variation are keys to success.

statistical tolerance interval—An interval estimator determined from a random sample so as to provide a specified level of confidence that the interval covers at least a specified proportion of the sampled *population*.

storage life—The length of time an item can be stored under specified conditions and still meet specified requirements.

storyboarding—A technique that visually displays thoughts and ideas and groups them into categories, making all aspects of a process visible at once. Often used to communicate to others the activities performed by a team as they improve a process.

stratified sampling—A *sampling* such that portions of the *sample* are drawn from different strata and each stratum is sampled with at least one sampling *unit*.

Note 1: In some cases, the portions are specified proportions determined in advance, however, in post-stratified sampling the specified proportions would not be known in advance.

Note 2: Items from each stratum are often selected by random sampling.

subgroup—(control chart usage) A group of data plotted as a single point on a control chart. See *rational subgroup*.

subsystem—A combination of sets, groups, and so on, that performs an operational function within a system and is a major subdivision of the system, for example, data processing subsystem, guidance subsystem.

supplier—Any provider whose goods and services may be used at any stage in the production, design, delivery, and use of another company's products and services. Suppliers include businesses, such as distributors, dealers, warranty repair services, transportation contractors, and franchises, and service suppliers, such as healthcare, training, and education providers. Internal suppliers provide materials or services to internal customers.

supply chain—The series of processes and/or organizations that are involved in producing and delivering a product to the final user.

supply chain management—The process of effectively integrating and managing components of the supply chain.

survey—An examination for some specific purpose; to inspect or consider carefully; to review in detail (survey implies the inclusion of matters not covered by agreed-upon criteria). Also, a structured series of questions designed to elicit a predetermined range of responses covering a preselected area of interest. May be administered orally by a survey-taker, by paper and pencil, or by computer. Responses are tabulated and analyzed to surface significant areas for change.

switching rules—Instruction within an *acceptance sampling scheme* for changing from one *acceptance sampling plan* to another of greater or lesser severity of sampling based on demonstrated quality level. *Normal, tightened, reduced inspection*, or discontinuation of inspection are examples of severity of sampling.

symptom—An indication of a problem or opportunity.

system—A network of interdependent processes that work together to accomplish a common mission.

system, general—A composite of equipment, skills, and techniques capable of performing or supporting an operation role, or both. A complete system includes all equipment, related facilities, material, software, services, and personnel required for its operation and support to the degree that it can be considered self-sufficient in its intended operational environment.

system effectiveness—The result of a combination of availability, dependability, and capability. The definition of *system effectiveness* is: a measure of the degree to which an item or system can be expected to achieve a set of specific mission requirements, and which may be expressed as a function of availability, dependability, and capability.

availability—A measure of the degree to which an item or system is in the operable and commitable state at the start of the mission, when the mission is called for at an unknown point in time. The point in time is a random variable.

dependability—A measure of the item or system operating condition at one or more points during the mission. It may be stated as the probability that an item will (a) enter or occupy any one of its required operational modes during a specified mission and (b) perform the functions associated with those operational modes.

capability—A measure of the ability of an item or system to achieve mission objectives given the conditions during the mission.

T

T^2 **(Hotelling's T^2)**—A multivariate test that is a generalization of the *t-test*.

t distribution—A theoretical distribution widely used in practice to evaluate the *sample mean* when the *population standard deviation* is estimated from the data. Also known as Student's distribution. It is similar in shape to the *normal distribution* with slightly longer tails. See *t-test*.

Taguchi design—See *robust parameter design*.

Taguchi loss function—Pertains to where product characteristics deviate from the normal aim and losses increase according to a parabolic function; by merely attempting to produce a product within specifications doesn't prevent loss (loss is that inflicted on society after shipment of a product).

takt time—The available production time divided by the rate of customer demand. Operating to takt time sets the production pace to customer demand.

tally sheet—Another name for check sheet.

target value—The preferred reference value of a *characteristic* stated in a specification.

t-confidence interval for means (one-sample)—When there is an unknown *mean* μ and unknown *variance* σ^2. For a *two-tailed test*, a $100(1 - \alpha)\%$ two-sided confidence interval on the true mean:

$$\overline{x} - t_{\alpha/2, n-1} \frac{s}{\sqrt{n}} \leq \mu \leq \overline{x} + t_{\alpha/2, n-1} \frac{s}{\sqrt{n}}$$

where $t_{\alpha/2, n-1}$ denotes the percentage point of the *t distribution* with $n - 1$ *degrees of freedom* such that

$$P\left(t_{n-1} \leq t_{\alpha/2, n-1}\right) = \alpha/2$$

See *confidence interval*.

t-confidence interval for means (two-sample)—When there is a difference in *means* (μ_1 and μ_2) and the *variances* are unknown. The combined or pooled estimate of the common variance, s_p is:

$$s_p^2 = \frac{(n_1 - 1)s_1^2 + (n_2 - 1)s_2^2}{n_1 + n_2 - 2}$$

The $100(1 - \alpha)\%$ two-sided confidence interval on μ_1 and μ_2 is

$$\left(\overline{x}_1 - \overline{x}_2\right) - t_{\alpha/2, n_1 + n_2 - 2} s_p \sqrt{\frac{1}{n_1} + \frac{1}{n_2}} \leq \mu_1 + \mu_2 \leq \left(\overline{x}_1 - \overline{x}_2\right) + t_{\alpha/2, n_1 + n_2 - 2} s_p \sqrt{\frac{1}{n_1} + \frac{1}{n_2}}$$

where $\overline{x}_1$ and $\overline{x}_2$ are the sample means $t_{\alpha/2}$ is the value from a *t-distribution* table where α is $1 - confidence\ level/100$. See *confidence interval*.

team—A group of two or more people who are equally accountable for the accomplishment of a task and specific performance goals; it is also defined as a small number of people with complementary skills who are committed to a common purpose.

team building/development—The process of transforming a group of people into a team and developing the team to achieve its purpose.

team dynamics—The interactions that occur among team members under different conditions.

team growth, stages of—Refers to the four development stages through which groups typically progress: forming, storming, norming, and performing. Knowledge of the stages helps team members accept the normal problems that occur on the path from forming a group to becoming a team.

team leader—A person designated to be responsible for the ongoing success of the team and keeping the team focused on the task assigned.

team-based structure—Describes an organizational structure in which team members are organized around performing a specific function of the business, such as handling customer complaints or assembling an engine.

temporary/ad hoc team—A team, usaually small, formed to address a short-term mission or emergency situation.

testing—A means of determining the ability of an item to meet specified requirements by subjecting the item to a set of physical, chemical, environmental, or operating actions and conditions.

tests for means—See *means, tests for.*

tests for proportions—See *proportions, tests for.*

tests for regression coefficients—See *regression coefficients, tests for.*

tests for significance—See *significance test.*

test for variances—See *variances, tests for.*

threshold control—Threshold control applies to *process control* where the primary interest is in events near *control limits.* Typical *control charts* used for this purpose are h, g, c, u, np, p, x (individuals), MR, $\bar{x}$, R, and s charts. This type of control is most often used in the early stages of a quality improvement process. See also *deviation control.*

throughput time—The total time required (processing + queue) from concept to launch or from order received to delivery, or raw materials received to delivery to customer.

tightened inspection—An *inspection* more severe than *normal inspection,* to which the latter is switched when inspection results of a predetermined number of *lots* indicate that the quality level achieved by the *process* is poorer than that specified.

time, active—That time during which an item is in an operational inventory.

time, checkout—That element of *maintenance time* during which performance of an item is verified to be in a specified condition.

time, delay—That element of downtime during which no maintenance is being accomplished on the item because of either supply or administrative delay.

time, down (downtime)—That element of active time during which an item is not in condition to perform its required function. (Reduces *availability* and *dependability.*)

time, supply delay—That element of *delay time* during which a needed replacement item is being obtained.

time, up (uptime)—That element of *active time* during which an item is in condition to perform its required functions. (Increases *availability* and *dependability.*)

time series—The sequence of successive time intervals.

tolerance—The difference between upper and lower *specification limits*.

tolerance limit—See *specification limit*.

top management commitment—Participation of the highest-level officials in their organization's quality improvement efforts. Their participation includes establishing and serving on a quality committee, establishing quality policies and goals, deploying those goals to lower levels of the organization, providing the resources and training that the lower levels need to achieve the goals, participating in quality improvement teams, reviewing progress organizationwide, recognizing those who have performed well, and revising the current reward system to reflect the importance of achieving the quality goals. Commitment is top management's visible, personal involvement as seen by others in the organization.

total productive maintenance (TPM)—Reducing and eventually eliminating equipment failure, setup and adjustment, minor stops, reduced speed, product rework, and scrap.

total quality management (TQM)—A term initially coined by the Naval Air Systems Command to describe its management approach to quality improvement. Total quality management (TQM) has taken on many meanings. Simply put, TQM is a management approach to long-term success through customer satisfaction. TQM is based on the participation of all members of an organization in improving processes, products, services, and the culture they work in. TQM benefits all organization members and society. The methods for implementing this approach are found in the teachings of such quality leaders as Philip B. Crosby, W. Edwards Deming, Armand V. Feigenbaum, Kaoru Ishikawa, J. M. Juran, and others.

training—Refers to the skills that employees need to learn in order to perform or improve the performances of their current job or tasks, or the process of providing those skills.

training evaluation—The techniques and tools used and the process of evaluating the effectiveness of training.

training needs assessment—The techniques and tools used and the process of determining an organization's training needs.

transformation—A reexpression of data aimed toward achieving *normality*.

treatment—The specific setting of *factor levels* for an *experimental unit*.

tree diagram—A management and planning tool that shows the complete range of subtasks required to achieve an objective. A problem-solving method can be identified from this analysis.

trimodal—A *probability distribution* having three distinct statistical modes.

TRIZ—(Russian) The theory of the inventive solution of problems. A set of analytical and knowledge-based tools that are typically hidden in the subconscious minds of creative inventors.

true value—A value for a *quantitative characteristic* that does not contain any sampling or measurement variability. (The true value is never exactly known; it is a hypothetical concept.)

***t*-test**—A *test for significance* that uses the *t distribution* to compare a *sample statistic* to a hypothesized *population mean* or to compare two means. See *t-test (one sample), t-test (two-sample), t-test (paired data)*.

Note: Testing the equality of the means of two normal populations with unknown but equal variances can be extended to the comparison of *k* population means. This test procedure is called *analysis of variance (ANOVA)*.

***t*-test (one-sample)**—

$$t = \frac{\bar{x} - \mu_0}{s / \sqrt{n}}$$

where $\bar{x}$ is the mean of the data, μ_0 is the hypothesized *population mean*, *s* is the *sample standard deviation*, and *n* is the *sample* size. The degrees of freedom are $n - 1$.16

***t*-test (paired data)**—Samples are paired to eliminate differences between specimens. The test could involve two machines, two test methods, two treatments, and so on. Observations are the pairs of the two machines, tests, treatments, and so on. The differences of the pairs of observations on each of the *n* specimens: $d_j = x_{1j} - x_{2j}$, $j = 1$,

$$t = \frac{\bar{d}}{s_d / \sqrt{n}}$$

where

$$\bar{d} = \frac{1}{n} \sum_{j=1}^{n} d_j$$

and

$$s_d^2 = \frac{\sum_{j=1}^{n} d_j^2 - \frac{\left(\sum_{j=1}^{n} d_j \right)^2}{n}}{n-1}$$

***t*-test (two-sample, equal variances)**—When there are two *populations* with unknown *means* and unknown *variances* that are assumed to be equal, the variances can be pooled to determine a pooled variance:

$$s_p^2 = \frac{(n_1 - 1)s_1^2 + (n_2 - 1)s_2^2}{n_1 + n_2 - 2}$$

where s_1 and s_2 are the individual sample variances and n_1 and n_2 are the respective *sample sizes.*

The *t*-test is:

$$t = \frac{\bar{x}_1 - \bar{x}_2}{s_p \sqrt{\dfrac{1}{n_1} + \dfrac{1}{n_2}}}$$

where s_p is the pooled variance calculated above, $\bar{x}_1$ and $\bar{x}_2$ are the means of the respective populations, and n_1 and n_2 are the respective *sample* sizes.

The degrees of freedom, ν: $\nu = n_1 + n_2 - 2$.

t-test (two-sample, unequal variances)—When there are two *populations* with unknown *means* and unknown variances that are assumed to be unequal, the sample standard deviation of $(\bar{x}_1 - \bar{x}_2)$ is:

$$s = \sqrt{\frac{s_1^2}{n_1} + \frac{s_2^2}{n_2}}$$

The degrees of freedom are:

$$\nu = \frac{\left(s_1^2 / n_1 + s_2^2 / n_2\right)^2}{\left[\left(s_1^2 / n_1\right)^2 / (n_1 - 1)\right] + \left[\left(s_2^2 / n_2\right)^2 / (n_2 - 1)\right]} - 2$$

two-tailed test—A *hypothesis* test that involves tails of a distribution. Example: We wish to reject the *null hypothesis*, H_0, if the true *mean* is within minimum and maximum (two tails) limits.

$$H_0: \mu = \mu_0$$

$$H_1: \mu \le \mu_0$$

Type I error—The *probability* or risk of rejecting a *hypothesis* that is true. This probability is represented by α *(alpha)*. See diagram below. See *operating characteristic curve* and *producer's risk.*

	H_0 true	H_0 false
Do not reject H_0	Correct decision	Error Type II
Reject H_0	Error Type I	Correct decision

Type II error—The *probability* or risk of accepting a *hypothesis* that is false. This probability is represented by β *(beta)*. See *power curve* and *consumer's risk*.

U

***u* (count per unit)**—The *events* or events per *unit* where the opportunity is variable. More than one event may occur in a unit.

Note: For *u*, the opportunity is variable; for *c*, the opportunity for occurrence is fixed.

U—See *upper specification limit*.

***u* chart**—An *attributes control chart* for number of *events* per unit where the opportunity is variable.

Note: Events of a particular type, for example, number of absentees or number of sales leads, form the count. In the quality field, events are often expressed as nonconformities and the variable opportunity relates to *subgroups* of variable size or variable amounts of material.

$$\text{Central line: } \bar{u}$$

$$\text{Control limits: } \bar{u} \pm 3\sqrt{\bar{u}/n}$$

where $\bar{u}$ is the average number of events per unit and n is the total number of samples. $\bar{u}$ is calculated as $\bar{u} = \bar{c}/n$.

Note: If the *lower control limit* (LCL) calculates ≤ 0 there is no LCL.

UCL—See *upper control limit*.

unacceptable quality level (UQL)—See *limiting quality level*.

uncertainty—A *parameter* that characterizes the *dispersion* of the values that could reasonably be attributed to the particular quantity subject to measurement or *characteristic*. Uncertainty indicates the variability of the measured value or characteristic that considers two major components of error: (1) *bias* and (2) the random error from the *imprecision* of the measurement process.

unconditional guarantee—An organizational policy of providing customers unquestioned remedy for any real or perceived product or service deficiency.

unique lot—A *lot* formed under conditions peculiar to that lot and not part of a routine sequence.

unit—A quantity of product, material, or service forming a cohesive entity on which a measurement or observation can be made.

universe—A group of *populations*, often reflecting different *characteristics* of the items or material under consideration.

upper control limit (UCL)—The *control limit* on a *control chart* that defines the upper control boundary.

upper specification limit or **upper tolerance limit (U)**—The *specification limit* that defines the upper limiting value.

UQL—See *limiting quality level.*

URPL—See *rejectable process level.*

USDA—U.S. Department of Agriculture.

useful life—The number of life units from manufacture to when an item has an unrepairable failure or unacceptable failure rate.

V

validation—Confirmation by examination of objective evidence that specific requirements and/or a specified intended use are met.

value chain—See *supply chain.*

value stream—The primary actions required to bring a product from concept to placing the product in the hands of the end user.

value stream mapping—The technique for mapping the value stream.

value-added—Refers to tasks or activities that convert resources into products or services consistent with customer requirements. The customer can be internal or external to the organization.

variable (control chart usage)—A quality *characteristic* that is from a *continuous scale* and is *quantitative* in nature.

variables, inspection by—See *inspection by variables.*

variables control chart—A *Shewhart control chart* where the measure plotted represents data on a *continuous scale.*

variance—A measure of the *variation* in the data. When working with the entire *population,* the population variance is used; when working with a *sample,* the sample variance is used. The population variance is based on the mean of the squared deviations from the *arithmetic mean* and is given by

$$\sigma^2 = \frac{1}{n}\sum(x-\mu)^2$$

The sample variance is based on the squared deviations from the *arithmetic mean* divided by $n-1$ and is given by

$$s^2 = \frac{1}{n-1}\sum(x-\bar{x})^2$$

where n is the number of data points in the sample or population, μ is the *population mean*, $\bar{x}$ is the *observed value* of the quality characteristic, and $\bar{x}$ is the sample mean. See *standard deviation*.

variances, tests for—A formal statistical test based on the *null hypothesis* that the *variances* of different groups are equal. Many times in *regression analysis* a formal test of variances is not done. Instead, *residual analysis* checks the assumption of equal variance across the values of the *response variable* in the *model*. For two variances, see *F-test*.

variation—The difference between values of a *characteristic*. Variation can be measured and calculated in different ways, such as *range, standard deviation,* or *variance*. Also known as *dispersion* or *spread*.

verification—The act of reviewing, inspecting, testing, checking, auditing, or otherwise establishing and documenting whether items, processes, services, or documents conform to specified requirements.

virtual team—A boundaryless team functioning without a commonly shared physical structure or physical contact, using technology to link the team members.

visual control—A technique of positioning all tools, parts, production activities, and performance indicators so that the status of a process can be understood at a glance by everyone; provides visual clues to aid the performer in correctly processing a step or series of steps, to reduce cycle time, to cut costs, to smooth flow of work, to improve quality.

vital few, useful many—A term used by J. M. Juran to describe his use of the Pareto principle, which he first defined in 1950. (The principle was used much earlier in economics and inventory control methodologies.) The principle suggests that most effects come from relatively few causes; that is, 80 percent of the effects come from 20 percent of the possible causes. The 20 percent of the possible causes are referred to as the vital few; the remaining causes are referred to as the useful many. When Juran first defined this principle, he referred to the remaining causes as the trivial many, but realizing that no problems are trivial in quality assurance, he changed it to useful many.

voice of the customer—An organization's efforts to understand the customers' needs and expectations (voice) and to provide products and services that truly meet such needs and expectations.

W

w—The span of the *moving average*.

w_t—The *exponentially weighted moving average (EWMA)* at the present time t.

w_{t-1}—The *exponentially weighted moving average (EWMA)* at the immediately preceding time interval.

warning limits—There is a high *probability* that the *statistic* under consideration is in a *state of statistical control* when it is within the warning limits (generally two standard deviations) of a *control chart*. See *Shewhart control limits*.

Note: When the value of the statistic plotted lies outside a warning limit but within the *action limit*, increased supervision of the *process* to prespecified rules is generally required.

waste—Activities that consume resources but add no value; visible waste (for example, scrap, rework, downtime) and invisible waste (for example, inefficient setups, wait times of people and machines, inventory).

WBS—See *work breakdown structure*

wearout—The process that results in an increase of the *failure rate* or probability of failure with increasing number of *life units*.

Weibull distribution—A distribution of continuous data that can take on many different shapes and is used to describe a variety of patterns; used to define when the "infant mortality" rate has ended and a steady state has been reached (decreasing failure rate); relates to the bathtub curve.

work analysis—The analysis, classification, and study of the way work is done. Work may be categorized as value-added (necessary work), or non-value-added (rework, unnecessary work, idle). Collected data may be summarized on a Pareto chart, showing how people within the studied population work. The need for and value of all work is then questioned and opportunities for improvement identified. A time use analysis may also be included in the study.

work breakdown structure (WBS)—A project management technique by which a project is divided into tasks, subtasks, and units of work to be performed.

work group—A group composed of people from one functional area who work together on a daily basis and whose goal is to improve the processes of their function.

work instruction—A document that answers the question: how is the work to be done?

workbook—A collection of exercises, questions, or problems to be solved during training; a participant's repository for documents used in training (for example, handouts).

X

x—The observed value of a quality characteristic; specific observed values are designated x_1, x_2, x_3, and so on. *x* is also used as a predictor value. See *input variable* or *predictor variable*.

x **chart**—See *individuals chart*.

$\bar{x}$ **(pronounced x-bar)**—The *average*, or *arithmetic mean*. The average of a set of *n* observed values is the sum of the observed values divided by *n*:

$$\bar{x} = \frac{x_1 + x_2 + \ldots + x_n}{n}$$

$\bar{x}$ **chart (averages control chart)**—A *variables control chart* for evaluating the *process* level in terms of *subgroup averages*. A variables control chart for evaluating the *process* level in terms of *subgroup averages*.

Central line: $\bar{\bar{x}}$ (if standard given, $\bar{x}_0$)

Control limits: $\bar{\bar{x}} \pm A_3\bar{s}$ or $\bar{\bar{x}} \pm A_2\bar{R}$ (if standard given, $\bar{x}_0 \pm A\sigma_0$ or $\bar{x}_0 \pm A_2R_0$)

where $\bar{\bar{x}}$ is the average of the subgroup values; $\bar{s}$ is the sample *standard deviation*; A, A_2, and A_3 are *control chart factors* (see Appendix D); σ_0 is the standard value of the population standard deviation; and R_0 is the standard value of the *range*. (Use the formula with $\bar{R}$ when the sample size is small, the formula with $\bar{s}$ when the sample is larger—generally > 10 to 12.)

$\bar{x}_0$—See *control chart, standard given*.

$\bar{x}_i$—The average of the ith subgroup (when $n = 1$, $\bar{x}_i = x = x_i$).

$\bar{\bar{x}}$ **(pronounced x-double-bar)**—The average for the set under consideration of the subgroup values of $\bar{x}$.

Y

y—y is sometimes used as an alternate to x as an observation. In such cases, y_0, and should be substituted where appropriate. See *output variable* or *response variable*.

$\bar{y}$ **(pronounced y-bar)**—The *average*, or *arithmetic mean*, of y. It is calculated exactly the same as $\bar{x}$.

$\bar{y}_0$—See y and *control charts, standard given*.

$\bar{\bar{y}}$ **(pronounced y-double-bar)**—The average for the set under consideration of the subgroup values of $\bar{y}$. See y.

Z

z-confidence interval for means (one-sample)—Given an unknown *mean* μ and known *variance* σ^2, then the $100(1 - \alpha)\%$ *two-sided* confidence interval on μ is:

$$\bar{x} - Z_{\alpha/2}\frac{\sigma}{\sqrt{n}} \leq \mu \leq \bar{x} + Z_{\alpha/2}\frac{\sigma}{\sqrt{n}}$$

where $\bar{x}$ is the *mean* of the data, σ is the *population standard deviation*, and n is the *sample* size.

zero defects—A performance standard popularized by Philip B. Crosby to address a dual attitude in the workplace: people are willing to accept imperfection in some areas, while in other areas they expect the number of defects to be zero. This dual attitude has developed because of the conditioning that people are human and humans make mistakes. The zero-defects methodology states, however, that if people commit themselves to watching details and avoiding errors, they can move closer to the goal of perfection.

zero investment improvement—Another name for a kaizen blitz.

z-test (one-sample)—Given an unknown *mean μ*, and known *variance σ²*:

$$z = \frac{\bar{x} - \mu_0}{\sigma / \sqrt{n}}$$

where $\bar{x}$ is the mean of the data, μ_0 is the standard value of the *population mean*, σ is the *population standard deviation*, and n is the *sample* size.

Note: If the variance is unknown, then the *t-test* applies.

z-test (two proportions)—To test if two binomial parameters are equal, the null hypothesis is H$_0$: $p_1 = p_2$ and the alternate hypothesis is H$_1$: $p_1 \neq p_2$. If the null hypothesis is true, then $p_1 = p_2 = p$, and

$$\hat{p} = \frac{n_1 \hat{p}_1 + n_2 \hat{p}_2}{n_1 + n_2}$$

The test is then:

$$z = \frac{\hat{p}_1 - \hat{p}_2}{\sqrt{\hat{p}(1-\hat{p})\left(\dfrac{1}{n_1} + \dfrac{1}{n_2}\right)}}$$

and H$_0$ should be rejected if $|z| > z_{\alpha/2}$.

z-test (two-sample)—Given unknown *means μ$_1$* and *μ$_2$*, and known *variances σ$_1^2$* and *σ$_2^2$*:

$$z = \frac{\bar{x}_1 - \bar{x}_2}{\sqrt{\dfrac{\sigma_1^2}{n_1} + \dfrac{\sigma_2^2}{n_2}}}$$

where $\bar{x}_1$ is the mean of the first *population* and $\bar{x}_2$ is the mean of the second population, σ_1 is the *standard deviation* of the first population and σ_2 is the standard deviation of the second population, and n_1 and n_2 are the respective *sample* sizes.

Note: If the variance is unknown, then the *t-test* applies.

References

American Society for Quality. *ASQ's Foundations in Quality Self-Directed Learning Series, Certified Quality Engineer*. Milwaukee: ASQ Quality Press, 2000.

———. *ASQ's Foundations in Quality Self-Directed Learning Series*, Certified Quality Manager. Milwaukee: ASQ Quality Press, 2001.

———. *Glossary and Tables for Statistical Quality Control*, 4th ed. Milwaukee: ASQ Quality Press, 2005.

Babbie, E. *The Practice of Social Research*, 10th ed. Belmont, CA: Thomson/Wadsworth, 2003.

Bauer, J. E., G. L. Duffy, and R. T. Westcott. *The Quality Improvement Handbook*, 2nd ed. Milwaukee: ASQ Quality Press, 2006.

Bayers, P. "Apply Poka-Yoke Devices Now to Eliminate Defects," in *51st Annual Quality Congress Proceedings* (Orlando, FL: 1997).

Benbow, D. W., A. K. Elshennawy, and H. F. Walker. *The Certified Quality Technician Handbook*. Milwaukee: ASQ Quality Press, 2003.

Benbow, D. W., and T. M. Kubiak. *The Certified Six Sigma Black Belt Handbook*. Milwaukee: ASQ Quality Press, 2005.

Benbow, D. W., R. W. Berger, A. K. Elshennawy, and H. F. Walker. *The Certified Quality Engineer Handbook*. Milwaukee: ASQ Quality Press, 2002.

Brassard, M., and D. Ritter. *The Memory Jogger II*. Methuen, MA: Goal QPC Press, 1994.

Brassard, M., L. Finn, D. Ginn, and D. Ritter. *The Six Sigma Memory Jogger II*. Salem, NH: Goal/QPC, 2002.

Chapman, C. D. "Clean House with Lean 5S," *Quality Progress* 38, no. 6 (June 2005).

Cheser, R. "Kaizen Is More Than Continuous Improvement," *Quality Progress* 27, no. 4 (April 1994).

Cooper, R. G. *Winning at New Products*, 3rd ed. Cambridge, MA: Perseus Publishing, 2001.

Douglas, A. "Improving Manufacturing Performance," in *56th Annual Quality Congress Proceedings* (Denver, CO: 2002).

Dovich, R. A. *Reliability Statistics*. Milwaukee: ASQ Quality Press, 1990.

Feigenbaum, A. V. *Total Quality Control*, 3rd ed., revised. New York: McGraw-Hill, 1991.

Feigenbaum, A. V., and D. S. Feigenbaum. "The Power of Management Capital" *Quality Progress* 37, no. 11 (November 2004).

Fisher, C., and J. Schutta. *Developing New Services*. Milwaukee: ASQ Quality Press, 2003.

Fisher, R. A. *Statistical Methods and Scientific Inference*. Edinburgh: Oliver and Boyd, 1956.

Freund, J. E. *Mathematical Statistics*. Englewood Cliffs, NJ: Prentice Hall, 1962.

Griffin, J. *Customer Loyalty*. New York: Lexington Books, 1995.

Hacking, I. *Logic of Statistical Inference*. Cambridge: Cambridge University Press, 1965.

Hradesky, J. L. *Total Quality Management Handbook*. New York, NY: McGraw-Hill, 1995.

Juran, J. M. *Juran on Planning for Quality*. New York: The Free Press, 1988.

Juran, J. M., and B. Godfrey. *Juran's Quality Handbook,* 5th ed. New York: McGraw-Hill, 1999.

Keeping, E. S. *Introduction to Statistical Inference.* Princeton, NJ: D. Van Nostrand, 1962.

Kiefer, J. *Journal of the American Statistical Association* 72 (1977).

King, B. M., and E. W. Minium. *Statistical Reasoning in Psychology and Education,* 4th ed. Hoboken, New Jersey: John Wiley & Sons, 2003.

Lawton, R. "8 Dimensions of Excellence," *Quality Progress* 39, no. 4 (April 2006).

Lindman, H. R. *Analysis of Variance in Complex Experimental Designs.* San Francisco: W. H. Freeman & Co, 1974.

Lowenstein, M. *Customer Retention.* Milwaukee: ASQC Quality Press, 1995.

MacInnes, R. *The Lean Enterprise Memory Jogger,* 1st ed. Salem, NH: Goal/QPC, 2002.

Manz, C. C., and H. P. Sims, Jr. *Business without Bosses: How Self-Managed Teams Are Building High-Performance Companies.* New York: John Wiley & Sons, 1993.

Michell, J. "Measurement Scales and Statistics: A Clash of Paradigms." *Psychological Bulletin* 3 (1986).

Neyman, J. *Philosophical Transactions of the Royal Society of London* (London: University College, 1937). (Seminal work.)

Ott, E. R., E. G. Schilling, and D. V. Neubauer. *Process Quality Control: Troubleshooting and Interpretation of Data,* 4th ed. Milwaukee: ASQ Quality Press, 2005.

Palmes, P. *An Introduction to the PDCA Cycle and Its Application.* MP3 Audiocasts. http://www.asq.org/learn-about-quality/project-planningtools/links-resources/audiocasts.html.

Pande, P., R. Neuman, and R. Cavanagh. *The Six Sigma Way.* New York: McGraw-Hill, 2000.

Pyzdek, T. *The Complete Guide to the CQE.* Tucson, AZ: Quality Publishing, 1996.

Robinson, G. K. "Some Counterexamples to the Theory of Confidence Intervals." *Biometrika* 62, no. 1 (1975).

Scholtes, P. R., B. L. Joiner, and B. J. Streibel. *The Team Handbook,* 3rd ed. Madison, WI: Joiner Associates Consulting Group, 2003.

Schultz, G. *The Customer Care and Contact Center Handbook.* Milwaukee: ASQ Quality Press, 2003.

Scriabina, N., and S. Fomichov. "Six Ways to Benefit from Customer Complaints." *Quality Progress* 38, no. 7 (September 2006).

Shingo, S. *A Revolution in Manufacturing: The SMED System.* Cambridge, MA: Productivity Press, 1985.

Shuker, T. J. "The Leap to Lean," in *54th Annual Quality Congress Proceedings.* Indianapolis, IN: 2000.

Sower, V. E., and F. K. Fair. "There Is More to Quality Than Continuous Improvement: Listening to Plato." *Quality Management Journal* 12, no. 1 (January 2005).

Stevens, S. S. "On the Theory of Scales of Measurement." *Science* 103 (1946).

"Mathematics, Measurement, and Psychophysics." In S. S. Stevens (ed.), *Handbook of Experimental Psychology.* New York: John Wiley & Sons, 1951.

Tague, N. R. *The Quality Toolbox,* 2nd ed. Milwaukee: ASQ Quality Press, 2004.

Van Patten, J. "A Second Look at 5S." *Quality Progress* 39, no. 10 (October 2006).

Velleman, P. F., and L. Wilkinson. "Nominal, Ordinal, Interval, and Ratio Typologies are Misleading." *The American Statistician* 47, no. 1 (1993). http://www.spss.com/research/wilkinson/Publications/Stevens.pdf.

Westcott, R. T. *The Certified Manager of Quality/Organizational Excellence Handbook,* 3rd ed. Milwaukee: ASQ Quality Press, 2006.

———. "Quality Level Agreements for Clarity of Expectations," *Stepping Up to ISO 9004:2000.* Chico, CA: Paton Press, 2003.

Westcott, R. T., and D. Okes. *The Certified Quality Manager Handbook,* 2nd ed. Milwaukee: ASQ Quality Press, 2006.

Zar, J. H. *Biostatistical Analysis.* Englewood Cliffs, NJ: Prentice Hall International, 1984.

Web Sites

http://en.wikipedia.org/wiki/Analysis_of_variance
http://en.wikipedia.org/wiki/Confidence_interval
http://en.wikipedia.org/wiki/Just_In_Time
http://en.wikipedia.org/wiki/Six_Sigma#The_.2B.2F-1.5_Sigma_Drift
http://en.wikipedia.org/wiki/SMED
http://en.wikipedia.org/wiki/Weibull_distribution
http://www.asq.org
http://www.isixsigma.com
http://www.lean.org
http://www.qms-ltd.com/elearning/resources/SixSigma/QFD.pdf
http://www.qualitydigest.com/oct96/godfrey.html
http://www.sevenrings.co.uk/SMED/HowtodoSMED.asp
http://www.themanagementor.com/EnlightenmentorAreas/mfg/QM/
    pokayoke.htm, copyright 2003, C & K Management Limited.
http://www.weibull.com/hotwire/issue21/hottopics21.htm
http://www.weibull.com/hotwire/issue22/hottopics22.htm
http://www.weibull.com/SystemRelWeb/availability.htm

Index

A

acceptable quality level (AQL), 138
acceptance number, in sampling, 140–42
acceptance sampling, 135–36
    by attributes, 137–42
acceptance stamp, 279
accuracy. in measurement, 169
achieved availability, 269
activity diagram, 80
activity network diagram (AND), 80
ad hoc teams, 30
affinity diagram, 75
alpha testing, 262
analysis of variance (ANOVA), 229–32
ANSI/ASQ Z1.4-2003 standard, for attributes
    sampling, 144–48
    tables, 154–59
ANSI/ASQ Z1.9-2003 standard, for variables
    sampling, 149–53
    tables, 160–63
appraisal costs, 11
areas under the standard normal curve,
    126–29
    Appendix B, 290–91
arrow diagram, 80
ASQ Code of Ethics, 2–4
assignable cause variation, 196
assumptions, in ANOVA, 230–31
attributes charts, 182–88
attributes data, 132
attributes sampling plans, types, 142–44
audit(s), 20–26
    process, 22–24
    roles and responsibilities, 25–26
    types, 20–22
audit process, 22–24
audit team member, 23
audit team leader, responsibilities, 25
auditee, responsibilities, 26
auditor, responsibilities, 26
automatic gauging, 133
availability, in reliability, 269
average outgoing quality limit (AOQL),
    139–40

average outgoing quality (AOQ), 138–39
average quality protection sampling, versus
    lot-by-lot, 136
averages and range ($\bar{X}$ and R) chart, 174
averages ($\bar{X}$) chart, 172–74

B

basic quality management tools, 74–81
basic quality tools, 54–65
bathtub curve, 264–66
benchmarking, 103–7
benefits-to-cost ratio (BCR), for training
    evaluation, 50
best practices, 103, 104–5
beta testing, 262
bias, in measurement, 171
bimodal distribution, 62–63
binomial distribution, 115–17
blocking, in DOE, 220, 226
Body of Knowledge, Quality Process Analyst
    (Appendix A), 284–89
breakthrough improvement, versus
    incremental improvement, 69

C

c charts, 184
capability, 190–95
categorical data, 132
cause-and-effect diagram, 63–64
cellular teams, 32
central tendency, measures of, 110–11
check sheets, 55–56
Code of Ethics, ASQ, 2–4
combinations, probability, 122–24
common cause variation, 196
companywide quality control, 13
    Ishikawa's definition, 6
complaint forms, 249–52
complaint management system (CMS), 250
complementation rule, probability, 118
conditional probability, 120–21

P

p charts, control limits for, 185–87
Pareto charts, 56–57
Pareto principle, 56
performance benchmarking, 106
permutations, probability, 124–25
philosophy, Six Sigma, 70
pilot run, 262
plan, organizational, requirements, 5
plan–do–check–act methodology, 13
  in complaint handling, 250–52
  steps, 66–67
planning, quality, 5–9
point estimate, 205
Poisson distribution, 117
poka-yoke, 100–102
population parameter, versus sample
  statistic, 205
p_p, calculating, 195
P_{pk}, calculating, 194
precision, in measurement, 169
pre-control chart, 195–96
predictor variables, 201
prevention costs, 11
preventive action, 273–74
preventive maintenance, 268
prioritization matrix, 79–80
probability, terms and definitions, 118–29
  areas under a normal curve, 126–29
  basic probability rules, 118–19
  combinations, 122–24
  conditional probability, 120–21
  contingency tables, 119–20
  general multiplication rule, 121
  independence and the special
    multiplication rule, 121–22
  permutations, 124–25
probability density functions, 264
problem identification, 55
problem solving and improvement, 53–107
  basic quality management tools, 74–81
  basic quality tools, 54–65
  benchmarking, 103–7
  continuous improvement models, 66–73
  lean, 95–102
  project management tools, 82–90
  Taguchi concepts, 91–94
problems, team process, 40–41
process approval systems, 259–62
process audit, 21–22
process average estimation, and sample size,
  210–11
process benchmarking, 106
process capability, measures, 190–95
process control, 225
process decision program chart, 80–81

process improvement teams, 28
producer's risk (α), 138
product approval systems, 259–62
product audit, 21
production part approval process (PPAP), 262
program evaluation and review technique
    (PERT), 80, 86
project benchmarking, 106–7
project budget, 88–89
project charter, 84
project constraints, 83
project management, principles of, 82–84
project management tools, 82–90
  Gantt chart, 85–86
  network diagram, 86–88
  principles of project management, 82–84
  project budget, 88–89
  project charter, 84
  project progress review, 90
  project variance analysis, 90
  responsibility assignment matrix (RAM),
    88
  work breakdown structure (WBS), 84
project progress review, 90
project variance analysis, 90
pull system, 97, 99
p-value, misunderstandings about use of, 219

Q

qualification, in product/process approval,
    261–62
qualitative data, 132
quality, basics, 1–51
  ASQ Code of Ethics, 2–4
  audits, 20–26
  cost of quality (COQ), 10–12
  documentation systems, 15–19
  quality planning, 5–9
  quality standards, requirements, and
    specifications, 13–14
  teams, 27–42
  training components, 43–51
quality assurance (QA), 14
  function, 13
quality control, companywide, Ishikawa's
    definition, 6
quality control, total, Feigenbaum's
    definition, 5–6
quality council, 6
quality function deployment, 253–58
  benefits, 258
  steps, 256–58
quality level agreements, 236–37
quality management tools, seven basic, 74–81
  activity diagram (arrow diagram), 80